Nanotechnology and Photovoltaic Devices

Nanotechnology and Photovoltaic Devices

Light Energy Harvesting with Group IV Nanostructures

edited by

Jan Valenta and Salvo Mirabella

CRC Press
Taylor & Francis Group
Boca Raton London New York

CRC Press is an imprint of the
Taylor & Francis Group, an **informa** business

Nanotechnology and Photovoltaic Devices: Light Energy Harvesting with Group IV Nanostructures

First published 2015 by Pan Stanford Publishing Pte. Ltd.

Published 2022 by CRC Press
Taylor & Francis Group
6000 Broken Sound Parkway NW, Suite 300 Boca Raton, FL 33487-2742

ISBN-13: 978-981-4463-63-8 (hbk)
ISBN-13: 978-981-4463-64-5 (ebk)

Visit the Taylor & Francis Web site at
http://www.taylorandfrancis.com

and the CRC Press Web site at
http://www.crcpress.com

British Library Cataloguing-in-Publication Data
A catalogue record for this book is available from the British Library.

Contents

Preface

The increasing energy demand of humankind on the Earth cannot be reasonably sustained by prolonged exploitation of fossil fuels. Therefore we have to turn toward efficient usage of the most abundant renewable supply of energy—it means the Sun. When considering Photovoltaics' aim, the direct transformation of solar photon flux into electrical energy, the most practical materials for this transformation are semiconductors whose absorption matches quite well solar photons' energy and whose conductivity can be adjusted so that photogenerated charge carriers are separated and directed to make useful work in an external circuit. Fortunately, some of these materials are very abundant, especially silicon, but other elements from group IV of the periodic table of elements are also extremely interesting. However, the maximum efficiency in energy conversion of the solar spectrum by a single semiconductor material is limited, as described by the famous Shockley–Queisser limit. To overcome this constraint, most of the proposed ideas, commonly labeled as third-generation Photovoltaics, are based on Nanotechnology employing materials whose energy scheme is more complex and variable. There are such materials, namely, semiconductor nanostructures, that enable us to tune their energy levels, density of electronic states, transition probabilities, etc., with large potential benefits for light energy conversion.

The purpose of this book is to summarize the knowledge and current advances of group IV semiconductor nanostructures potentially applicable in the next generations of solar cells. Considering the increasing research efforts devoted to nanostructure applications in Photovoltaics, our intention was to provide a clear background to students and newcomer researchers as well as to point out some open questions and promising directions of future development.

The book presents a broad overview on group IV nanostructures in Photovoltaics, beginning with a theoretical background, presentation of main solar cell principles, technological aspects, and nanostructure characterization techniques and finishing with the design and testing of prototype devices. The limited space of one book did not allow us to include some special nanostructure-related subjects, such as nanocrystal-sensitized solar cells (Grätzel cells or polymer cells), microcrystalline and amorphous silicon materials, rare-earth-doped nanostructures, plasmonic structures, etc. It is not intended to be just a review of the most up-to-date literature, but the contributing authors' ambition was to provide an educative background of the field. In view of the harsh economic competition in the solar cell business it might be that nanostructures will never be a commonly used material in Photovoltaics' massive production; still the solid background knowledge gained by researchers and summarized in this book will help in applying nanostructures to this and other fields.

The idea to compile this book was born in 2012 within the framework of a successful European research project (NASCEnT, Silicon nanodots for solar cell tandem, 2010–2013, 7FP project contract 245997), and in fact, many authors of the book participated in that project. Therefore we shall thank the European Commission for the support and Pan Stanford Publishing for its effort and helpful cooperation. The main acknowledgment goes to all chapter authors, who invested a lot of time and effort into the success of this book.

Jan Valenta and Salvo Mirabella
Prague and Catania
January 2015

Chapter 1

Introduction to Photovoltaics and Potential Applications of Group IV Nanostructures

Jan Valenta[a] **and Salvo Mirabella**[b]

[a] *Department of Chemical Physics and Optics, Faculty of Mathematics and Physics, Charles University, Prague, Czechia*
[b] *Institute for Microelectronics and Microsystems, Consiglio Nazionale delle Ricerche, via Santa Sofia 64, 95123 Catania, Italy*
jan.valenta@mff.cuni.cz, salvatore.mirabella@ct.infn.it

The human population increase along with the raise of living standards is about to cause doubling of the global primary energy consumption in less than 50 years [1]. Such continuous increase of energy demand will soon become unsustainable when considering that most of the currently exploited energy comes from fossil fuels whose resources are, obviously, limited. Moreover, burning of fossil fuels by the humankind in the past 250 years released such a quantity of carbon (in the form of CO_2—an important greenhouse gas) that it took our planet about 250 million of years to sequester [2]. An increasing awareness comes up on the energy demand issue and new pressing challenges arise to provide people with enough energy within a sustainable development scenario.

Nanotechnology and Photovoltaic Devices: Light Energy Harvesting with Group IV Nanostructures
Edited by Jan Valenta and Salvo Mirabella

ISBN 978-981-4463-63-8 (Hardcover), 978-981-4463-64-5 (eBook)
www.panstanford.com

Among other means, the research and development of new technologies and materials for energy is particularly important. In this context, a wide and exciting range of possible solutions is provided by nanotechnologies offering innovative materials with unique properties exploitable for energy production, distribution and saving.

In this book we deal with nanostructures based on group IV elements (e.g., Si quantum dots, C nanotubes, Ge nanowires, etc.) which attracted great attention during the last two decades. Their main advantages are abundance, nontoxicity, high attainable purity, and mature technology, which promise effective exploitation of these nanomaterials in advanced photovoltaic devices.

1.1 Energy from the Sun

The solar photon flux is the only sustainable source of energy for the earth (the current knowledge predicts that this flux will be slowly increasing during next billion years (Gyr) for which the life can survive on the earth[a]) [3]. On the other hand electricity is currently the most versatile form of energy used by human civilization. Therefore the direct transformation of photon energy into electricity in devices called solar cells (SCs) attracts still more interest and motivates effort of scientific research and industry.

The photon emission comes from the solar outer shell called photosphere, which has temperature around 5800 K and the spectrum corresponds to the thermal radiation of the black body with this temperature (Fig. 1.1a). On the surface of the earth the sunshine spectrum is modified by absorption in the earth atmosphere; the main absorption occurring in the ultraviolet (UV) and infrared (IR) spectral regions.

[a]The sun is now about 4.6 Gyr old and will remain in the main sequence of star evolution in total for about 10 Gyr. Then, after ~5 Gyr, it will enter the red giant stage (expanding and cooling down). However, luminosity of the sun is—in the current main sequence state—increasing by about 10% per billion years. This will probably disable the life on the earth in about 1 Gyr [3]. The possible lifetime of humankind is another question.

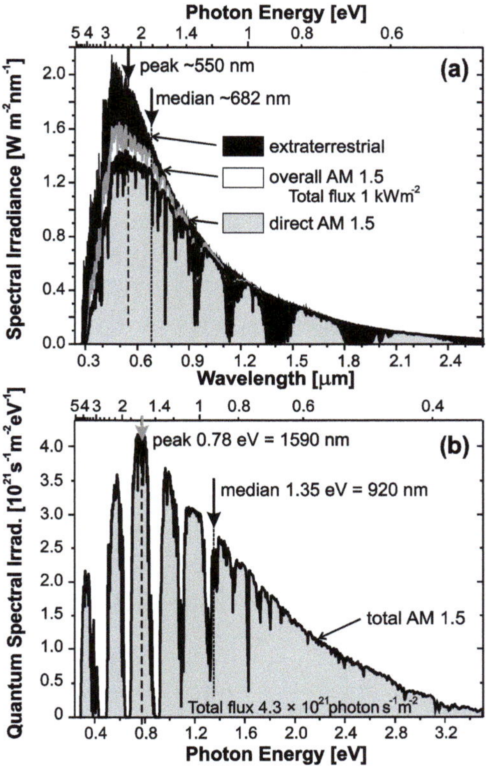

Figure 1.1 Solar spectrum given by the standard ASTM G173-03. (a) The extraterrestrial (AM 0) and AM 1.5 spectra (overall and direct radiation) expressed as spectral irradiance vs. wavelength. (b) The overall AM 1.5 spectrum expressed as quantum spectral irradiance vs. photon energy.

Throughout the photovoltaic (PV) research and development (R&D) the calculations are mostly done using the standard spectrum known as the "AM 1.5" (abbreviation for air mass 1.5—which means that the solar rays traverse the atmosphere at a tilt angle of 48°, so the apparent thickness of the atmosphere is 1.5 times the perpendicular thickness)[a] [4]. There are two spectra in this standard: the first one for the direct radiation and the second for

[a]The AM 1.5 spectrum was defined by the PV industry and the US government laboratories in conjunction with the American Society for Testing and Materials (ASTM). The standard ASTM G173-03 replaced the previous G159 in January 2003

total (direct and scattered) radiation from the whole hemispherical 2π solid angle (Fig. 1.1a).

The AM 1.5 spectrum is represented in spectral irradiance [$W{\cdot}m^{-2}{\cdot}\ nm^{-1}$], so it gives the energy which hits area of 1 m^2 per second and per the spectral interval of 1 nm. When integrating the whole spectrum we obtain the total power received by 1 m^2 and it is 900.1 and 1000.4 $W{\cdot}m^{-2}$ for direct and total AM 1.5 radiation, respectively. This means that the ratio of direct vs. indirect radiation power is about 9, in other words, 90% of radiation comes from the sun within the narrow viewing angle of $\alpha_s = 32'$ or solid angle $\Omega_s = 6.8 \times 10^{-5}$ sr (steradian). But under cloudy sky conditions the radiation is much reduced and distributed more evenly over the hemisphere. The peak of this AM 1.5 spectral irradiance curve is observed at 550 nm, that is, green light.

The interaction of light with a semiconductor on the microscopic level should be described from the quantum physics point of view. Therefore, we will need to describe the incoming radiation as flux of photons. Photon frequency ν, wavelength λ, and energy E are related through the well-known relation

$$E = h\nu = \frac{hc}{\lambda}, \quad E\,[eV] = \frac{1239.5}{\lambda\,[nm]} \tag{1.1}$$

where h and c are the Planck constant and the speed of light, respectively. The right-hand side shows the numerical relation for E and λ expressed in the common units of electron volts and nanometers, respectively (taking the value of c for air under normal conditions). Let us now transform the spectral irradiance into quantum spectral irradiance, it means the number of photons incident on 1 m^2 area per second per energy interval of 1 eV. The first step is the transformation of irradiance from wavelength density into energy density. The relation between infinitesimal steps of wavelength $\mathrm{d}\lambda$ and energy $\mathrm{d}E$ is found by differentiation of Eq. 1.1

$$\mathrm{d}E = -\frac{hc}{\lambda^2}\mathrm{d}\lambda \tag{1.2}$$

where the negative sign—causing flipping the spectral scale—is due to reciprocal relation of E and λ. Then we shall transform

(last approval in 2012) and the corresponding international standard is ISO 9845-1:1992.

the spectrum from $d\lambda$ to dE by multiplying with $\lambda^2/(hc)$ (for more details see Ref. [5]). Subsequently, we shall transform energy irradiance into number of photons by dividing with photon energy $(hc)/\lambda$ (so the total multiplication factor is $\lambda^3/(hc)^2$). The result is shown in Fig. 1.1b. One can see that the most frequent are IR photons with energy 0.78 eV (i.e., 1590 nm). By dividing the total power (1 $kJ \cdot s^{-1} \cdot m^{-2}$) with the total photon flux (4.3×10^{21} $s^{-1} \cdot m^{-2}$) we obtain the average energy carried by one solar photon 1.45 eV.

1.2 The Basic Principles of Photovoltaic Solar Cells

Any exploitation of solar photon energy consists of two basic steps: *Absorption* (when absorbed photons excite electrons of the absorbing material into higher states) and *conversion* (when the electronic excitation is converted into usable form of energy, for example, heat, electrical current or chemical energy). The core physical principle exploited in PV cells is the *internal photoeffect*—excitation of electron from the valence band of a semiconductor material into the conduction band, leaving behind an unoccupied state, called hole (and behaving like positively charged "quasiparticle"). This electron–hole pair created by absorption must be separated before its recombination (radiative or nonradiative) can happen. For this an internal potential must be created, which drives negatively charged electrons and positively charged holes into opposite directions. A *p–n* junction (connection of *p*- and *n*-type doped semiconductor regions) creates such internal potential in PV devices (eventually, the Schottky contact between metal and semiconductor can induce the charge separation). We can say that it is a compositional gradient in the material which drives the charge separation.

1.2.1 *Energy Balance*

The first silicon *p–n* junction SC introduced by Chappin, Fuller, and Pearson at Bell Telephone Laboratories in 1954 [6] gave power of 60 $W \cdot m^{-2}$, that is, power efficiency of 6%. These authors provided the first simple estimation of maximum conversion efficiency to be about 22%. Then the question of maximum attainable power efficiency attracted considerable attention as it would give a technological target for the PV R&D.

Already since 1960, *thermodynamics* has been applied as the most general approach to calculate the PV conversion efficiency [7]. If we treat a PV cell as a heat engine working between the temperature of sun T_S (5770 K) and ambient T_A (300 K), then the maximum efficiency η is given by the *second law of thermodynamics* and described as the *Carnot cycle* (consisting of two isothermal and two adiabatic steps) to be $\eta = 1-(T_A/T_S) = 0.95$. This hypothetic cycle produces no entropy; it is reversible and infinitely slow, therefore having infinitely low power. The theoretical treatment by Landsberg and Tonge [8] gives more precise thermodynamic limit for solar energy converters $\eta = 0.933$ (often called Landsberg limit), which is just slightly lower than the Carnot limit. However, if we want to maximize power delivered by the Carnot machine the efficiency limit decreases significantly to $\eta = 1 - \sqrt{T_A/T_S} = 0.77$ [9].

Under constant cell illumination and constant ambient temperature (the ambient thermal capacity is considered infinite) the cell is in the steady state condition and all incoming and outcoming energy fluxes must compensate together. Such system in equilibrium must fulfill the *principle of detailed balance*—each elementary process in the system must be microscopically reversible (which is valid for any closed system in nature). The detailed balance principle was first used by Shockley and Queisser in their seminal paper in 1961 [10] and the derived efficiency limit is commonly called the *Shockley–Queisser (SQ) limit*. This limit is derived for a single *p–n* junction cell made of a semiconductor with the energy band gap E_g which is supposed to absorb all photons with energy above E_g and transmit all photons with energy below E_g (no losses by reflection). The created electrons and holes relax to the band minima (thermalization by emission of phonons—quanta of lattice vibration) and are collected to external circuit without losses except the unavoidable radiative recombination. We can illustrate the detailed balance energy fluxes by Fig. 1.2. Thermal energy released during relaxation of hot electrons and holes (excited into higher states by absorption of high energy photons) is radiated to the environment (1). We illustrate thermal radiation on the right panel of Fig. 1.2. supposing that SC is heated from ambient $T = 300$ K to 334 K. In dark, the thermal radiation emitted and absorbed by

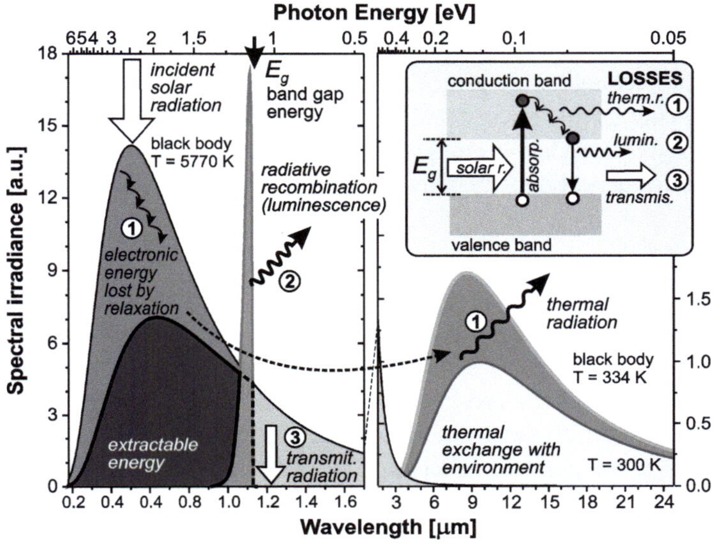

Figure 1.2 The spectral representation of the detailed balance energy fluxes in a single-junction PV cell with $E_g = 1.1$ eV (Si). The black-body radiation spectrum ($T = 5770$ K) is used as an approximate representation of the solar spectrum. Perfect absorption of all photons above E_g and no reflection losses are considered. Thermalization of hot carriers is supposed to increase the temperature of the PV cell by 34°C. The intensity scale of the right-hand side panel is expanded and the radiative emission is not to scale. The inset shows the energetic scheme for the three mechanisms of losses.

a PV cell are in equilibrium (white area in Fig. 1.2). The radiative recombination (luminescence) of electron-hole pairs (2) appears in the spectral region just below the semiconductor band gap E_g.

1.2.2 *Energy Conversion: Efficiency and Limits*

The SQ approach was later extended by C. H. Henry [11] and followers to multijunction cells. Henry also proposed very instructive graphical representation of the detailed balance calculation which we are going to exploit here. First, we make a cumulative summing up of number of photons N_p in the AM 1.5 solar spectrum from the highest to the lowest photon energies (solid line in Fig. 1.3a). Then we take a semiconductor with certain E_g (here 1.1 eV as for Si) and make a vertical line at this energy. The intersection with N_p is number of absorbed photons with energy above E_g and all these photons relax to the band minima and contribute with energy just equal to E_g. It means that the available energy is represented by area of the rectangle $N_p(E_g) \times E_g$. The role of radiative recombination is reducing usable energy and it is represented by shifting the curve $N_p(E_g)$ toward lower energy (dashed line in Fig. 1.3a).

The current density delivered to the load is

$$J = J_{sc} - J_S \left[\exp\left(\frac{eV}{k_B T_c} \right) - 1 \right] \quad (1.3)$$

where the *short-circuit* ($V = 0$) *current* $J_{sc} = eN_p$ is determined only by the number of absorbed photons (multiplied by the elementary charge e).[a] The second term characterizes the reduction of current due to the radiative recombination, where J_S is the reverse saturation current in dark [12] (k_B is the Boltzmann constant and T_c is temperature of a cell). Equation 1.3 describes the current–voltage (J–V) characteristic of an ideal PV cell (diode). It is shown in the inset of Fig. 1.3a, plotted with parameters of bulk Si ($E_g = 1.1$ eV, refractive index $n = 3.5$) and the cell temperature $T_c = 300$ K.

The open-circuit voltage V_{oc} ($J = 0$), which gives the separation of quasi-Fermi energies at which e–h recombination and generation are in equilibrium is [11, 12]

$$V_{oc} = \frac{k_B T}{e} \ln\left(\frac{J_{sc}}{J_S} + 1 \right) \quad (1.4)$$

The maximum power point (J_m,V_m) is then found as the peak of the $J \cdot V$ function (see inset in Fig. 1.3a) [11]:

$$V_m = V_{oc} - \frac{1}{e} \left[k_B T_c \ln\left(1 + \frac{eV_m}{k_B T_c} \right) \right], \; J_m = \frac{eN_p}{1 + k_B T_c / eV_m} \quad (1.5)$$

The work performed per absorbed photon is limited to

$$W = \frac{J_m V_m}{N_p} = \frac{eV_m}{1 + k_B T_c / eV_m} \cong eV_m - k_B T_c \quad (1.6)$$

where $eV_m \gg k_B T_c$ at room temperature ($k_B T_c \cong 26$ meV), and power efficiency of the PV cell is

$$\eta = \frac{J_m V_m}{\int_{\text{spectr.}} N_p(E)\, dE} \quad (1.7)$$

[a] Note that we adopted the common convention: *Current driven by photovoltage conversion in an illuminated PV cell has a positive sign.* It is opposite to the normal definition of current in electronics.

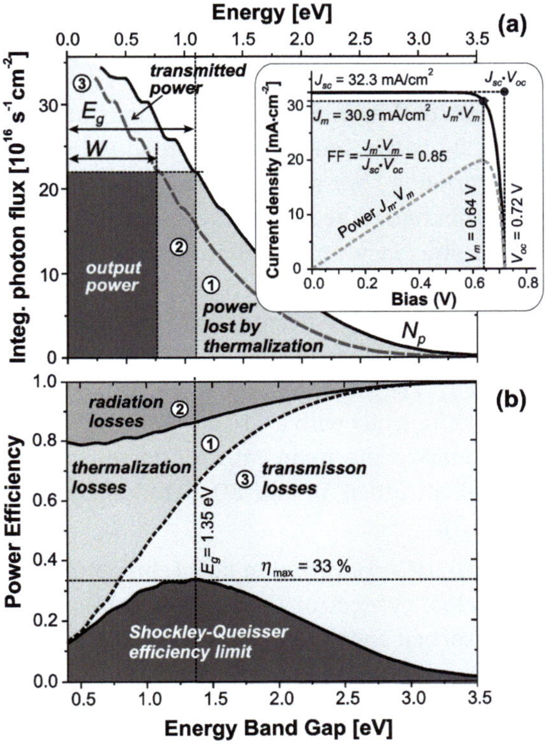

Figure 1.3 (a) Representation of the detailed balance treatment of a single-band PV cell according to Henry [11]. The work W delivered by a photon is reduced from the band-gap energy E_g due to radiative recombination. The inset shows the J–V characteristic of an ideal PV cell made of bulk Si; FF is the fill factor. (b) The SQ efficiency limit as a function of E_g plotted along with distribution of intrinsic losses due to the hot-carrier thermalization, nonabsorption of below-gap photons, and radiative recombination (the numerical labels correspond to Fig. 1.2).

where the denominator is the integral power of the solar spectrum, that is, the area under the curve in Fig. 1.3a. The task to find the semiconductor material for a single p–n junction PV cell which should have maximum efficiency is then identical to finding the rectangle with the maximal area—this takes place for $E_g = 1.35$ eV and limiting efficiency is ~33%. We plot the SQ limit efficiency as a function of E_g in Fig. 1.3b along with losses due to hot-carrier

(HC) relaxation, nonabsorbed below-gap photons and radiative recombination. All these calculations are done for nonconcentrated illumination by 1 sun AM 1.5.

The solar concentration by ratio C (the maximum C is $\pi/\Omega_S = 46200$, where Ω_S is the solid angle subtended by the sun) could increase the work done per photon to $W + k_B T_c \cdot \ln(C)$ (if the effects connected with increased heating of the cell are neglected) [11]. The single junction efficiency limit is then increased to about 37% at concentration of $C = 1000$ and to 41% at full concentration [13]. The reason for increase of efficiency with increasing C can be seen from Eq. 1.7 where both the incident power density (denominator) and the current J_m are linear functions of C while the voltage V_m is increasing logarithmically with C [14].

Let us summarize the main *intrinsic losses* of the single *p–n* junction PV cell included in the SQ detailed balance treatment (Figs. 1.2 and 1.3):

- Energy of HCs converted to heat (entropy)
- Energy lost by electron–hole radiative recombination
- Nonabsorbed energy of below-band-gap photons

There are many more sources of losses in real PV devices. Between these *extrinsic losses* we can find:

- Reflection on interfaces
- Absorption by inactive layers
- Light shadowing by electrical contacts
- Nonradiative recombination of photogenerated carriers
- Incomplete collection of carriers
- Series resistance (I^2R losses)
- Heating of the PV cell above ambient temperature

For a more detailed description of physics behind the *p–n* junction SCs we refer readers to the various textbooks, for example, Refs. [12, 14].

1.3 Advanced Concepts for Photovoltaics

The following three generations of PV devices are commonly distinguished [15] (Fig. 1.4a):

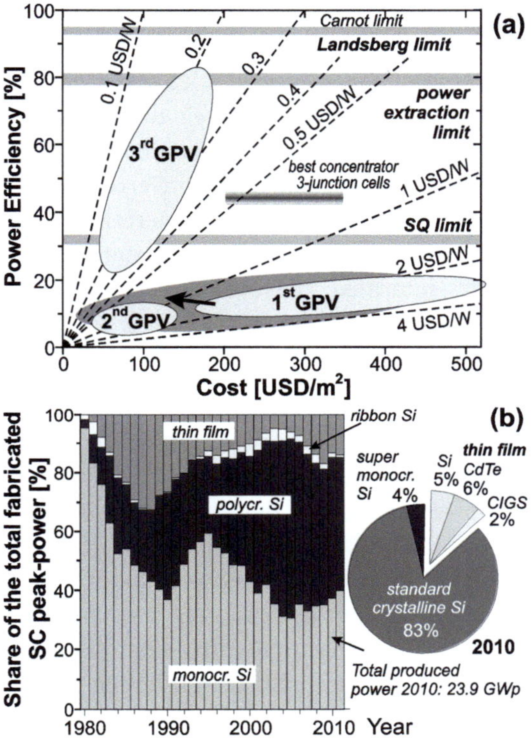

Figure 1.4 (a) The three PV generations plotted in coordinates of price/m^2 and energy efficiency, as proposed by M. A. Green [15] (reprinted with permission). The tilted dashed lines indicate price per power (taking solar power of 1 kW/m^2). Different theoretical limits (mentioned in Section 1.3.1) are indicated. Light gray ellipsoids show the position of PV generations according to Green. The dark gray area marks recent evolution of the first generation toward low price and improved efficiency (15%–20%). (b) Share of the PV production by material—thin-film PV is mostly limited between 10% and 20%, except the a-Si boom around 1986. Data from Fraunhofer ISE, Photovoltaics report 2012.

- **First generation**: Based on a bulk material *p–n* junction is represented by the silicon PV cells made of mono- or polycrystalline wafers.
- **Second generation**: Based on thin-film technology and possibly using various materials (a-Si, CdS, CdTe, copper indium/gallium diselenide [CIGS], etc.).

- **Third generation**: High-efficiency PV cells which overcome the SQ limit by reducing losses due to nonabsorbed photons and thermalized HCs. It is using advanced concepts employing several energy levels to harvest the solar spectrum more efficiently. The main concepts include tandem cells (TCs), light-frequency converters (up- and down-conversion), multiple-exciton generation (MEG) or carrier multiplication (CM), HC exploitation through selective contacts, intermediate band (IB) cells, etc.

The first-generation cells (first produced in 1954, as mentioned above) are still representing the major part of the PV production (Fig. 1.4b). The producers of bulk Si SCs continue to decrease price per power beyond limits which were anticipated when the second generation was developed for production. This strong competition limits the market share for thin-film PV cells, which remains between 10% and 20% already for about two decades. Therefore the plot in Fig. 1.4a evolved from the original publication by M. A. Green in 2001, so the first-generation cells merged with the second generation. One can suppose that different "generations" and concepts will coexist and possibly find different (niche) application fields under the severe economical conditions.

The third-generation PV cells are still under development, mostly in the stage of different prototypes or even in the proof-of-concept stage. The only type of commercially produced "beyond SQ limit" cells are multijunction III–V semiconductor cells for use with concentrators or in space. They are defining the current record efficiency of about 43.5% a for GaInP/GaInAs/Ge triple cell under the concentration ~480 suns [16] and even 44% at ~950 suns [17].

There are three basic routes to be followed to improve solar spectrum harvesting:

- Increasing the number of energy bands which can separately absorb different parts of the spectrum (multijunction approach)
- Creating more low-potential electron–hole pairs from one absorbed photon of high energy or inversely, creating one high-potential electron–hole pair from several low-energy photons (up and down conversion)

- Capturing highly excited (hot) photoexcited carriers before their thermalization (relaxation) to the band minima (HC extraction)

Let us look at these concepts closely.

1.3.1 *The Multijunction Approach*

The idea to split the solar spectrum between several cells made from semiconductor of different band-gap width appeared shortly after the first PV device was fabricated. TC attracted much attention due to potentially high increase of efficiency. The detailed balance limit for optimized TC under maximum concentration shifts from the SQ limit to 42% (55%), 49% (63%), and 53% (68%) for two, three, and four cells, respectively (the lower values are for 1 sun illumination and values in parentheses are for maximum concentration). For the hypothetical system with infinite number of cells the limit efficiency is 68.2% and 86.8% for 1 sun and the maximum concentration, respectively [18].

There are two main types of TCs (Fig. 1.5a), (i) *stacked cells* and (ii) *separated cells*.

The stacked TCs are produced by a thin-film deposition technique one over another and interconnected in series by a tunnel contact (having only two external contacts). The short-wavelength part of the solar spectrum is absorbed by the upper cells, while the long-wavelength part is transmitted to the lower cell. The bias of both cells is summed up but the current is determined by the worst cell. Therefore the current generated in all cells must be matched together (see the chapter by Janz et al.).

The separated TCs are electronically independent and the spectrum is split between them using optical filters or dispersive elements. The disadvantage of this approach consists in a more complicated (expensive) optical, mechanical, and electronic assembly.

1.3.2 *Up- and Down-Conversion*

1.3.2.1 Wavelength conversion

These processes, as shown in Fig. 1.5b, alternate the spectrum of solar radiation by converting several low-energy photons into one

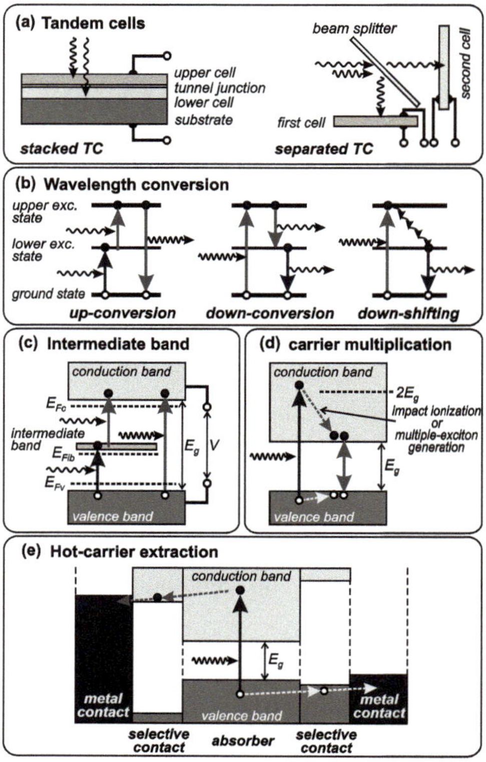

Figure 1.5 Schematic illustration of the main approaches to the third-generation PV cells. (a) Tandem cells in the stack (left) of separated (right) configuration, (b) wavelength conversion for "shaping" of the incoming solar radiation, (c) intermediate-band SC, (d) SC with carrier multiplication (called impact ionization in bulk semiconductor), and (e) hot-carrier-extraction SC with energy-selective contacts.

high-energy photon (*up-conversion*) or one high-energy photon into several low-energy photons (*down-conversion*, also called *quantum cutting* [19]).[a] This enables to reduce losses due to transmitted light or HC relaxation, respectively. The incident spectrum can also be modified by changing wavelength of one photon to longer value

[a]Up- and down-conversion are processes known from nonlinear optics. But such nonlinear phenomena cannot be exploited in photovoltaics as the typical power density (0.1–100 W/cm^2 under 1–1000 suns) is at least 2 orders of magnitude lower than thresholds for nonlinear optical phenomena.

(lower energy)—then we speak about *down-shifting* (which can be easily obtained as luminescence Stokes shift [5]). Up- or down-conversion is most often achieved using special dopants, usually rare-earth ions (lanthanides) like Sm^{3+}, Eu^{3+}, Dy^{3+}, Ho^{3+}, Er^{3+}, Tm^{3+}, or Yb^{3+}. However, their absorption is characterized by very narrow bands, therefore combination with sensitizers (like dye molecules) is advantageous [20]. The mechanism of up-conversion can be based on the excited-state absorption, energy transfer, or migration [21]. Down-conversion can be realized by a kind of cascaded photon emission.

1.3.2.2 Intermediate-band SCs

This proposed concept makes use of an electronic state (or a narrow band of states) within the host semiconductor band gap, termed "*intermediate band*." The IB enables excitation of electron into the conduction band not only directly by absorption of one photon with energy $>E_g$ but also by sequential excitation into IB and then to conduction band (absorbing two low-energy photons, Fig. 1.5c) [22]. Thus, the below band-gap photons can be used to increase the photocurrent. However, to improve power efficiency, IB must be electronically separated from the host semiconductor. It means that it has its separate (quasi-) Fermi levels and no current can be extracted directly from IB. Otherwise the photovoltage (and so power of SC), would be reduced [23]. Three main approaches to realization of IB SC have been followed: (i) *impurity levels* (e.g., due to deep levels produced by implanting In [24] or Ti [25] in Si), (ii) *quantum dots* [26], or (iii) *highly mismatched semiconductor alloys* [27]; but many problems remain to be solved. The attractive feature of IB cells is the theoretical efficiency limit significantly exceeding an ideal tandem SC, that is, 63% vs. 55% under full concentration [23].

1.3.2.3 Carrier multiplication

The excess energy of a carrier excited highly above the bottom of a respective energy band (called *HC*) can be, in principle, transformed into excitation of another electron–hole pair. This is de facto an inverse process to Auger recombination (when energy released by annihilation of an electron–hole pair is used for excitation of a remaining carrier). Such process is called *CM* or *MEG*. In bulk semiconductors this is known as *impact ionization* (II). The influence of II on the PV internal quantum efficiency in Si was demonstrated by Kolodinski et al. [28]. Recent effort by many research groups has led to the conclusion that CM can be much more efficient in quantum-confined semiconductor structures than II in bulk and therefore considerably increase power efficiency of SCs [29, 30]. (For more details see the chapter by Marri et al.)

1.3.3 *Hot-Carrier Extraction*

The energy of HCs could be exploited if they are extracted by special contacts before losing potential energy by thermalization via inelastic carrier–phonon scattering. This requires slowing down the thermalization rate (i.e., reducing electron–phonon interaction in absorber material), which is usually very fast (tens of picosecond time scale), and realization of energy-selective contacts [31]. The former may be achieved via modifying phonon dispersion in special materials and the later by using resonant tunneling barrier contacts [32, 33]. The problem of slowing down the HC relaxation is related to the so called *phonon bottleneck*—splitting of semiconductor electronic bands into discrete levels which takes place in nanostructures could make distance between electronic excited states which does not match with any of optical phonon energies. Then multiphonon relaxation takes place which is much less probable and therefore slower [34]. (For details on HC relaxation, see the chapter by Trojanek et al.)

1.4 Group IV Nanostructures

The group IV A (or 14 [35])[a] of the periodic table of elements (Fig. 1.6a) contains carbon, silicon, germanium, tin, and lead (the

[a] Different notations for groups in the periodic table of elements have been used in history. The current notation using Arabic instead of Roman numerals was proposed by the International Union of Pure and Applied Chemistry (IUPAC) in 1988 and should be preferred [35].

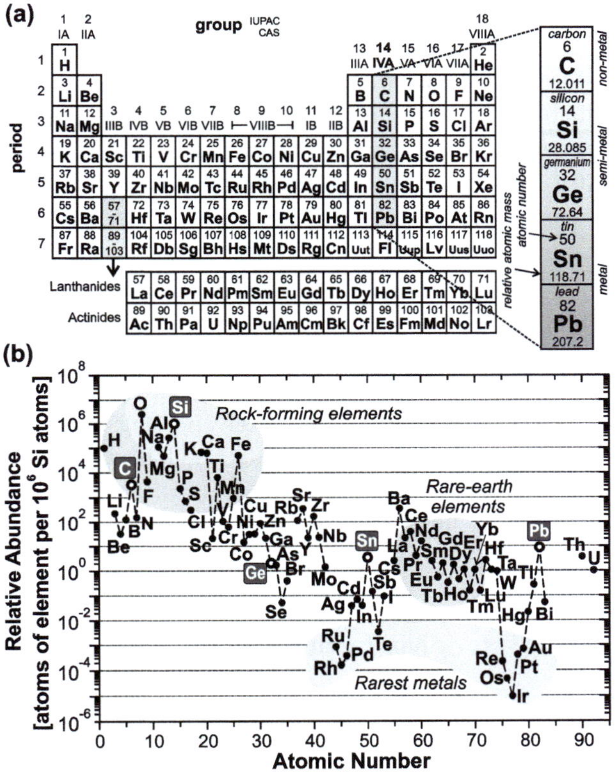

Figure 1.6 (a) The position of group IVA in the periodic table of elements. The properties of C, Si, Ge, Sn, and Pb evolve from nonmetallic (or very-large-band semiconductor—diamond) through semiconductor to metal. (b) Abundance of elements in the earth's upper continental crust relative to Si. Group IVA elements are highlighted by dark boxes. Data from Ref. [36].

latest member of this group is the element 114 flerovium—a superheavy radioactive element). The first three members (C, Si, Ge) are semiconductors with decreasing band gap (diamond has extremely wide gap of about 5.5 eV), while Sn and Pb are metals.

Great advantage for large-scale applications is that these elements are very abundant (Fig. 1.6b). Especially, Si is actually the second (after oxygen) most abundant element in the upper continental crust of the earth (with mass concentration of 28.2% [36]).

Silicon has exceptional position among man-exploited materials as it became the basis of integrated microelectronic technology. Due to the large R&D effort and continuously increasing volume of production silicon is available at exceptionally low price and perfectly controlled quality (crystallinity, purity etc.). As mentioned above, it is also the main element for production of the first- and second-generation PV cells. Therefore, it would be the most advantageous material for the third-generation PV as well. Carbon is not only the basis of organic materials and life, but also very attractive element for material science. Especially, carbon nanostructures—like fullerenes, nanotubes, graphene, or nanodiamond—provide some unique properties which promise variety of applications, including PVs (see the chapter by Svrcek and Mariotti).

While the group IV materials in their bulk form have been thoroughly studied and exploited for PV applications, new challenging perspectives arise for the nanostructural forms of these materials [37].

1.4.1 *Prospects of Nanomaterials in Photovoltaics*

What is the advantage of using nanostructured materials instead of their bulk forms in PV cells? The main reason is the *quantum confinement* (QC) *effect* [38], that is, the modification of material energy states (and consequently the physical and chemical properties) when the characteristic size of structures is reduced down to the dimension of nanometers or tens of nm. More precisely, the QC effect becomes significant if the nanostructure size is comparable or smaller than the exciton radius (about 5 nm for Si and 18 nm for Ge [39]). Under strong confinement the energy gap opens up with decreasing size of structure which enables tuning the absorption spectrum and possibly achieving better harvesting of the solar spectrum. In addition, the increased overlap of electron and hole wavefunctions in the confined structure enhances the oscillator strength of optical transitions, as it is proportional to the square of normalized overlap integral of electron and hole states (see the chapter by Guerra and Ossicini).

Considering ensembles of closely spaced nanostructures, additional effects can arise due to collective phenomena. When a photon is absorbed in a quantum dot and part of its energy is exploited to create an electron–hole pair, long-range interactions (mainly the dipole–dipole interaction) can promote energy transfer between nanocrystals, eventually a superlattice with energy minibands can arise from periodically arranged nanocrystals (see the chapter by Garrido et al.).

On the other side, nanostructure shaping can also improve optical properties of SCs through the so-called *light management* (LM) by reducing reflection losses and increasing the light path through active layers. This is extremely important as the nanostructured SCs cannot be built optically, i.e., only small fraction of incident light is absorbed during perpendicular transmission through an active layer.

1.4.2 *Light Management in Solar Cells*

The term "light management" is commonly used for various approaches to optimize light path through a PV cell. We shall mention only the two most important tasks of LM:

- *Reduce losses by reflection* on the front surface of a SC: Reflection of light from a semiconductor surface is considerably high. The simplest estimate using Fresnel equation for perpendicular reflectivity $R = (n_{sc} - 1)^2/(n_{sc} + 1)^2$, where n_{sc} is refractive index of the SC material (the ambient medium, air, is supposed to have $n = 1$), gives $R = 31\%$ for $n = 3.5$ (bulk Si). The traditional way to reduce reflection losses (proposed already by Lord Rayleigh in 1886) is fabrication of *antireflective coatings* (ARCs) by deposition of one or multiple layers of appropriate materials on the front interface of a SC. In case of a single-layer coating, ARC can be optimized only for certain wavelength λ by depositing a layer of material having refractive index $n_c(\lambda) = [n_{sc}(\lambda)]^{1/2}$ and the quarter-wave thickness $d = \lambda/(4n_{sc})$. A common material for ARC on silicon is SiN [40]. Some nanostructures (e.g., random or periodic pores, rods, cones or spheres) fabricated on the front interface of SC possess reduced "effective medium" refractive index and can act as an efficient ARC [41, 42].

- *Couple and confine light* in the active layers of a SC: In a plan-parallel SC with a perfect ARC (Fig. 1.7a) the rays entering in a cell pass the active layer only once (exiting the back interface) or twice (when reflected on the back interface). If the active material is "optically thin" (thinner than the effective light penetration length $1/\alpha$, where α is the absorption coefficient) large part of incident light is not absorbed. However, we can increase the light path through the active layer or even couple light into the guided modes by structuring one or both SC interfaces (Fig. 1.7b,c), eventually various scattering centers can be introduced directly in the active layer (Fig. 1.7d) [43, 44]. Efficient light trapping can optimize absorption and consequently the short-circuit current I_{sc} of an SC.

The case of randomly textured surface was described in 1982 by Eli Yablonovitch in his seminal paper [45] where the limit for absorption enhancement of $4n^2$ times is derived and nowadays

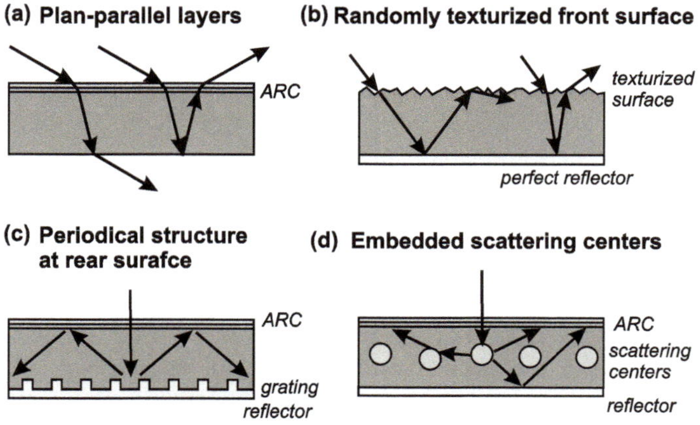

Figure 1.7 Light management in solar cells. (a) External light cannot be coupled into a perfect plan-parallel layer—in the case of a perfect ARC coating photons can pass once or twice when reflected on the rear interface. (b) Trapping of light by scattering, refraction, and reflection on a randomly structured interface. (c) Coupling of light into guided modes by diffraction on a grating. (d) Scattering of light on centers placed within the active layer.

called the *Yablonovitch limit* (or the ergodic ray limit). We have to stress that the Yablonovitch theory is based on the ray optics and for more precise description (especially for nanostructured materials) the full electromagnetic approach must be adopted [46].

Obvious disadvantage of textured surfaces is not only the increased fabrication cost but also larger interface surface area and related increase of surface recombination probability. However, the problem of surface recombination can be solved as demonstrated, for example, by the record-breaking 24% efficient bulk Si SC with the inverted-pyramid structuring of front surfaces fabricated by M.A. Green with colleagues in 1994 [47].

Recently, the design of an optimal surface morphology attracts considerable attention and research effort. The periodically textured surfaces (gratings, Fig. 1.7c) can avoid light escape from the front surface (and go beyond the Yablonovitch limit) but only for certain wavelengths, while the random structures are less efficient but effective in a wide spectral and angular range [48, 49]. Therefore, some more complex approaches were proposed, for example the supercell structure with nonperiodic grating pattern (supercell) which is periodically repeated [50]. Another example of the beyond-Yablonovitch scheme uses random texturing with a fluorescent layer which shifts frequency of light [51]. D.M. Callahan et al. showed that the increased (compare to bulk) local density of optical states in absorbing layer is the key criterion for exceeding the ray optic light trapping limit (in combination with efficient "incoupling" to many optical modes of the absorber layer) and can be achieved, for example, by plasmonic structures or photonic crystals [52].

In the case of SCs based on nanostructures (like the structures described in this book) the efficient LM must be included into the design of the cell structure in order to optimize simultaneously electrical and optical performance of SCs [53].

1.5 Conclusions

In this introductory chapter we went back to basics of PV devices. By considering the limits of solar spectrum conversion efficiency in simple bulk semiconductor SCs we demonstrated the needs for novel approaches often covered by the term "third-generation photovoltaics." Finally, we summarized advantages of the group IV nanostructures in PV concepts for better exploitation of the solar spectrum. These materials and concepts will be the subject of the following 10 chapters dealing with various silicon, germanium, and carbon nanomaterials from all relevant points of view: theory, material growth, characterization, and devices.

There are, however, other important topics which are also related to group IV nanostructures in PVs but cannot be treated within the limited space of one book. Among them we should mention *plasmonic metal nanostructures* which can substantially increase the light-material interaction [54] or the waste field of *organic SCs* which can also be combined with group IV materials in hybrid organic/inorganic SC (for recent review see Ref. [55]).

Acknowledgments

Part of this work was supported within the framework of the EU project NASCEnT (FP7-245977). One of the authors (JV) is grateful to Prof. K. Matsuda for invitation to his group in the Institute of Advanced Energy of Kyoto University, where most of this chapter was written down in a peaceful and creative atmosphere.

References

1. D.S. Ginley and D. Cahen (2012). *Fundamentals of Materials for Energy and Environmental Sustainability*, 1st Ed. (Cambridge University Press, USA).
2. Green, M. L., Espinal, L., Traversa, E., and Amis, E. J. (2012). Materials for sustainable development, *MRS Bull.*, **37**, pp. 303–308.
3. K.-P. Schröder and R.C. Smith (2008). *Mon. Not. R. Astron. Soc.*, **386**, 155–163.
4. ASTM website: http://www.astm.org/Standards/G173.htm (as for 2013-02-22).
5. I. Pelant and J. Valenta (2012). *Luminescence Spectroscopy of Semiconductors* (Oxford University Press).

6. D. M. Chapin, C. S. Fuller, and G. L. Pearson (1954). *J. Appl. Phys.*, **25**, 676.
7. A. Rose (1960). *J. Appl. Phys.*, **31**, 1640–1641.
8. P.T. Landsberg and G. Tonge (1980). Thermodynamic energy conversion efficiencies, *J. Appl. Phys.*, **51**(7), R1.
9. C. Van den Broeck (2005). Thermodynamic efficiency at maximum power, *Phys. Rev. Lett.*, **95**(19), 190602.
10. W. Shockley and H. J. Queisser (1961). *J. Appl. Phys.*, **32**, 510–519.
11. C. H. Henry (1980). *J. Appl. Phys.*, **51**, 4494.
12. J. Nelson (2003). *The Physics of Solar Cells* (Imperial College Press, London).
13. G.L. Araújo and A. Martí (1994). Absolute limiting efficiencies for photovoltaic energy conversion, *Sol. Cell. Mater. Sol. Cells*, **33**, 213.
14. P. Würfel (2009). *Physics of Solar Cells: From Basic Principles to Advanced Concepts*, 2nd Ed. (Willey-VCH, Weinheim).
15. M. A. Green (2001). Third-generation photovoltaics: ultra-high conversion efficiency at low cost, *Prog. Photovolt: Res. Appl.*, **9**, 123.
16. R. R. King et al. (2012). Solar cell generations over 40 % efficiency, *Prog. Photovolt: Res. Appl.*, **20**, 801.
17. http://www.sj-solar.com.
18. A. de Vos (1980). Detailed balance limit of the efficiency of tandem solar cells, *J. Phys. D: Appl. Phys.*, **13**, 893.
19. R.T. Wegh et al. (1999). Visible quantum cutting in $LiGdF_4{:}Eu^{3+}$ through downconversion, *Science*, **283**, 663–666.
20. X. Xie and X. Liu (2012). Upconversion goes broadband, *Nat. Mater.*, **11**, 842–843.
21. F. Wang et al. (2011). Tuning upconversion through energy migration in core-shell nanoparticles, *Nat. Mater.*, **10**, 968–973.
22. M. Wolf (1960). Limitations and possibilities for improvement of photovoltaic solar energy converters, part I., *Proc. Inst. Radio Eng.*, **48**, 1246–1263.
23. A. Luque and A. Martí (1997). Increasing the efficiency of ideal solar cells by photon induced transitions at intermediate levels, *Phys. Rev. Lett.*, **78**(26), 5014.
24. H. Kasai and H. Matsumura (1997). Study for improvement of solar cell efficiency by impurity photovoltaic effect, *Sol. En. Mater. Sol. Cells*, **48**, 93–100.
25. H. Castán et al. (2013). Experimental verification of intermediate band formation on titanium-implanted silicon, *J. Appl. Phys.*, **113**, 024104.

26. Y. Okada et al. (2011). Increase in photocurrent by optical transitions via intermediate quantum states in direct-doped InAs/GaNAs strain-compensated quantum dot solar cell, *J. Appl. Phys.*, **109**, 024301.
27. A. Luque and A. Martí (2011). Towards the intermediate band, *Nat. Photon.*, **5**, 137–138.
28. S. Kolodinski, J. H. Werner, T. Wittchen, H. J. Queisser (1993). Quantum efficiencies exceeding unity due to impact ionization in silicon solar cells, *Appl. Phys. Lett.*, **63**(17), 240–242.
29. M.C. Beard et al. (2010). Comparing multiple exciton generation in quantum dots to impact ionization in bulk semiconductors: Implications for enhancement of solar energy conversion, *Nano Lett.*, **10**, 3019–3027.
30. M.C. Beard (2011). Multiple exciton generation in semiconductor quantum dots, *J. Phys. Chem. Lett.*, **2**, 1282–1288.
31. P. Würfel (1997). Solar energy conversion with hot electrons from impact ionization, *Sol. En. Mater. Sol. Cells*, **46**, 43–52.
32. D. König et al. (2010). Hot carrier solar cells: Principles, materials and design, *Phys. E*, **42**, 2862–2866.
33. D. König et al. (2012). Lattice-matched hot carrier solar cell with energy selectivity integrated into hot carrier absorber, *Jpn. J. Appl. Phys.*, **51**, 10ND02.
34. H. Benisty, C.M. Sottomayor-Torres, and C. Weisbuch (1991). Intrincic mechanism for the poor luminescence properties of quantum-box systems, *Phys. Rev. B*, **44**(19), 10945–10948.
35. E. Fluck (1988). New notations in the periodic table, *Pure Appl. Chem.*, **60**(3), 431–436.
36. D.R. Lide (Editor-in-Chief) (1999). *CRC Handbook of Chemistry and Physics*, 40th ed. (Boca Raton, London, New York, Washighton DC), section 14, p. 14.
37. A.J. Nozik (2010). Nanoscience and nanostructures for photovoltaics and solar fuels, *Nano Lett.*, **10**, 2735.
38. J.H. Davies (1998). *The Physics of Low-Dimensional Semiconductors: An Introduction* (Cambridge University Press).
39. A.G. Cullis, L.T. Canham, and P.D.J. Calcott (1997). The structural and luminescence properties of porous silicon, *J. Appl. Phys.*, **82**(3), 909–965.
40. H. Nagel, A.G. Aberle, and R. Hezel (2009). Optimised antireflection coatings for planar silicon solar cells using remote PECVD silicon nitride and porous silicon dioxide, *Prog. Photovolt: Res. Appl.* **7**, 245.
41. M. Cao et al. (2006). Fabrication of highly antireflective silicon surfaces with superhydrophobicity, *J. Phys. Chem.*, **110**, 13072.

42. S. Jeong, S. Wang, and Y. Cui (2012). Nanoscale photon management in silicon solar cells, *J. Vac. Sci. Technol. A*, **30**(6), 060801.
43. J.R. Nagel and M.A. Scarpulla (2010). Enhanced absorption in optically thin solar cells by scattering from embedded dielectric nanoparticles, *Opt. Express*, **18**, A139.
44. K. Vynck, M. Burresi, F. Riboli, and D.S. Wiersma (2012). Photon management in two-dimensional disordered media, *Nat. Mater.*, **11**, 1017.
45. E. Yablonovitch (1982). Statistical ray optics, *J. Opt. Soc. Am.*, **72**, 899.
46. Z. Yu, A. Raman, and S. Fan (2010). Fundamental limit of light trapping in grating structures, *Opt. Express*, **18**, A366.
47. J. Zhao et al. (1995). Twenty-four percent efficient silicon solar cells with double layer antireflection coatings and reduced resistance loss, *Appl. Phys. Lett.*, **66**, 3636.
48. S. Mokkapati and K.R. Catchpole (2012). Nanophotonic light trapping in solar cells, *J. Appl. Phys.*, **112**, 101101.
49. C. Battaglia et al. (2012). Light trapping in solar cells: can periodic beat random?, *ACS Nano*, **6**(3), 2790.
50. E. R. Martins et al. (2012). Engineering gratings for light trapping in photovoltaics: the supercell concept, *Phys. Rev. B*, **86**, 041404.
51. T. Markvart (2011). Beyond the Yablonovitch limit: trapping light by frequency shift, *Appl. Phys. Lett.*, **98**, 071107.
52. D.M. Callahan, J.N. Munday, and H.A. Atwater (2012). Solar cell light trapping beyond the ray optic limit, *Nano Lett.*, **12**, 214.
53. M.G. Deceglie, V.E. Ferry, A.P. Alivisatos, and H.A. Atwater (2012). Design of nanostructured solar cells using coupled optical and electrical modeling, *Nano Lett.*, **12**, 2894.
54. P. Spinelli et al. (2012). Plasmonic light trapping in thin-film Si solar cells, *J. Opt.*, **14**, 024002.
55. T. Song, S.T. Lee, B. Sun (2012). Prospects and challenges of organic/group IV nanomaterial solar cells, *J. Mater. Chem.*, **22**, 4216.

Chapter 2

The Dielectric Function and Spectrophotometry: From Bulk to Nanostructures

Caterina Summonte

Consiglio Nazionale delle Ricerche, Istituto per la Microelettronica e i Microsistemi, via Gobetti 101, 40129 Bologna, Italy

caterina.summonte@cnr.it

2.1 Introduction

The world of group IV nanomaterials for applications in photovoltaics is vast and heterogeneous. It includes silicon and germanium quantum dots and nanowires (NWs) and their combinations; quantum wells; graphene and carbon nanotubes.

Silicon and germanium nanoparticles are candidates in tunable band-gap absorbers in third-generation multijunction solar cells [1]; NWs exhibit remarkable scattering and are used to enhance the absorption well beyond the value of the corresponding bulk materials [2]; if fabricated in quantum dimensions, they combine a tunable band gap with electrical transport properties [3]; graphene

Nanotechnology and Photovoltaic Devices: Light Energy Harvesting with Group IV Nanostructures
Edited by Jan Valenta and Salvo Mirabella

ISBN 978-981-4463-63-8 (Hardcover), 978-981-4463-64-5 (eBook)
www.panstanford.com

sheets have been proposed as transparent conducting material in organic photovoltaic devices [4]; and Ge nanoparticles are used to enhance photocurrent in dye-sensitized solar cells [5].

The different materials cover different roles, have different experimental characteristics, and are treated in different ways. For materials of optical quality, spectroscopic ellipsometry (SE) and reflectance and transmittance (R&T) spectroscopy can be used to retrieve the dielectric function (DF), detect features such as the crystallized fraction, or investigate the surface quality. SE can be applied to nondepolarizing materials, whereas R&T spectroscopy is the only option for depolarizing, highly scattering materials, such as NWs or structured surfaces.

In photovoltaics, the relevant spectral range is determined by the range of highest intensity of the solar spectrum, that is, photon energies from 0.8 to less than 4 eV, as can be seen in Fig. 2.1 (yellow pattern), which is described in Section 2.3.1. For bulk silicon, most of this region corresponds to the range of medium-low absorption (Fig. 2.1). This region is best analyzed by R&T rather than SE, which is only moderately sensitive to low absorption. In contrast, in the opaque range where $T = 0$ and R&T bears limited information, SE performs at best. SE detects the spectral shape of the DF around the critical points, and gives a fundamental insight into the material. R&T spectroscopy is the best choice to determine the absorption edge of materials, the optical gap, and its direct or indirect character. R, T are parameters of direct interest for those devices whose performance is related to absorption or transparency.

SE and R&T do not detect the sub-band-gap absorption related to defect states. Knowledge of such parameter allows us to gain an insight into material quality rather than being of interest in photovoltaic conversion, and will not be discussed in this text. This chapter is mainly focussed on R&T spectroscopy. Reviews on the state of the art of SE and polarimetry applied to the nanoscale can be found in Refs. [6, 7].

When speaking of nanoparticles applied to photovoltaics, an emerging topic is the application of metal nanoparticles for plasmon induced light trapping. This exciting topic is however out of the scope of this review.

2.2 The Dielectric Function: Why do we Need an Approximation?

The DF represents the proportional factor between the macroscopic displacement field and the externally applied electric field. This quantity always exists, irrespective of the nature of the material under examination. Its detection at the nanoscale is not straightforward. Going to the nanoscale means that if the sample size is at least of the order of the wavelength, (also if we do not insist for a practical sample size) we deal with an aggregation of several nanoparticles, which by definition must be separated by one another. This implies that a surrounding medium always exists or that we are dealing with an aggregation of phases, at least one being not spatially interconnected. Hence, we do not have a material at the nanoscale, but a composition of at least two materials, that are probed together with the same experiment. For this reason, to analyze experimental data we need a theoretical description of how the DF of the mixture is related to those of the components. This situation has been treated using quite a number of approaches, none of them appearing to be universal. An exhaustive review on existing mixing formulas can be found in Ref. [8]. In the next section, a brief summary is given.

2.2.1 *Electromagnetic Mixing Formulas*

The most widely used mixing formulas are the Maxwell Garnett [9] (MG) approximation and the Bruggeman effective medium approximation (BEMA) [10]. The MG formula reads:

$$\frac{\varepsilon_{\mathrm{eff}} - \varepsilon_{\mathrm{m}}}{\varepsilon_{\mathrm{eff}} + 2\varepsilon_{\mathrm{m}}} = f_i \left(\frac{\varepsilon_{\mathrm{i}} - \varepsilon_{\mathrm{m}}}{\varepsilon_{\mathrm{i}} + 2\varepsilon_{\mathrm{m}}} \right) \tag{2.1}$$

where ε_{m} is the complex DF of the hosting medium, ε_{i} is the DF of the inclusions, and $\varepsilon_{\mathrm{eff}}$ is the DF of the effective medium. It derives from the Clausius Mossotti (CM) or Lorenz–Lorentz equation taken in the dilute-mixture limit, which relates the dielectric constant of a medium to density and polarizability of the constituent molecules. If in the CM equation we replace the polarizability of molecules with that of spherical inclusions, approximated by dipoles, of a host material, the MG formula is obtained by assuming that (1) the effective medium is a composition of host material with dielectric constant ε_m, and spherical inclusions of a second material, with dielectric constant ε_I; (2) the inclusions occupy a limited fraction f_i of the volume in order for them to be separated one from another; (3) the inclusions are spherical, with monodisperse radius; and (4) the radius and interparticle distance are smaller than the photon wavelength λ.

It is recognized that the validity of such ideal model is limited [8, 11, 12]. For instance, if the inclusions are not spherical, the introduction of a shape-dependent depolarization factor is needed, which may not be analytical apart from simple geometries, with the additional difficulty that not always the shape is known, or it is the same for all inclusions [8].

The MG formula implicitly depends on the size of inclusions through volume fraction and polarizability of the particles, because the polarizability increases with volume. Typically, the MG formula is used to describe metal particles dispersed in a medium with slightly varying ε_m. However, it was also applied to the DF of Si or Ge nanoparticles [13–15] although in this case its applicability is limited to nanoparticles not smaller than 1 nm diameter, and to energies below the first main absorption peak of the matrix, due to modifications of the matrix properties that are not predicted by MG [11]. A recent derivation of MG that takes into account the density fluctuations of particles and presence of multiple scattering has shown a range of applicability extended to a higher density of scattering centers [16]. The dispersion in radius is explicitly considered in Refs. [17, 18], and it has been shown that if this feature is neglected, an incorrect prediction of the DF is obtained.

If the fractions are similar, it may not be straightforward to identify the host medium. If it is assumed that the host medium is actually the effective medium in which the components are embedded, the BEMA formula is obtained:

$$\sum_j f_j \frac{\varepsilon_j - \varepsilon_\mathrm{eff}}{\varepsilon_j + 2\varepsilon_\mathrm{eff}} = 0 \tag{2.2}$$

where the sum extends over the components. The factors in the quotient of Eq. 2.2) reflect the assumption of spherical inclusions [8].

This formula has the advantage of being insensitive to the exchange of components, and performs at best for similar volumes of the components. By definition, the dielectric constant resulting from BEMA is between the extremes of the components; it is in fact a sort of an "average" of the DFs of the components. This is not always the case with the MG formula. In fact, if the inclusions are lossy, that is, have the imaginary part ε'' of the DF larger than the real part ε', then MG predicts an $\varepsilon'_{\text{eff}}$ larger than that of the components. The BEMA model is recognized to supply adequate prediction of the effective DF in the case of mixtures of dielectric and semiconductor materials [19], but it fails in predicting the absorption peak due to plasma resonance of metal nanoparticles embedded in a dielectric, nonlossy medium [20].

Both MG and BEMA functions are widely used to describe the DF of mixtures of optical materials. BEMA was used also to describe the DF even below the nanoscale. In Ref. [21], the amorphous SiC_xH_y compound was considered to be decomposed into tetrahedra representing the Si and C bonding configurations (Si–Si_4, Si–Si_3C, Si–Si_3H, and so on) then optically modeled as components of an effective medium approximation (EMA) mixture. The results, applied to the statistics of the network, allowed the authors of Ref. [21] to draw conclusions on the microscopical structure of the material. The approach was also applied by different authors, in particular to the determination of the DF of nc-Si [22].

2.3 The Dielectric Function at the Nanoscale

Much work exists on the theoretical prediction of the optical properties of nanomaterials. Experiments do not always confirm the expectations. There are several reasons for this. First, theoretical predictions are almost generally based on the assumption that atoms are located at fixed crystallographic positions, which is equivalent to assume 0 K temperature [23]. Second, it is really hard to reduce the intricacy of actual systems to a model. Third, the approximations in the elaboration of experimental data may play a fundamental role. In this section, we briefly review some results on optical properties of nanomaterials, and relate them to theoretical expectations, with particular attention to the nc-Si case.

2.3.1 *Silicon Nanoparticles*

The idea behind the use of nc-Si in third-generation solar cells is that the quantum confinement deriving from the nanometric dimensions produces an increase of the effective band gap, with advantages described in detail in Chapters 1 and 11 of this book. Besides a general agreement on such band-gap opening, widely proven by photoluminescence (PL) as also described in Chapter 4 of this book, it is not clear which optical properties, and in particular which change in the absorption edge, we should expect. A compilation of literature results is reported in Fig. 2.1a. Let's first focus on monocrystalline (c-Si) [24] and microcrystalline silicon (μc-Si). In c-Si, (the gray bold line in Fig. 2.1a), the imaginary part of the DF ε'', shows sharp features, with two main critical points, CP1 and CP2, at 3.4 and 4.25 eV, related to direct interband transitions. For interconnected, fine-grained (FG) crystalline silicon (FG in the following, the generally accepted reference for μc-Si [25], cyan in the figure), ε'' shows fairly smeared CP1 and CP2 structures. CP2 is slightly red shifted, and ε'' at CP2 remains higher than at CP1. The smearing of CP1 implies increased absorption on the red side of CP1 (energies ~3 eV, circled area in Fig. 2.1a), and consequently a red shift of the absorption edge. Amorphous silicon (a-Si [31], light gray) shows a single smeared band, with remarkable absorption in the region 1.7 to 3.5 eV.

What one should expect from isolated silicon nanocrystals (nc-Si), is less clear. The finite dimensions, the structural stress, the transition region, that is, the terminations of Si bonds of edge atoms and therefore the origin of the surrounding matrix, produce a modification of the Brillouin zone diagram. Ab initio calculations of nc-Si predict a survival of the CP2 and not CP1 signal [33]. Experimentally, a variety of results is reported in the literature, as is apparent at a first glance from Fig. 2.1a.

We leave the analysis for a different occasion. Here, we only note that in some cases a reminiscence of the CP1, CP2 features is maintained; in other cases a single band is detected. This case is the

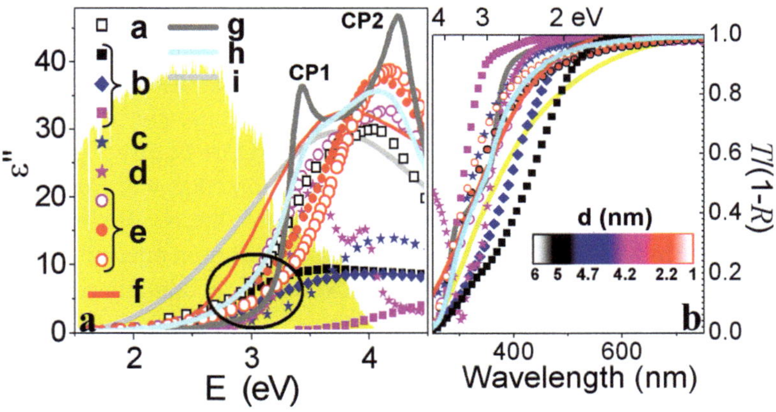

Figure 2.1 (a) Imaginary part of the dielectric constant of nc-Si. **a**: [26], **b**: [27], **c**: [28], **d**: [29], **e**: [30], **f**: [22], **g**: c-Si [24], **h**: μc-Si [25], and **i**: a-Si [31]. Other results in Refs. [13, 18, 32]. Yellow pattern: solar spectrum (a.u.). (b) Reduced transmittance $T/(1-R)$ for a hypothetical 350 nm SiO_2 film containing 10% nc-Si on fused silica, computed introducing in the BEMA model the functions as in panel (a). $T/(1-R)$ is preferred over T because it does not contain oscillating terms (see Section 2.4.3.1).

most similar to theoretical expectations. Such a single band is (very approximately) located around CP2, with remarkable differences in intensity and central energy, which may arise from differences in grain size but also misestimation of volume fractions of nc-Si, residual amorphous phases or voids, or different contribution of the matrix and of the transition region. For the surviving CP1–CP2 structure, we may suppose an effect of undetected interconnection between grains, actual grain size and size distribution, surface roughness relatively to the reduction of CP2 with respect of CP1 [19].

The broadening of the absorption structure has the effect of producing increased absorption on the red side of CP1 (~3 eV, again the circled area in Fig. 2.1a), with a red shift of the absorption edge. Or, if only CP2 survives, the absorption edge will undergo a blue shift. The impact of such differences on transmittance is illustrated in Fig. 2.1b (the details are given in the caption). Transmittance is essentially dominated by the rise of ε'' at ~3 eV. This is also the most important region for photovoltaics, because the solar spectrum is still intense (see the yellow pattern in Fig. 2.1a), before declining at higher energies. High absorption in this region is desirable, as it directly reflects in efficient absorption of solar radiation. Not too high, though, because this would reduce the absorption depth, which is detrimental because, roughly speaking, surface regions are defective and electrically poor. It is therefore of direct practical importance to learn if for nc-Si the absorption edge, with respect to c-Si or μc-Si, is red or blue shifted, a feature that is not clear from Fig. 2.1b. (We note that we are dealing with photon energies higher than the region involved in the indirect band-gap opening related to quantum confinement).

Such a picture leaves room for intensive investigation through R&T spectroscopy. Qualitative considerations on R&T spectra, and the information behind, will be given in Section 2.4.3.

2.3.2 *Germanium Nanoparticles*

The optical properties of Ge nanoparticles (np-Ge) have been theoretically investigated by several authors [3, 34–38]. In general, due to the increased contribution of the lowest-allowed highest occupied molecular orbital (HOMO)–lowest unoccupied molecular orbital (LUMO) transition, the decreasing size of np-Ge is predicted to produce a transition to a direct band gap, with band-gap opening and increase of the absorption threshold associated to an increase of the oscillator strength, and to a blue shift of the CP1, CP1 $+\Delta 1$ critical points.

Some features are experimentally confirmed, such as the band gap opening, detected on the basis of the shape or blue shift of the absorption edge [39, 40], or of PL [11, 38, 41]. A blue shift of the CPs was also detected [42–45]. However, in contrast to expectations, in some cases a decrease of the absorption cross section was observed [15, 46]. Evidence itself of quantum confinement was not always detected [15]. Some authors have shown that the blue shift of absorption is related to interface states [47]. The CP structure itself is not always evident. The role of the matrix and of trapped carriers was also pointed out to explain why the experimental band-gap opening is less pronounced than expected [34]. A review on optical results on Ge nanoparticles can be found in Ref. [15]. The interested reader will find results on ellipsometric detection of the DF of

Ge nanoparticles in Refs. [43–45], and spectrophotometric results in Refs. [39, 40, 42, 46–50]. Extensive information on quantum confinement effects in Ge nanostructures for optical applications is given in Chapter 9 of this book.

2.3.3 *Nanowires*

NWs are characterized by high aspect ratios, typically a factor 10^2 or 10^3, which implies optical anisotropy [51]; with diameter and spacing of the order of wavelength range, they do not meet the requirements needed to treat them as a homogeneous material characterized by an analytical DF.

The absorption edge of NWs is predicted to shift to a higher energy for decreasing diameter [52]. SiGe NWs are predicted to show a lower band gap than the separate materials, which is an important property in photovoltaics as it is expected to result in efficient charge separation [3]. Scattering plays a dominant role—in fact, the interest in optical properties of NWs resides in the unique properties of light trapping and very large specific surface, which enhance the effective absorption beyond the value of the corresponding bulk material. In photovoltaics, this property directly translates into enhanced carrier generation [35, 45, 53–54].

R&T spectra acquired with an integrating sphere are often evaluated as they are, and associated to the geometrical parameters of the NW array [54, 55]. In some cases, the optical response is analytically predicted [45, 56], and experimentally verified [56].

If the NWs are etched from a c-Si wafer and do not have a quantum sized diameter, the optical properties of the material do not differ from that of the bulk, and the composite material can be treated as a photonic crystal. The optical properties can be analytically obtained by solving the Maxwell equations for the ideal structure [57–59].

Arrays of disordered NWs require a different approach, as even the most accurate mathematical description implies strong simplifications. For purely disordered NWs, a semianalytical approach was obtained through the description of the scattering properties of the material associated to a partially reflecting substrate [2, 60]. Scattering was treated using the Rayleigh–Mie scattering theory

[61]. An extensive description of R&T theory and results on Si NWs can be found in Ref. [62]. Results on the application of Si NWs in photovoltaics can be found in Refs. [35, 58, 63].

2.3.4 *Graphene*

Graphene has the unique feature of optical properties being directly related to the fine structure constant [64–65]. Graphene is a nanomaterial that shows at best its peculiarities when at monoatomic thicknesses, which has stimulated the use of specific investigation techniques. Transmittance acquired through the gray scale of a photograph [64] or microellipsometry [66–68] are some examples. As predicted by the Maxwell equations [69], the DF of graphene differs from that of graphite and depends on the substrate [65, 68]. Optical properties of graphene based on SE [70] or R&T spectroscopy [71–72] have been reported. The number of graphene layers is readily detected by transmittance [64], which is also a directly relevant parameter in device applications. A review on graphene as electronic and optical material in photovoltaic devices can be found in Ref. [4].

2.4 Measurements and Elaboration

In this part of the chapter, we will focus on handling of R&T spectra. Scope of the section is to discuss some aspects of data elaboration. The known equations $\varepsilon = \varepsilon' + i\varepsilon''$; $\varepsilon' = n^2 - k^2$; $\varepsilon'' = 2nk$; $\alpha = 4\pi k/\lambda$ relating the DF to a complex refractive index $n + ik$ and absorption coefficient, α, are used.

2.4.1 *Volume Fractions of Composite Materials*

Whatever the procedure used to determine the effective DF, an assumption on the volume fraction V of the components is required. By focusing on the case of nc-Si obtained by Si precipitation from a Si-rich compound such as silicon-rich carbide, oxide, nitride, or oxynitride, as also described in Chapter 4 of this book, the expected V can be determined on the basis of the molecular weights and densities of the final compounds [73]. The formula and its derivation are reported in Fig. 2.2. The formula can be applied to any combination of the four elements, just setting to zero the concentrations of the missing elements. In presence of shrinkage after thermal treatment, the shrunk thickness must be used for consistency with the use of nominal densities.

Knowledge of nominal volume fractions allows us to reduce the number of free variables, or to check the results if fractions are left as free parameters.

Nominal volume fractions

Suppose to have the mixture $Si_wC_xO_yN_z$, where $w = 1-x-y-z$, and want to know the nominal volume fractions of SiC, SiO_2, and Si_3N_4 and residual Si under the hypothesis of complete phase separation. By setting:

Index	i	*0*	*1*	*2*	*3*
Compound		*Si*	*SiC*	*SiO_2*	*Si_3N_4*
*Element other than Si in the compound**		*Si*	*C*	*O*	*N*
Concentration of the element	E	w	x	y	z
*Si-to-other element ratio in the formula unit**	r	*0**	*1*	*0.5*	*0.75*
Atomic mass of the element	m	*28*	*12*	*16*	*14*
Nominal density of the compound (g/cm³)	ρ	*2.32*	*3.17*	*2.62*	*3.30*

* *In precipitated Si, silicon is assumed to represent the 'other element'. For compactness of formulations.*

It is obtained:

Concentration of excess silicon atoms that are supposed to precipitate into *Si*:

$$Si_{exc} = w - x - \frac{y}{2} - \frac{3z}{4} = w - \sum r_i E_i = E_0$$

Mass M_i of the compound i, obtained from the mass of Si, plus the mass of the other atom divided by r_i, times the concentration of Si atoms involved in that compound, in turn given by $E_i r_i$:

$$M_i = \left(m_0 + \frac{m_i}{r_i}\right) \cdot E_i r_i = (m_0 r_i + m_i) E_i$$

Volume fraction of the i compound, (or of Si by setting i=0), obtained by dividing by ρ_i and normalizing by the sum over the compounds:

$$V_i = \frac{\dfrac{E_i (m_0 r_i + m_i)}{\rho_i}}{\sum_i \dfrac{E_i (m_0 r_i + m_i)}{\rho_i}}$$

Overall volume of precipitated Si for a multilayer that alternates a compound barrier layer of thickness d_B^+, and a well layer of thickness d_W^+ that contains the same compound plus a volume fraction $V_{Si,W}$ of Si precipitates:

$$V_{Si} = V_{Si,W} \frac{d_W}{d_B + d_W}$$

$^+$ *Shrunk values. For consistency with the use of nominal densities.*

Figure 2.2 Nominal volume fractions.

2.4.2 *R&T Spectroscopy Experimental Setup*

R&T spectroscopy requires a conceptually simple apparatus. The main issue is the difficulty in performing accurate intensity measurements.

Possible arrangements are illustrated in Fig. 2.3. Briefly, the intensity of monochromatic light (typically 200 nm to 1 μm, or

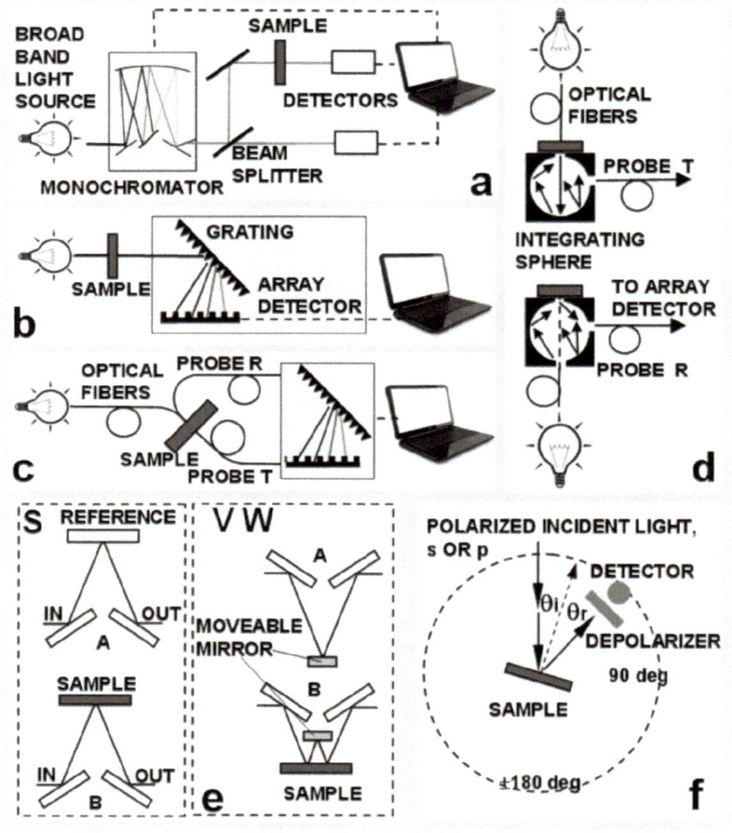

Figure 2.3 Possible arrangements for spectrophotometric measurements. (a) Double beam; (b) with parallel acquisition; (c) with optical fibers; (d) with integrating sphere and optical fibers; (e) configuration for *R*. S: standard, VW: named after the V- and W-shaped light path in reference and sample configuration; and (f) with rotating detector; elaborated from Ref. [74].

to 2.5 μm using a specific near-infrared [NIR] detector) reflected or transmitted by the sample, is detected and normalized to the incident reference intensity. If the spectral analysis occurs after the interaction with the sample, parallel acquisition is used, with elimination of the control of the λ scan. The reference is acquired either separately, or through a double beam set up. Detection of R requires either a calibrated mirror, or the so-called VW configuration (Fig. 2.3e) [75]. Recently, an instrument which permits independent control of sample and detector rotation and no need of a calibrated reflector has become commercially available [74]. If a reference sample is used, the expensive calibrated mirror with unavoidably limited shelf life [75] can be replaced by other optical materials, provided that their R spectrum is well known. A good option is single polished c-Si, which is a superior-quality, extremely well-characterized optical material, readily available at the best standard and low cost, and easily restorable to the original quality. An integrating sphere is needed for not shiny etched samples. If a fiber probe is used, R is measured at virtually normal angle of incidence ϑ. With mirror optics, ϑ is typically 8°. Fortunately, the Fresnel coefficients exhibit negligible dependence on ϑ for small ϑ [76], and $\vartheta = 0$ can always be assumed in the calculations.

2.4.3 *Elaboration of R&T Spectra*

2.4.3.1 Determination of absorption

R&T is the most suitable technique to determine absorption, and the relative procedures have received great attention in the past. In spite of absorption being simply the complement to 1 of $R + T$, retrieving the absorption coefficient α is complicated by the interference occurring between transmitted and reflected waves at the interfaces within the film, and by the difficulty in directly inverting the complete formulas. Although the use of computers has virtually eliminated this complication, it is not infrequent that R&T data are still inverted through simple but approximated analysis. In this section, a few formulations are reviewed, chosen among the most simple and with no claim of completeness.

Given R, T, the spectral specular reflectance and transmittance of an optical system, in absence of scattering and of a diffused component due to unsmooth surfaces, the absorption of the system is given by

$$A = l - R - T \tag{2.3}$$

If the optical system can be modeled by a single layer of absorption coefficient α and thickness d on a transparent, infinite substrate, then absorption only occurs in the film, according to

$$A = (1 - R)(1 - e^{-\alpha d}) \tag{2.4}$$

Equation 2.4 is often used to determine α from the equation

$$\frac{T}{1 - R} = e^{-\alpha d} \tag{2.5}$$

Equation 2.5 neglects the multiple reflections at the substrate/ambient interface of practical samples, and is rigorously valid only for nonsupported films. The approximation is valid for materials with refractive index $n \leq 2$ and leads to overestimated absorption for higher n (3.4 or higher, such as Si, Ge) [77]. If R is not experimentally available, its average value can be used, determined as

$$R_{\mathrm{AVE}} = \frac{(n-1)^2+k^2}{(n+1)^2+k^2} \tag{2.6}$$

where k is virtually not known but is small compared to n in the range of applicability of (Eq. 2.5) where $T \neq 0$. [78]. As T is an oscillating quantity, oscillations will appear as an artifact in the resulting α. For thin films, on which the fringe pattern is not evident, one may be induced to draw incorrect conclusions on absorption [79].

As R and T are opposite in phase and with the same amplitude, the quantity $T/(1 - R)$ does not oscillate [80] (see, for instance, Fig. 2.1b). On the basis of this, an interference-free form for α that takes into account the substrate/ambient interface can be derived [77]:

$$\alpha = \frac{1}{d} \cdot \ln \frac{D + \sqrt{D^2 - 4(C_9 - C_A - R_S C_B)(C_{10} - C_B - R_S C_A)}}{2(C_9 - C_A - R_S C_B)} \tag{2.7}$$

n = ambient
n_2 = substrate
n = film
d = film thickness

$$T_i = \frac{T}{1-R} \qquad T_s = \frac{4n_2n_0}{(n_2+n_0)^2}$$

$$C_c = 2(n_0^2 - n^2)(n^2 - n_2^2)$$
$$C_A = (n_0 - n)^2(n + n_2)^2$$
$$C_B = (n_0^2 + n)^2(n - n_2)^2$$
$$R_s = \frac{(n_2 - n_0)^2}{(n_2 + n_0)^2}$$
$$T_{NUM} = 16n_0n_2n^2$$

$$C_9 = (n_0 + n)^2(n + n_2)^2$$
$$C_{10} = (n_0 - n)^2(n - n_2)^2$$
$$Y = 2C_C(C_A + C_B)$$
$$Z = (C_A + C_B)^2 - T_{NUM}^2 + C_C^2$$
$$D = \frac{T_S T_{NUM}}{T_i} - R_S\frac{ZC_9 - YC_C}{C_9^2 - C_C^2}$$

$$\alpha d = \ln\left[\frac{D + \sqrt{D^2 - 4(C_9 - C_A - R_SC_B)(C_{10} - C_B - R_SC_A)}}{2(C_9 - C_A - R_SC_B)}\right]$$

Figure 2.4 Direct determination of α through Eq. 2.7: value of all coefficients [77]. It can been shown that the imaginary part of n_f can be safely neglected in all coefficients [77].

where the coefficients depend on the real part of the refractive indexes of the optical system, which are normally wavelength dependent, with moderate deviations if a fixed approximated value is used [77]. The coefficients in Eq. 2.7 are explained in Fig. 2.4.

An iterative procedure to extract α from T only, especially well performing for low dispersive media, was recently proposed [15]. Other iterative methods can be found in the literature (see, for instance, Refs. [80, 81]).

We note that α determined by R&T is reliable down to 10^3 cm^{-1}. Lower values are of the order of the noise of the measurement, mainly caused by internal scattering in the materials.

The determination of the spectral dependence of α leads to the determination of the optical gap, as illustrated in the next section.

2.4.3.2 Determination of the optical gap

The determination of the optical band gap from absorption coefficient is reported in fundamental books and will be only briefly reviewed here. The interested reader will find detailed information in Refs. [78, 82–84].

From the spectral dependence of α, the optical gap of the materials, and its direct or indirect character, can be determined. For a direct semiconductor, a dependence of the form [82]

$$\varepsilon''^2 = A_d(E - E_g) \tag{2.8}$$

is expected for the imaginary part ε'' of the DF, which is linked to α by

$$\frac{\alpha}{E} = 1.61 \cdot 10^{4} \frac{\pi}{n} \cdot \varepsilon''^{2} \tag{2.9}$$

where E is the photon energy in eV, α is given in cm^{-1}, and n is the refractive index of the material. For an indirect-gap semiconductor the following form holds [82]:

$$\varepsilon''^{1/2} = A_{\mathrm{i}}(E - E_{\mathrm{g}}) \tag{2.10}$$

The case of direct forbidden transitions, for which the exponent 2/3 holds, applies to materials extraneous to this chapter [82] and is not discussed here.

Equations 2.8–2.10 are often given by simple replacement of ε'' with α [78]. In this case, the coefficients A_{d}, A_{i}, depend on energy [82], and are considered approximately constant for $E \approx E_{\mathrm{g}}$.

For amorphous semiconductors, due to relaxation of the k selection rules, and under the hypothesis of parabolic band edges and constant momentum in the matrix element, the following dependence holds [84]:

$$(\alpha E)^{1/2} = B(E - E_{\mathrm{g}}) \tag{2.11}$$

where E_{g} is the Tauc gap of the material. The constant B contains information on the band edges and on the matrix element of the optical transitions [85]. A steeper increase of absorption, or higher B, is characteristic of a more ordered material [86]. The inverse of B was taken as a measure of the breadth of the conduction band tail [84, 87]. Equations 2.10 and 2.11 have formally the same behavior, essentially because they are both based on the consideration that k is not conserved.

Intercept with zero ordinate of the term on the left of Eqs. 2.8–2.11 plotted against energy gives the appropriate band gap for the material under examination, whereas the linearity of either of the related plots gives indication on the band edge of the material. In general, the linearity of Tauc plot declines with alloying and with crystallization, which makes the determination of E_{g} uncertain. E_{04}, defined as $\alpha(E_{04}) = 10^{4}\ \mathrm{cm}^{-1}$, is often used as a term of comparison between samples. Moderate linearity was also shown to lead to a thickness dependent determination of E_{g} [88]: in fact, for thicker samples, $T \neq 0$ in an energy range more compressed to low energies; for an upward curvature of the Tauc plot, this leads to an apparent larger E_g for thinner samples.

The hypothesis of constant momentum matrix element is subject of debate. The constant dipole matrix element was proposed to hold instead, which leads to an energy dependence of the form [89]

$$(\alpha/E)^{1/2} = C(E - E_C) \quad (2.12)$$

where E_C in this case is the Cody gap, and C contains information on the optical matrix element, the ratio of the conduction- and valence-band effective masses and the number of valence electrons per atom [89]. The Cody equation (Eq. 2.12) was shown to describe accurately the energy dependence of α in amorphous silicon, and was introduced in combination with the Lorentz oscillator to design an analytical function for amorphous semiconductors [90].

In the case of nanoparticles embedded in a transparent matrix, the plot of the left term of Eqs. 2.8–2.12 determined from the overall composite material will give information on E_g and gap character of the sole absorbing component, also in lack of knowledge of the thickness of the material, whereas the slope will depend on the relative volume fractions.

In general, when applied to nanocrystals, none of the relations (Eqs. 2.8–2.12) is linear over an extended energy range. A difficulty in identifying a linear region in the Tauc plot in nc-Si was encountered in Ref. [91]. The lack of linearity was interpreted as indicative of the presence of a mixture of two separate components [92] or, based on a dependence such as in Eqs. 2.8 and 2.10, in terms of direct and indirect band gap [93]. A double slope was also observed for 2 nm Si nanoparticles, whereas an indirect band gap for larger diameters was obtained using constant energy in the matrix element in Eq. 2.10 [94], or based on the observation of a behavior of the type of Eq. 2.10 [95], yet on a rather limited energy range. For self-organized germanium quantum dots, based on an energy dependence of type Eq. 2.8, a direct band gap was deduced [40].

2.4.3.3 Qualitative evaluation of R&T spectra

Let's consider a film with thickness d and refractive index $n + ik$ on a transparent substrate with refractive index n_2. In the low-absorption

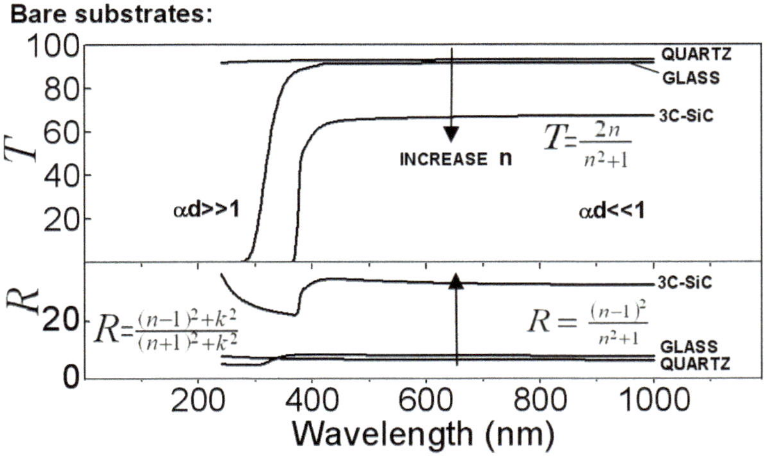

Figure 2.5 R&T spectra of bare substrates (1 mm). Above the absorption edge, R probes the front surface only. Below, R increases, due to multiple incoherent reflections at the back surface. Formulas valid in the respective ranges ($\alpha d>>1, \alpha d<<1$), in air, for normal incidence.

region (AR) ($\alpha < d^{-1}$, with $\alpha = 4\pi k/\lambda$) then the R&T spectra will exhibit an interference pattern with argument $4\pi nd/\lambda$. The pattern is intuitively evident only for d large enough that at least a few periods occur in the investigated range of λ. A wealth of information can be deduced by simply observing R&T spectra [75]. To this aim, some basic features and formulas are summarized in Figs. 2.5 and 2.6, which the reader is encouraged to analyze in detail. The figures were produced using the open access code Optical, which can be downloaded at the link in Ref. [96].

R&T of a bare substrate in its region of transparency are simply linked to n_2 (Fig. 2.5). The two curves $T(\lambda)$, $R(\lambda)$ of the bare substrate will represent the point of tangency of interference fringes if a transparent film is introduced, upper (lower) for T and lower (upper) for R if $n>n_2$ ($n<n_2$). This is illustrated in Fig. 2.6 for the case of quartz substrate and $n>n_2$. This implies that if tangency is not observed, either the film is not transparent ($k \neq 0$), or an interface layer exists. It is important to realize that, when performing a simulation (Section 2.4.4), lack of tangency cannot be simulated by varying n. Also, if tangency is observed for T, and not for R (or the opposite), this is an indication that the measurements must be checked for some anomaly. In general, if $k = 0$, then the entire R and T spectra, and not only the points of tangency, are symmetrical with respect to 50%, and oscillate opposite in phase with the same amplitude. Departure from this may be caused by rough surfaces. This occurrence is normally visible as a bright spot on the sample during the measurement. The amplitude of fringes increases for increasing contrast between n and n_2; moreover, fringes are damped by absorption. The oscillating pattern is then enveloped by simple functions of n_2, n, k (Fig. 2.6). The functions, as formalized by Swanepoel [97], are reported in Fig. 2.6, and can be used to retrieve the three parameters. This is called the envelope method. The argument of R&T oscillation is $\beta = 4\pi nd/\lambda$, that is, fringes travel toward long λ for increasing n and/or d, but it is easy to distinguish, because the increase of n also causes an increase of fringe contrast (Fig. 2.6), whereas the increase of d does not. For low d, it may occur that in the investigated range a minimum (maximum) of T is not present [argument valid if $n>n_2$ ($n<n_2$)]. In this case, multiple solutions are found, because, as said above, the maxima (minima) of T only depend on the substrate. The correct solution may be individuated by considerations of continuity of n between the high AR ($\alpha d>>1$, $T = 0$) and the low AR, where fringes are visible. In the high AR only R is available, and the determination of n, k is subordinated to assumptions on the n, k spectral form (Section 2.4.3.5), and on a possible presence of a low density surface layer (Section 2.5.1.4).

2.4.3.4 Single layer on a transparent substrate

When applied to nanoparticles, the special case of a single layer on a transparent substrate is only interesting for the determination of thickness and DF of the effective medium, from which, on the basis of mixing formulas (Section 2.2.1) and volume fractions (Section 2.4.1), the DF of nanoparticles can in principle be retrieved. In practice, the much more versatile generalized transfer matrix (GTM) method (Section 2.4.4) is preferred.

Direct retrieval of the DF from R&T is based on the formulations for a single film on a transparent substrate. The formulations contain

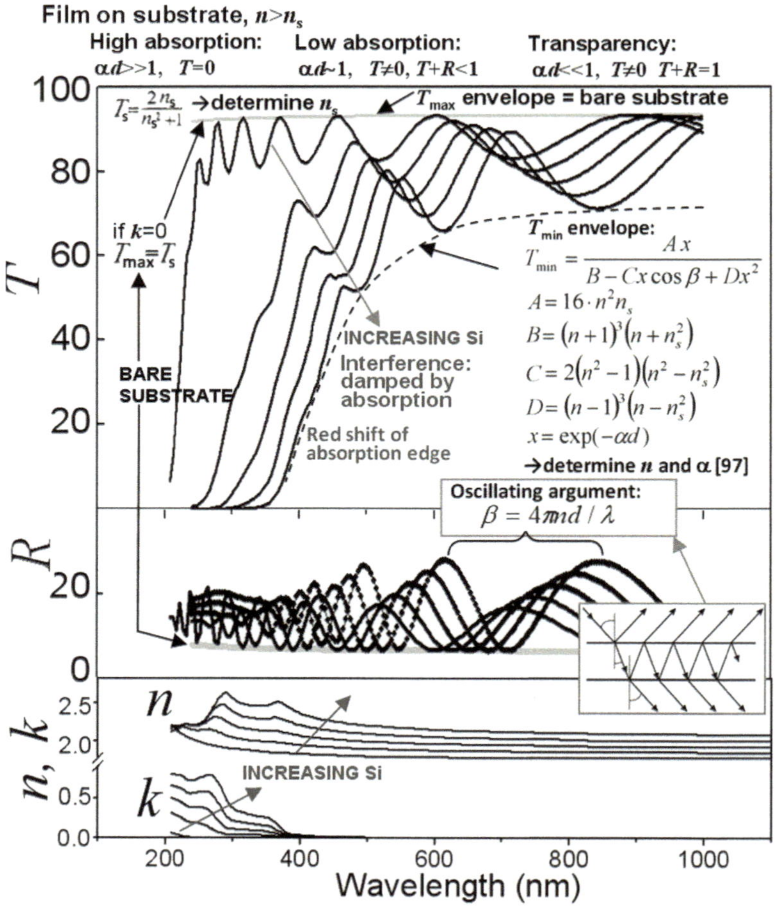

Figure 2.6 R&T spectra of 500 nm thick composite films Si_3N_4+ Si (0% to 20%) on quartz. Relevant features are indicated in the figure, for instance, the red shift of the absorption edge at increasing Si content in film composition, the T lower envelope carrying information on n and α, allowing for direct retrieval of n irrespective of thickness if $k = 0$; the upper (lower) tangency of T (R) with the curve of bare substrate for $k = 0$, providing direct evidence of $k \neq 0$.

the spectral n, k, which represent the unknowns. Typically, R&T are computed using an initial guess on n, k, which are determined, after some iterations, through a fit to the experimental data. The initial guess is an appropriate function of λ (see 2.4.3.5). A review on methods for DF retrieval can be found in Ref. [75]. One method is the envelope method described in Section 2.4.3.3. We can mention direct inversion [98] or numerical inversion based on the genetic algorithm [99]. The effect of roughness is accounted for in Ref. [100]. A numerical inversion allows us to avoid artifacts derived from forcing spectral features, whereas an analytical fitting allows relating of fitting parameters to physical quantities.

2.4.3.5 Spectral forms for the DF

Spectral forms for the DF are used not only in the just described context, but also in the more general GTM method, mainly for the elaboration of SE data. Some spectral forms are reviewed in Refs. [75, 90]. In the region of transparency, the semi-empirical equations of Cauchy or Sellmeier can be used to approximate n. To describe the overall absorption spectrum, a Lorentz oscillator (LO) is required, and indeed most spectral forms are based on it. One LO is sufficient for amorphous materials; a sum of LOs is required for the articulated features of crystalline materials. In the LO ε'', or k, never vanishes, and the low AR is poorly described. A modified LO model that forces $k = 0$ at E_g, known as the Forouhi and Bloomer (FB) model, is indeed widely used in SE [101]. FB does not describe correctly the sub-band-gap region. This is a problem for R&T, due to its sensitivity to low absorption. Improved description of the low AR is obtained by including the Tauc region in the LO (Tauc–Lorentz model [102]). Further modifications account for tail state absorption: the Ferlauto model [90], the damped LO [73] or the Jellison-Tauc-Lorentz with Gaussian Band (JTL-GB) method [103]. An open source code based on these last two models is accessible in Ref. [104].

2.4.4 *The Generalized Transfer Matrix Approach*

SE and R&T share most of the mathematics: both techniques are based on the Fresnel coefficients at interfaces, and can be treated using the scattering matrix on the basis of the Jones vectors/Müller matrix approach [76]. The scattering matrix follows the light path from source to detector, and is the product of the Müller matrices of all optical elements in the light path, including all interfaces and layers of a sample. In the generalized version (GTM), the treatment is extended to partially coherent and incoherent layers [96, 105]. In this mathematics, the expected R&T (or SE) spectra are computed on the basis of a hypothesis on the parameters of the optical system: number and thickness of layers, DF of each layer, parameters of the effective medium if any. The correct parameters are then determined by fitting the experimental spectra. The advantage of the GTM is its versatility. Surface and interface layers, multilayers, composition profiles, roughness at any interface, incoherent layers as substrates or superstrates of solar cells, can all be treated with the same mathematics. The mathematics of light scattering is also subject to continuous progress, so that R&T spectroscopy can now benefit from a considerable theoretical and numerical apparatus. In this section, only a limited number of cases is reviewed, directly related to semiconductor nanoparticles in photovoltaics.

2.5 R&T Spectroscopy Applied to Nanoparticles

In this section, some examples of applications of R&T spectroscopy to group IV nanoparticles are reported. In Section 2.5.1 the composite material is treated as a single EMA layer, whereas in Section 2.5.2 the multilayer approach is discussed.

2.5.1 *Single-Layer Approach*

2.5.1.1 Management of the unknown parameters

In a simulation, the free parameters depend on the system under examination. A case is the determination of the DF of a specific material. If the material is represented by nanoparticles embedded in a host medium, then the knowledge of the DF of the host medium and the relative volume fractions f are required. As the error on the DF of the unknown material increases for decreasing f, care must be taken in setting the constraints, especially for low f. A second case is the determination of fs of a mixture of given materials with known DF. In this case, the accuracy of the procedure depends on how well the DFs are known, or better, how well do the mixture components reproduce the quality of reference materials.

2.5.1.2 Determination of the dielectric function of nc-Si

An example of the determination of the DF for nc-Si in SiO_2 matrix obtained by precipitation of Si from a silicon-rich oxide (SRO) is reported in Ref. [106]. In this system, separately determined complete Si crystallization and Si–SiO_2 phase separation allow us to neglect a component of a-Si and of residual undissolved material. By setting the volume fractions at the nominal value estimated from composition (Section 2.4.1), the DF for nc-Si is determined (Fig. 2.7a). The measured and computed R&T spectra are reported (Fig. 2.7b). The accuracy of the result is strictly related to the accuracy with which the volume fractions are known, which stresses the importance of a precise knowledge of the composition.

2.5.1.3 Volume fractions and Si crystallized fractions

An example for the determination of the crystallized silicon fraction X_c of silicon nanoparticles by R&T is reported in Fig. 2.8 [73]. It is related to nc-Si in a SiC matrix, obtained through the fabrication of a multilayer with alternated SiC/silicon-rich SiC, followed by

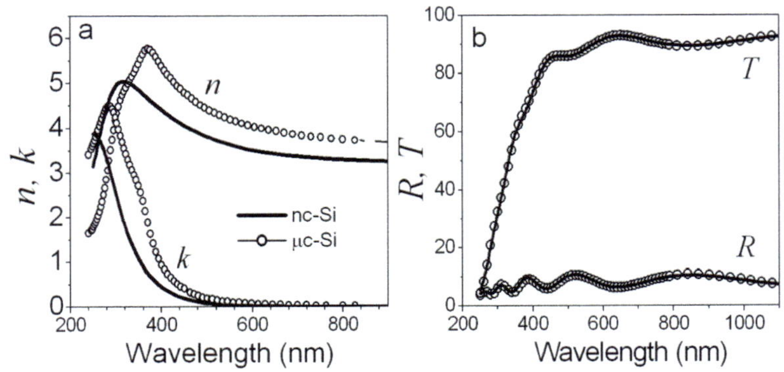

Figure 2.7 (a) n–k spectra of μc-Si (symbols) and of nc-Si (lines). (b) Measured (symbols) and simulated (lines) R&T spectra of nc-Si in SiO_2.

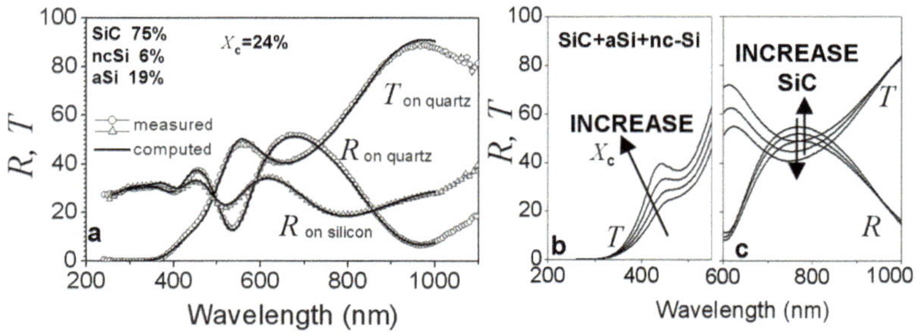

Figure 2.8 (a) Measured (symbols) and computed (lines) R&T spectra of annealed multilayers: R&T on quartz and R on c-Si. (b, c) Simulated R&T spectra for the mixture SiC + aSi + nc-Si, varying composition. The arrows illustrate the effect of the variation of the silicon-crystallized fraction X_c (b) and SiC fraction (c).

annealing (see Chapters 4 and 11 of this book). The figure reports the measured and simulated R&T (or R) spectra on fused silica (or on c-Si). The final material is composed by μc-SiC, nc-Si, and residual a-Si. In such a composite material, the Si/SiC ratio and the silicon-crystallized fraction X_c can be separately determined because they have different signature in the R&T spectra. In the low AR (700 to 1000 nm), the contrast of the interference pattern depends on the difference in refractive index n between layer and substrate, whereas the fringe period depends on nd. As for increasing Si concentration, n increases from the value of SiC (2.7 at 800 nm) to that of Si (3.7 at 800 nm), the Si/SiC ratio is univocally determined by fringe contrast, independently of X_c, because the n of a-Si and nc-Si are similar in that range. However, the absorption at the band edge (350–450 nm) is remarkably higher for a-Si. Thus the onset of T depends on X_c, which can therefore be determined. The features are illustrated in Fig. 2.8b,c where simulated spectra with varying X_c or SiC fractions are reported.

2.5.1.4 Detection of a low-density surface layer

The spectra reported in Fig. 2.8 also show the presence of a low-density, few-nanometer-thick surface layer, representing surface roughness and/or oxidation. The signature of this feature is located in the UV, where the absorption is high and the absorption depth

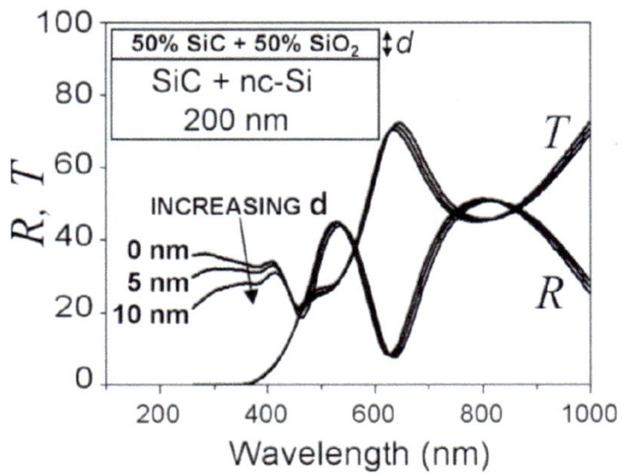

Figure 2.9 Simulated R&T spectra of a film composed by a SiC + nc-Si film, 200 nm, with increasing thickness *d* (0 to 20 nm) of a low-density surface layer, 50% SiC + 50% SiO_2.

is within few nm from the surface. The experimental *R* spectrum then probes the topmost region, and is independent of substrate. In fact, similar *R* in the ultraviolet (UV) is observed for the two substrates. The expected effect of the low-density surface layer is illustrated in Fig. 2.9. The figure shows the computed R&T spectra of an nc-Si + SiC film, covered by a 50% SiC + 50% SiO_2 layer. Note the damping in the UV for increasing thickness of the surface layer, and moderate effect at longer λ related to the slight increase of *d*. Even with no specific assumptions on sample composition, a surface layer can be detected because, if neglected, the *n* and *k* retrieved by direct inversion of the R&T spectra would not have an analytical form such as those illustrated in Section 2.4.3.5, and would not be consistent with Kramers–Kronig integration [90]. However, although the indication of the presence of a low-density surface layer is clear, the EMA mixture needed to obtain such layer is not univocally defined by R&T.

2.5.1.5 Phase separation in silicon-rich oxides

Simulation of R&T spectra was used to follow the phase separation occurring in Si-rich oxynitrides (SRONs) upon annealing, designed to fabricate nc-Si in oxynitride [107] (see Fig. 2.10).

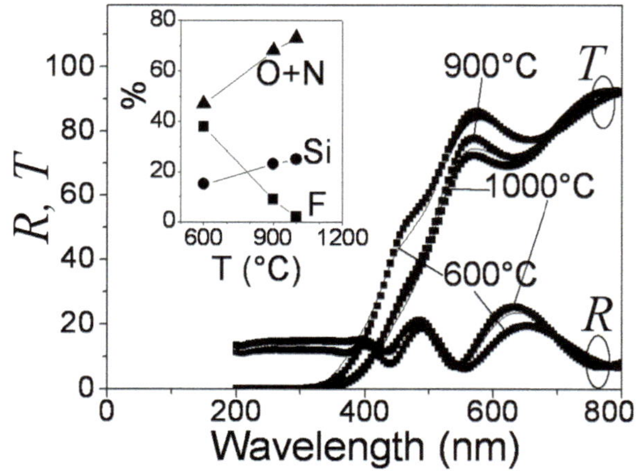

Figure 2.10 Measured (symbols) and computed (lines) R&T spectra of annealed samples. Inset: Volume fractions of Si, $SiO_2 + Si_3N_4$ and the residual F vs. annealing temperature [107]. Copyright © 2011 WILEY-VCH Verlag GmbH & Co. KGaA, Weinheim.

Phase separation represents the transition from random bonding to the random mixture configuration. The DF of the as-deposited material, modeled by a damped LO [108], is first obtained by fitting R&T by GTM. After annealing, the samples are supposed to be partially separated into Si, SiO_2 and Si_3N_4, with a residual unseparated fraction F. The R&T spectra were then simulated using a BEMA containing the DF of the as-deposited material, and the DF of the phase separated material, assumed to contain the same three compounds at a combination fixed by the nominal fractions. The only free parameter in the simulation is F. In spite of the approximations of the method, such as using the DF of the stoichiometric compounds, or inserting in the BEMA a material that is itself a mixture, the method supplies an evident trend of phase separation with annealing (inset of Fig. 2.10).

2.5.2 *Single Layers and Multilayers*

In Section 2.5.1, the superlattices used to fabricate the nc-Si were treated using single layers representing the bulk of the superlattices and simulated using the BEMA, with possible introduction of surface layers. However, if a well-defined multilayer structure is present, R&T spectroscopy is sensitive to alternating sequence of the topmost layers, and the simulation with a single layer may not be possible. This is because the light penetration at the absorption edge is of the order of sublayer spacing and varies rapidly with wavelength, so that adjacent spectral regions probe different thicknesses, and therefore different compositions, and the simulation with a unique effective medium is no longer possible. The situation is mitigated at longer wavelengths, where the absorption length is longer than the sample thickness.

The description of a multilayered structure through a single effective medium is not always an acceptable approximation. An EMA can be safely used only for thicknesses much lower than the wavelength, and also in this case the resulting effective absorption is overestimated with respect to that obtained using the superlattice structure. When applied to the retrieval of volume fractions, this peculiarity leads to an overestimation of the absorbing component, in the case given in Section 2.5.1.3 the residual amorphous fraction, with consequent underestimation of X_C. The refractive index n is also overestimated, and this leads to an underestimation of the fraction of the low-density component, SiC in the example above. The reason of the discrepancy is to be sought in the multiple reflections at the sublayer interfaces, and consequent redistribution of the energy flow, which, however, rigorously applies only in case of surfaces of optical quality. In practice, light propagation is also affected by nanometric inhomogeneities, scattering at interfaces, as well as reciprocal influence on polarizability of the nanostructured components, and the accuracy of the EMA description is only one of the approximations. A discussion of similar topics applied to SE can be found in Ref. [109]. In summary, the description of light propagation in nanostructured materials through the effective medium approach cannot be considered as consolidated, and room exists for improvements.

2.6 Conclusions

A brief review on spectrophotometry applied to semiconductor nanostructures is presented. The problem of mixing formulas has been addressed, and some peculiarities of the DF at the nanoscale have been mentioned. Some guidance has been given on an intuitive interpretation of the spectra, with the hopes of having helped to clarify that the information contained within the spectral data goes well beyond just absorption and band gap. Practical examples are also reported. Several topics were left behind or just rapidly reviewed, one for all the world of scattering measurements and the related mathematical apparatus. Yet, we hope to have stimulated the reader to further investigation of the topic.

Acknowledgments

I wish to acknowledge the invaluable support that I received from Prof. A. Desalvo while preparing this chapter. This work has been made possible thanks to funding from the European Community's Seventh Framework Programme (FP7/2007–2013) under grant agreement no. 245977.

References

1. Conibeer, G., Green, M., Cho, E. C., Konig, D., Cho, Y. H., Fangsuwannarak, T., Scardera, G., Pink, E., Huang, Y. D., Puzzer, T., Huang, S. J., Song, D. Y., Flynn, C., Park, S., Hao, X. J., and Mansfield, D. (2008). Silicon quantum dot nano-structures for tandem photovoltaic cells. *Thin Solid Films* **516**, 6748–6756.
2. Street, R. A., Wong, W. S., and Paulson, C. (2009). Analytic model for diffuse reflectivity of silicon nanowire mats. *Nano Lett.* **9**, 3494–3497.
3. Ossicini, S., Amato, M., Guerra, R., Palummo, M., and Pulci, O. (2010). Silicon and germanium nanostructures for photovoltaic applications: *ab-initio* results. *Nanoscale Res. Lett.* **5**, 1637–1649.
4. Pan, Z., Gu H., Wu M.-T., Li Y., and Chen, Y. (2012). Graphene-based functional materials for organic solar cells [invited]. *Opt. Mater. Express* **2**, 814–824.
5. Kim, C.-H., Ha E.-S., Baik, H., and Kim, K.-J. (2011). Treatment of TiO2 with COOH-functionalized germanium nanoparticles to enhance the photocurrent of dye-sensitized solar cells. *Chem. Asian* **6**, 850–855.

6. Losurdo, M., Bergmair, M., Bruno, G., Cattelan, D., Cobet, C., de Martino, A., Fleischer, K., Dohcevic-Mitrovic, Z., Esser, N., Galliet, M., Gajic, R., Hemzal, D., Hingerl, K., Humlicek, J., Ossikovski, R., Popovic, Z. V., and Saxl, O. (2009). Spectroscopic ellipsometry and polarimetry for materials and systems analysis at the nanometer scale: state-of-the-art, potential, and perspectives. *J. Nanopart. Res.* **11**, 1521–1554.
7. Losurdo, M. (2010) *Defining and Analysing the Optical Properties of Materials at the Nanoscale,* Series NanoCharM Publication, 2010, http://www.nanocharm.org/images/stories/ProjectDocs/dielectric-function-booklet.pdf.
8. Sihvola, A. H. (1999). *Electromagnetic Mixing Formulas and Application, Electromagnetic Waves Series,* Vol. 47. ISBN 0-85296-722-1 (IEE, London).
9. Maxwell Garnett, J. C. (1904). Colours in metal glasses and metal films. *Trans. R Soc.* **CCIII**, 385–420.
10. Bruggeman, D. A. G. (1935). Berechnung verschiedener physikalischer Konstanten von heterogenen Substanzen, I. Dielektrizitätskonstanten und Leitfähigkeiten der Mischkörper aus isotropen Substanzen, *Ann. der Physiks* **5**. Folge, Band 24, 636–664.
11. Weissker, H. C., Furthmüller, J., and Bechstedt, F. (2003). Validity of effective-medium theory for optical properties of embedded nanocrystallites from ab initio supercell calculations. *Phys. Rev. B* **67**, 165322.
12. Whites, K. W. (2000). Permittivity of a multiphase and isotropic lattice of spheres at low frequency. *J. Appl. Phys.* **88**, 1962–1970.
13. Zhang, R. J., Chen, Y. M., Lu W. J., Cai, Q. Y., Zheng, Y. X., and Chen, L. Y. (2009). Influence of nanocrystal size on dielectric functions of Si nanocrystals embedded in SiO2 matrix. *Appl. Phys. Lett.* **95**, 161109.
14. Pavesi, L., Dal Negro, L., Mazzoleni, C., Franzò, G., and Priolo, F. (2000). Optical gain in silicon nanocrystals. *Nature* **408**, 440–444.
15. Uhrenfeldt, C., Chevallier, J., Larsen, A. N., and Nielsen, B. B. (2011). Near-infrared-ultraviolet absorption cross sections for Ge nanocrystals in SiO2 thin films: effects of shape and layer structure. *J. Appl. Phys.* **109**, 094314.
16. Mallet, P., Guerin, C. A. & Sentenac, A. (2005). Maxwell-Garnett mixing rule in the presence of multiple scattering: derivation and accuracy. *Phys. Rev. B* **72**, 014205 (1–9).
17. Bànyai, L., Koch, S. W. (1993). Chapter 2. In *Semiconductors Quantum Dots,* 20–29 (World Scientific, Singapore).

18. Keita, A. S., Naciri, A. E., Delachat, F., Carrada, M., Ferblantier, G., and Slaoui, A. (2011). Ellipsometric demonstration of the existence of a strong correlation between size distribution and optical responses of silicon nanoclusters in a nitride matrix. *Appl. Phys. Lett.* **99**, 131903.
19. Aspnes, D. E., and Theeten, J. B. (1979). Investigation of effective-medium models of microscopic surface-roughness by spectroscopic ellipsometry. *Phys. Rev. B* **20**, 3292–3302.
20. Gittleman, J. I., and Abeles, B. (1977). Comparison of effective medium and Maxwell-Garnett predictions for dielectric-constants of granular metals. *Phys. Rev. B* **15**, 3273–3275.
21. Mui, K., and Smith, F. W. (1987). Tetrahedron model for the dielectric function of amorphous silicon-carbon alloys. *Phys. Rev. B* **35**, 8080–8088.
22. Losurdo, M., Giangregorio, M. M., Capezzuto, P., Bruno, G., Cerqueira, M. F., Alves, E., and Stepikhova, M. (2003). Dielectric function of nanocrystalline silicon with few nanometers (<3 nm) grain size. *Appl. Phys. Lett.* **82**, 2993–2995.
23. Cardona, M. (2010) Dependence of ellipsometric spectra and other optical properties on lattice vibrations: temperature and isotopic masses. In *Defining and Analysing the Optical Properties of Materials at the Nanoscale* (Losurdo, M., ed.), Series NanoCharM Publication, 2010, 62, http://www.nanocharm.org/images/stories/ProjectDocs/dielectric-function-booklet.pdf.
24. Palik, E. D. E. (1998). *Handbook of Optical Constants of Solids.* (Academic Press, San Diego).
25. Jellison, G. E., Chisholm, M. F., and Gorbatkin, S. M. (1993). Optical functions of chemical-vapor-deposited thin-film silicon determined by spectroscopic ellipsometry. *Appl. Phys. Lett.* **62**, 3348–3350.
26. Petrik, P., Fried, M., Vazsonyi, E., Basa, P., Lohner, T., Kozma, P., and Makkai, Z. (2009). Nanocrystal characterization by ellipsometry in porous silicon using model dielectric function. *J. Appl. Phys.* **105**, 024908 (1–6).
27. Alonso, M. I., Marcus, I. C., Garriga, M., Goni, A. R., Jedrzejewski, J., and Balberg, I. (2010). Evidence of quantum confinement effects on interband optical transitions in Si nanocrystals. *Phys. Rev. B* **82**, 8.
28. Moreno, J. A., Garrido, B., Pellegrino, P., Garcia, C., Arbiol, J., Morante, J. R., Marie, P., Gourbilleau, F., and Rizk, R. (2005). Size dependence of refractive index of Si nanoclusters embedded in SiO2. *J. Appl. Phys.* **98**, 013523.

29. Ding, L., Chen, T. P., Liu, Y., Ng C. Y., and Fung, S. (2005). Optical properties of silicon nanocrystals embedded in a SiO2 matrix. *Phys. Rev. B* **72**, 125419.

30. Stenger, I., Gallas, B., Siozade, L., Kao, C. C., Chenot, S., Fisson, S., Vuye, G., and Rivory, J. (2008). Evolution of the optical properties of Si nanoparticles embedded in SiO2 as function of annealing conditions. *J. Appl. Phys.* **103**, 114303.

31. Aspnes, D. E., Studna, A. A., and Kinsbron, E. (1984). Dielectric-properties of heavily doped crystalline and amorphous-silicon from 1.5 to 6.0 eV. *Phys. Rev. B* **29**, 768–779.

32. Charvet, S., Madelon, R., and Rizk, R. (2001). Structural, ellipsometry and photoluminescence spectroscopy studies of silicon nanograins embedded in a silica matrix. *Solid-State Electron.* **45**, 1505–1511.

33. Daldosso, N., Luppi, M., Ossicini, S., Degoli, E., Magri, R., Dalba, G., Fornasini, P., Grisenti, R., Rocca, F., Pavesi, L., Boninelli, S., Priolo, F., Spinella, C., and Iacona, F. (2003). Role of the interface region on the optoelectronic properties of silicon nanocrystals embedded in SiO2. *Phys. Rev. B* **68**, 085327.

34. Niquet, Y. M., Allan, G., Delerue, C., and Lannoo, M. (2000). Quantum confinement in germanium nanocrystals. *Appl. Phys. Lett.* **77**, 1182–1184.

35. Yu, L. W., O'Donnell, B., Foldyna, M., and Roca i Cabarrocas, P. (2012). Radial junction amorphous silicon solar cells on PECVD-grown silicon nanowires. *Nanotechnology* **23**, 194011.

36. Garoufalis, C. S. (2009). Optical gap and excitation energies of small Ge nanocrystals. *J. Math. Chem.* **46**, 934–941.

37. Palummo, M., Onida, G., Del Sole, R., Stella, A., Tognini, P., Cheyssac, P., and Kofman, R. (2001). Optical properties of germanium quantum dots. *Phys. Status Solidi B* **224**, 247–251.

38. Zdetsis, A. D., Garoufalis, C. S., and Koukaras, E. N. (2009). Mixed silicon-germanium nanocrystals: a detailed study of SixGe47-x:H. *J. Math. Chem.* **46**, 942–951.

39. Yang, H. Q., Wang, X. J., Shi, H. Z., Xie, S. H., Wang, F. J., Gu X. X., and Yao, X. (2002). Photoluminescence of Ge nanoparticles embedded in SiO2 glasses fabricated by a sol-gel method. *Appl. Phys. Lett.* **81**, 5144–5146.

40. Peng, C. S., Huang, Q., Cheng, W. Q., Zhou, J. M., Zhang, Y. H., Sheng, T. T., and Tung, C. H. (1998). Optical properties of Ge self-organized quantum dots in Si. *Phys. Rev. B* **57**, 8805–8808.

41. Wilcoxon, J. P., Provencio, P. P., and Samara, G. A. (2001). Synthesis and optical properties of colloidal germanium nanocrystals. *Phys. Rev. B* **64**, 035417.
42. Tognini, P., Andreani, L. C., Geddo, M., Stella, A., Cheyssac, P., and Kofman, R. (1996). Above the band gap structures in Ge nanoparticles: optical-absorption spectra. *Il Nuovo Cimento D* **18**, 865–872.
43. Gallas, B., and Rivory, J. (2003). Spectroscopic ellipsometry study of a self-organized Ge dot layer. *J. Appl. Phys.* **94**, 2248–2253.
44. Alonso, M. I., Garriga, M., Bernardi, A., Goni, A. R., Lopeandia, A. F., Garcia, G., Rodriguez-Viejo, J., and Labar, J. L. (2008). Ellipsometric measurements of quantum confinement effects on higher interband transitions of Ge nanocrystals. *Phys. Status Solidi A* **205**, 888–891.
45. Li, P. W., Kuo, D. M. T., Liao, W. M., and Tsai, M. J. (2004). Optical and electronic characteristics of germanium quantum dots formed by selective oxidation of SiGe/Si-on-insulator. *Jpn. J. Appl. Phys. 1* **43**, 7788–7792.
46. Mirabella, S., Cosentino, S., Failla, M., Miritello, M., Nicotra, G., Simone, F., Spinella, C., Franzò, G., and Terrasi, A. (2013). Light absorption enhancement in closely packed Ge quantum dots. *Appl. Phys. Lett.* **102**, 193105 (1–6).
47. Cosentino, S., Mirabella, S., Miritello, M., Nicotra, G., Lo Savio, R., Simone, F., Spinella, C., Terrasi, A. (2011). The role of the surfaces in the photon absorption in Ge nanoclusters embedded in silica. *Nanoscale Res. Lett.* **6**, 135 (1–7).
48. Dowd, A., Elliman, R. G., and Luther-Davies, B. (2001). Linear optical properties of Ge nanocrystals in silica. *Appl. Phys. Lett.* **79**, 2327–2329.
49. Zhao, X. Y., Wei, C. M., Yang, L., and Chou, M. Y. (2004). Quantum confinement and electronic properties of silicon nanowires. *Phys. Rev. Lett.* **92**, 236805 (1–4).
50. Xia, X., Meng, C., Mei-Yan, C., Xiao-Tao, Z., Sha, Z., and Lu-Min, W. (2010). Optical and structural properties of Ge-ion-implanted fused silica after annealing in different ambient conditions. *Chin. Phys. B* **19**, 018107 (1–6).
51. Qi, J. F., Belcher, A. M., and White, J. M. (2003). Spectroscopy of individual silicon nanowires. *Appl. Phys. Lett.* **82**, 2616–2618.
52. Zhao, X. Y., Wei, C. M., Yang, L., and Chou, M. Y. (2004). Quantum confinement and electronic properties of silicon nanowires. *Phys. Rev. Lett.* **92**, 236805.

53. Lin, C. X., and Povinelli, M. L. (2009). Optical absorption enhancement in silicon nanowire arrays with a large lattice constant for photovoltaic applications. *Opt. Express* **17**, 19371–19381.

54. Tsakalakos, L., Balch, J., Fronheiser, J., Shih, M. Y., LeBoeuf, S. F., Pietrzykowski, M., Codella, P. J., Korevaar, B. A., Sulima, O., Rand, J., Davuluru, A., and Rapol, U. (2007). Strong broadband optical absorption in silicon nanowire films. *J. Nanophoton.* **1**, 013552.

55. Tang, J., Shi, J. W., Zhou, L. L., and Ma Z. Q. (2011). Fabrication and Optical Properties of Silicon Nanowires Arrays by Electroless Ag-catalyzed Etching. *Nano-Micro Lett.* **3**, 129–134.

56. Cao, L. Y., White, J. S., Park, J. S., Schuller, J. A., Clemens, B. M., and Brongersma, M. L. (2009). Engineering light absorption in semiconductor nanowire devices. *Nat. Mater.* **8**, 643–647.

57. Muskens, O. L., Rivas, J. G., Algra, R. E., Bakkers, E., and Lagendijk, A. (2008). Design of light scattering in nanowire materials for photovoltaic applications. *Nano Lett.* **8**, 2638–2642.

58. Street, R. A., Qi P., Lujan, R., and Wong, W. S. (2008). Reflectivity of disordered silicon nanowires. *Appl. Phys. Lett.* **93**, 163109.

59. Zhu, J., Yu, Z. F., Burkhard, G. F., Hsu, C. M., Connor, S. T., Xu, Y. Q., Wang, Q., McGehee, M., Fan, S. H., and Cui, Y. (2009). Optical absorption enhancement in amorphous silicon nanowire and nanocone arrays. *Nano Lett.* **9**, 279–282.

60. Bao, H., Zhang, W. X., Chen, L. L., Huang, H. X., Yang, C., and Ruan, X. L. (2012). An investigation of the optical properties of disordered silicon nanowire mats. *J. Appl. Phys.* **112**, 124301.

61. Silverman, M. P. (1998). *Waves and Grains* (Princeton, NY).

62. Adachi, M. M., Khorasaninejad, M., Saini, S. S., and Karim, K. S. (2013) Optical properties of silicon nanowires. In *UV-VIS and Photoluminescence Spectroscopy for Nanomaterials Characterization* (CSSRK, ed.) (Springer-Verlag, Berlin, Heidelberg).

63. Tsakalakos, L., Balch, J., Fronheiser, J., Korevaar, B. A., Sulima, O., and Rand, J. (2008, 1–5 Sept). Silicon nanowire solar cells: device physics, fabrication, and optoelectronic properties. In *23rd European Photovoltaic Solar Energy Conference (EPVSEC)*, Valencia, 11–16.

64. Nair, R. R., Blake, P., Grigorenko, A. N., Novoselov, K. S., Booth, T. J., Stauber, T., Peres, N. M. R., and Geim, A. K. (2008). Fine structure constant defines visual transparency of graphene. *Science* **320**, 1308–1308.

65. Kravets, V. G., Grigorenko, A. N., Nair, R. R., Blake, P., Anissimova, S., Novoselov, K. S., and Geim, A. K. (2010). Spectroscopic ellipsometry of graphene and an exciton-shifted van Hove peak in absorption. *Phys. Rev. B* **81**, 155413.
66. Gaskell, P. E., Skulason, H. S., Strupinski, W., and Szkopek, T. (2010). High spatial resolution ellipsometer for characterization of epitaxial graphene. *Opt. Lett.* **35**, 3336–3338.
67. Wurstbauer, U., Roling, C., Wegscheider, W., Vaupel, M., Thiesen, P. H., and Weiss, D. (2010). Imaging ellipsometry of graphene. *Appl. Phys. Lett.* **97**, 231901.
68. Weber, J. W., Calado, V. E., and van de Sanden, M. C. M. (2010). Optical constants of graphene measured by spectroscopic ellipsometry. *Appl. Phys. Lett.* **97**, 091904.
69. Meera, V., and Setlur, G. S. (2010). Ellipsometry of graphene on a substrate. *J. Appl. Phys.* **107**, 033525.
70. Skulason, H. S., Gaskell, P. E., and Szkopek, T. (2010). Optical reflection and transmission properties of exfoliated graphite from a graphene monolayer to several hundred graphene layers. *Nanotechnology* **21**, 295709 (1–8).
71. Bruna, M., and Borini, S. (2009). Optical constants of graphene layers in the visible range. *Appl. Phys. Lett.* **94**, 031901.
72. Ni, Z. H., Wang, H. M., Kasim, J., Fan, H. M., Yu T., Wu Y. H., Feng, Y. P., and Shen, Z. X. (2007). Graphene thickness determination using reflection and contrast spectroscopy. *Nano Lett.* **7**, 2758–2763.
73. Summonte, C., Canino, M., Allegrezza, M., Bellettato, M., Desalvo, A., Shukla, R., Jain, I. P., Crupi, I., Milita, S., Ortolani, L., Lopez-Conesa, L., Estrade, S., Peiro, F., and Garrido, B. (2013). Boron doping of silicon rich carbides: electrical properties. *Mater. Sci. Eng. B* **178**, 551–558.
74. Tikhonravov, A. V., Amotchkina, T. V., Trubetskov, M. K., Francis, R. J., Janicki, V., Sancho-Parramon, J., Zorc, H., and Pervak, V. (2012). Optical characterization and reverse engineering based on multiangle spectroscopy. *Appl. Opt.* **51**, 245–254, www.agilent.com/chem/cary7000UMS
75. Poelman, D., and Smet, P. F. (2003). Methods for the determination of the optical constants of thin films from single transmission measurements: a critical review. *J. Phys. D Appl. Phys.* **36**, 1850–1857.
76. Azzam, R. M. A., and Bashara, N. M. (1987). *Ellipsometry and Polarized Light* (Elsevier, North Holland, Amsterdam).

77. Summonte, C. (1993). An invertible equation for the absorption coefficient from reflectance and transmittance. *Mater. Res. Soc. Symp. Proc.* **297**, 395–400, www.bo.imm.cnr.it/users/summonte/LaFormula.
78. Pankove, J. I. (1971). *Optical Processes in Semiconductors* (General, Toronto).
79. Elliman, R.G., Lederer, M.J., and Luther-Davies, B. (2002). Optical absorption measurements of silica containing Si nanocrystals produced by ion implantation and thermal annealing. *Appl. Phys. Lett.* **80**, 1325–1327.
80. Hishikawa, Y., Nakamura, N., Tsuda, S., Nakano, S., Kishi, Y., and Kuwano, Y. (1991). Interference-free determination of the optical-absorption coefficient and the optical gap of amorphous-silicon thin-films. *Jpn. J. Appl. Phys. 1* **30**, 1008–1014.
81. Demichelis, F., Kaniadakis, G., Tagliaferro, A., and Tresso, E. (1987). New approach to optical analysis of absorbing thin solid films. *Appl. Opt.* **26**, 1737–1740.
82. Yu, P. Y., and Cardona, M. (1996) *Fundamentals of Semiconductors* (Springer-Verlag, Berlin).
83. Adachi, S. (1999) Chapter 4. In *Optical Properties of Crystalline and Amorphous Semiconductors* (Kluwer Academic, Norwell).
84. Mott, N. F., and Davis, E. A. (1970) *Electronic Processes in Non-Crystalline Materials* (Clarendon Press, Oxford).
85. Tauc, J., Grigorovici, R., and Vancu, A. (1966). Optical properties and electronic structure of amorphous germanium. *Phys. Status Solidi* **15**, 627.
86. Chu, V., Conde, J. P., Jarego, J., Brogueira, P., Rodriguez, J., Barradas, N., and Soares, J. C. (1995). Transport and photoluminescence of hydrogenated amorphous silicon-carbon alloys. *J. Appl. Phys.* **78**, 3164–3173.
87. Conde, J. P., Chu, V., da Silva, M. F., Kling, A., Dai, Z., Soares, J. C., Arekat, S., Fedorov, A., Berberan-Santos, M. N., Giorgis, F., and Pirri, C. F. (1999). Optoelectronic and structural properties of amorphous silicon-carbon alloys deposited by low-power electron-cyclotron resonance plasma-enhanced chemical-vapor deposition. *J. Appl. Phys.* **85**, 3327–3338.
88. Mok, T. M., and O'Leary, S. K. (2007). The dependence of the Tauc and Cody optical gaps associated with hydrogenated amorphous silicon on the film thickness: αl experimental limitations and the impact of curvature in the Tauc and Cody plots. *J. Appl. Phys.* **102**, 113525.

89. Cody, G. D., Brooks, B. G., and Abeles, B. (1982). Optical-absorption above the optical gap of amorphous-silicon hydride. *Sol. En. Mater.* **8**, 231–240.
90. Ferlauto, A. S., Ferreira, G. M., Pearce, J. M., Wronski, C. R., Collins, R. W., Deng, X. M., and Ganguly, G. (2002). Analytical model for the optical functions of amorphous semiconductors from the near-infrared to ultraviolet: applications in thin film photovoltaics. *J. Appl. Phys.* **92**, 2424–2436.
91. Hao, X., Perez-Wurfl, I., Cho, E., Gentle, A., Shen, Y., Park, S., Conibeer, G., and Green, M. A. (2008). Study on quantum size effect in silicon quantum dot superlattices by structural and optical properties, *European Photovoltaic Solar Energy Conference*, Valencia Sept. 1–5, 186.
92. Perez-Wurfl, I., Ma L., Lin, D., Hao, X., Green, M. A., and Conibeer, G. (2012). Silicon nanocrystals in an oxide matrix for thin film solar cells with 492 mV open circuit voltage. *Sol. En. Mater. Sol. Cells* **100**, 65–68.
93. Hao, X. J., Cho, E. C., Flynn, C., Shen, Y. S., Conibeer, G., and Green, M. A. (2008). Effects of boron doping on the structural and optical properties of silicon nanocrystals in a silicon dioxide matrix. *Nanotechnology* **19**, 424019 (1–8).
94. Gardelis, S., Manousiadis, P., and Nassiopoulou, A. G. (2011). Lateral electrical transport, optical properties and photocurrent measurements in two-dimensional arrays of silicon nanocrystals embedded in SiO2. *Nanoscale Res. Lett.* **6**, 227.
95. Meier, C., Gondorf, A., Luttjohann, S., Lorke, A., and Wiggers, H. (2007). Silicon nanoparticles: Absorption, emission, and the nature of the electronic bandgap. *J. Appl. Phys.* **101**, 103112.
96. Centurioni, E. (2005). Generalized matrix method for calculation of internal light energy flux in mixed coherent and incoherent multi-layers. *Appl. Opt.* **44**, 7532–7539, http://www.bo.imm.cnr.it/users/centurioni/optical.html.
97. Swanepoel, R. (1983). Determination of the thickness and optical-constants of amorphous-silicon. *J. Phys. E: Sci. Instr.* **16**, 1214–1222.
98. Barybin, A., and Shapovalov, V. (2010). Substrate effect on the optical reflectance and transmittance of thin-film structures. *Int. J. Opt.* **2010**, 137572 (1–18).
99. Gungor, T., and Saka, B. (2004). Calculation of the optical constants of a thin layer upon a transparent substrate from the reflection spectrum using a genetic algorithm. *Thin Solid Films* **467**, 319–325.

100. Nowak, M. (1995). Determination of optical-constants and average thickness of inhomogeneous-rough thin-films using spectral dependence of optical transmittance. *Thin Solid Films* **254**, 200–210.

101. Forouhi, A. R., and Bloomer, I. (1988). Optical-properties of crystalline semiconductors and dielectrics. *Phys. Rev. B* **38**, 1865–1874.

102. Jellison, G. E., and Modine, F. A. (1996). Parameterization of the optical functions of amorphous materials in the interband region. *Appl. Phys. Lett.* **69**, 371–373.

103. Allegrezza, M., Gaspari, F., Canino, M., Bellettato, M., Desalvo, A., and Summonte, C., (2014). Tail absorption in the determination of optical constants of silicon rich carbides, *Thin Solid Films* **556**, 105–111; ibid. **564**, 426.

104. http://www.bo.imm.cnr.it/users/allegrezza/minuit.

105. Katsidis, C. C., and Siapkas, D. I. (2002). General transfer-matrix method for optical multilayer systems with coherent, partially coherent, and incoherent interference. *Appl. Opt.* **41**, 3978–3987.

106. Allegrezza, M., Löper, P., Hiller, D., Summonte, C. (2013). Optical function and absorption edge of silicon nanocrystals, E-MRS, Strasbourg, symp. D.

107. Summonte, C., Centurioni, E., Canino, M., Allegrezza, M., Desalvo, A., Terrasi, A., Mirabella, S., Di Marco, S., Di Stefano, M. A., Miritello, M., Lo Savio, R., Simone, F., and Agosta, R. (2011) Optical properties of silicon rich oxides. *Phys. Status Solidi C* **8**, 996–1001.

108. Summonte, C., Allegrezza, M., Canino, M., Bellettato, M., and Desalvo, A. (2013). Analytical expression for the imaginary part of the dielectric constant of microcrystalline silicon. *Res. Appl. Mater.* 6–11.

109. Zeppenfeld, P. (2010) On the applicability of the effective medium approximation in ellipsometry. In *Defining and Analysing the Optical Properties of Materials at the Nanoscale* (Losurdo, M., ed.), Series NanoCharM Publication, 2010, 35, http://www.nanocharm.org/images/stories/ProjectDocs/dielectric-function-booklet.pdf.

Chapter 3

Ab initio Calculations of the Electronic and Optical Properties of Silicon Quantum Dots Embedded in Different Matrices

Roberto Guerra and Stefano Ossicini

Dipartimento di Scienze e Metodi dell'Ingegneria and Centro Interdipartimentale En&Tech, Università di Modena e Reggio Emilia, via Amendola 2 Pad. Morselli, I-42122 Reggio Emilia, Italy

guerra@sissa.it

3.1 Introduction

The use of Si in photonic applications is limited by the indirect gap of the Si band structure: radiative interband transitions from the conduction-band minimum (Δ-point) to the top of the valence band (Γ-point) require electron–phonon coupling in order to satisfy the momentum conservation rule. Such coupling is quite weak, and consequently the phonon-assisted emission of a photon results a very unfavorable process with respect to the direct no-phonon Γ-Γ radiative transitions.

Nanotechnology and Photovoltaic Devices: Light Energy Harvesting with Group IV Nanostructures
Edited by Jan Valenta and Salvo Mirabella

ISBN 978-981-4463-63-8 (Hardcover), 978-981-4463-64-5 (eBook)
www.panstanford.com

Some works in the early 1990s suggested that the problems related to the indirect band gap of bulk Si might be overcome in highly confined systems, like porous silicon (PS) [6, 13, 14] or Si quantum dots (Si QDs) [99], in which the exciton is constrained in a narrow region of space while the momentum distribution spreads due to the Heisenberg uncertainty relation. In this case the momentum conservation law is not violated, allowing the Δ-Γ radiative transitions even in the absence of phonons. More recent works demonstrated the possibility to achieve efficient photoluminescence (PL) and optical gain from Si QDs [82].

Theoretically, the optical emission has been attributed to transitions between states localized inside the QD [as a consequence of the so-called quantum confinement (QC) effect] [24, 26, 52, 77, 98], or between defect states [3, 27, 55, 58, 63, 74]. While there is still some debate on which of the above mechanisms primarily determines the emission energy, some recent works have proposed that a concomitance of both mechanisms is always present, favoring one or the other depending on the structural conditions [1, 30, 38, 39, 41, 49, 68, 71, 90, 110, 115]. In this picture, it was suggested that for QD diameters above a certain threshold (of about 2 nm) the emission peak should follow the QC model, while interface states would assume a crucial role only for small-sized QDs [50].

Embedding Si QDs in wide band-gap insulators is one way to obtain a strong QC. Si QDs embedded in a silica matrix have been obtained by several techniques as ion implantation [7, 55], chemical vapor deposition [35, 38, 51, 90], laser pyrolysis [23, 58], electron beam lithography [97], sputtering [2, 3], and some others. Experimentally, several factors contribute to make the interpretation of measurements on these systems a difficult task. First of all, independently on the fabrication technique, in experimental samples no two QDs are the same. For instance, samples show a strong dispersion in the QD size that is difficult to be determined. In this case it is possible that the observed quantity does not correspond exactly to the mean size but instead to the most responsive QDs [19]. Moreover, QDs synthesized by different techniques often show different properties in size, shape and in the interface structure [33, 76, 84]. Finally, in solid QD arrays some collective effects caused by electron, photon, and phonon transfer between the QDs can strongly influence the electron dynamics in comparison with the case of isolated QDs [55]. In practice, all the conditions remarked above lead to measurements of collective quantities, making the identification of the most active configurations at the experimental level a nontrivial task.

Previous works already highlighted the dramatic sensitivity of the optoelectronic properties to the interface configuration, especially for very small QDs (d $\lesssim$ 1 nm), where a large proportion of the atoms is localized at the interface. For these sizes, QD conditions such as passivation, symmetry, and strain considerably concur for the determination of the final optoelectronic response, producing sensible deviations from the QC model [41]. Moreover, many PL experiments demonstrated that only a very small fraction of the QDs in the samples contributes to the observed PL, enforcing the idea that precise structural conditions are required in order to achieve high absorption and emission rates. Finally, recent works reported especially high optical yields for smaller QDs [42, 75], upgrading their role in the observed response of real samples.

It is thus clear the importance of understanding the factors that, at these sizes, contribute to enhance (or reduce) the global optical response.

Besides the intense experimental work, devoted to the improvement of the nanostructures growth and characterization techniques and to the realization of nanodevices, an increasing number of theoretical works, based on empirical and on ab initio approaches, is now available in the literature [21, 80, 93]. The importance of the theoretical efforts lies not only in the interpretation of experimental results, but also in the possibility of predicting structural, electronic, optical, and transport properties aimed at the realization of more efficient devices.

From the theoretical side, the possibility of atomically manipulating the structures and of associating the selected configuration to the calculated response allows in principle to elucidate some of the fundamental aspects related to the physics of the QDs. Important progress in the theoretical description of the electronic properties of semiconductor nanostructures have been reported in the last decade, also thanks to the development of ab initio techniques like the density functional theory (DFT) [31], but an exhaustive understanding is still lacking. This is due, on one side to the not obvious transferability of the empirical parameters to low dimensional systems and, on the other side, to the deficiency of the DFT approach in the correct evaluation of the excitation energies [79]. In fact, due to their reduced dimensionality, the inclusion of many-body and excited-state calculations is mandatory for a proper validation of the results. In particular, the quasiparticle structure is a key for the calculation of the electronic gap and to the understanding of charge transport, while the inclusion of excitonic effects is necessary for the description of the optical properties.

However, it must be taken into account that the full ab initio approach limits nowadays the systems size to a few thousands of atoms in the case of DFT-based methods. In addition, the calculation of realistic optical absorption or emission spectra, involving excited states, requires refined treatments that dramatically increase the computational effort, further reducing the maximum manageable system size [81].

3.2 Structures

The possibility of producing structures representing real samples is a prerequisite of every theoretical work. Different methods have been developed to reproduce particular features (presence of defects, bond lengths and angles, etc.), and their selection is based on the specific needs of the researcher.

One important employment of theoretical models is the possibility of distinguishing between the properties that depend only on the QD from those that are instead influenced by the presence of the matrix. Also, the comparison of the results relative to different passivating species (e.g., H or OH) could give some insight on the role played by the interface region. Moreover, the fine control over the amorphization of the system can reveal the role of disorder in real samples. In the next, these and other aspects are discussed on the basis of theoretical results. For the sake of simplicity we will consider only SiC ($E_{gap}\simeq$2.4 eV), Si_3N_4 ($E_{gap}\simeq$5.0 eV), or SiO_2 ($E_{gap}\simeq$9.0 eV) as embedding materials.

All the results reported in the next have been obtained in the framework of DFT using norm-conserving pseudopotentials within the local-density approximation (LDA) [31]. The absorption spectra, represented by the imaginary part of the dielectric function, are then evaluated within the random-phase approximation (RPA), with and without the inclusion of local-field (LF) and many-body effects [79].

3.2.1 *Embedded Silicon Quantum Dots*

In theoretical modelling, the simplest way to obtain embedded structures from an hosting matrix is by removing (e.g., for SiO_2) or by replacing with Si (e.g., for SiC) the non-Si atoms included in a cutoff sphere, whose radius determines the size of the QD. By centering the cutoff sphere on one Si or in an interstitial position it is possible to obtain structures with different symmetries. This procedure can be applied only to matrices with T_d local symmetry. In other case (e.g., Si_3N_4) one must employ more complicated methods, often resuting in defected structures. Instead, the above procedure allows to completely avoid defects (dangling bonds) that may severely complicate the DFT convergency and the interpretation of the results (one usually wants to treat defects separately and in the most possible controllable way).

To model QD of increasing size, the hosting matrix must be enlarged so that the separation between QDs guarantees a correct description of the stress localized around the QD [25, 71, 72] and to avoid the overlapping of states belonging to the QD, due to the application of periodic boundary conditions [20]. The proper separation between the QD replica depends on the insulating capability of the hosting material, generally correlated to their gap, and can vary from about 1 nm for SiO_2 up to several nm for SiC.

The optimized structures are achieved by relaxing the total volume of the cell (see Fig. 3.1). In all cases, after the relaxation the embedding matrix is strongly distorted near the QD, and reduces progressively its stress far away from the interface [106]. The strained interface is due to the difference between the lattice constant of embedding and embedded materials [64, 111].

For small SiO_2-embedded QDs produced by the above method, the Si–Si distance results in about 2.43 Å, larger with respect to the

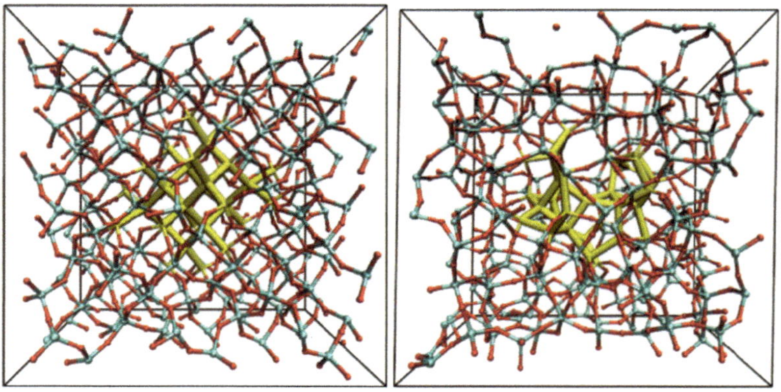

Figure 3.1 Si_{32} QD generated in a (left) betacristobalite SiO_2-3x3x3 supercell and (right) 192-atom SiO_2 glass (relaxed structures). Red spheres represent the O atoms, cyan spheres represent the Si of the matrix, and the yellow thick sticks represent the Si–Si bonds of the QD.

bulk value of 2.35 Å, in fair agreement with the outcomes of Yilmaz et al. [111].

Since real samples are always characterized by a certain amount of amorphization, in particular for QDs of small diameter [39, 89, 104], the complementary cases of crystalline and amorphous QDs should be considered in order to understand the effect of the amorphization.

With the above method it is possible to generate amorphous QDs by making use of amorphous hosting matrices. The glass models (a-SiO_2, a-Si_3N_4, a-SiC) can be generated using molecular dynamics (MD) simulations of quenching from a melt. In well prepared samples, the long-range order gets clearly broken, while the short-range order (determining the optical properties) remains substantially unvaried with respect to the crystalline case. Moreover, we have found that for a-SiO_2-embedded QDs the number of SiO–Si bonds (bridge bonds)[a] increases with the dimension of the QD, in nice agreement with structures obtained by different methods [48, 54].

[a]Such type of bond forms after the exposition of uncompletely passivated samples in air, and constitutes the most energetically favored (stable) bond type at the QD/SiO_2 interface.

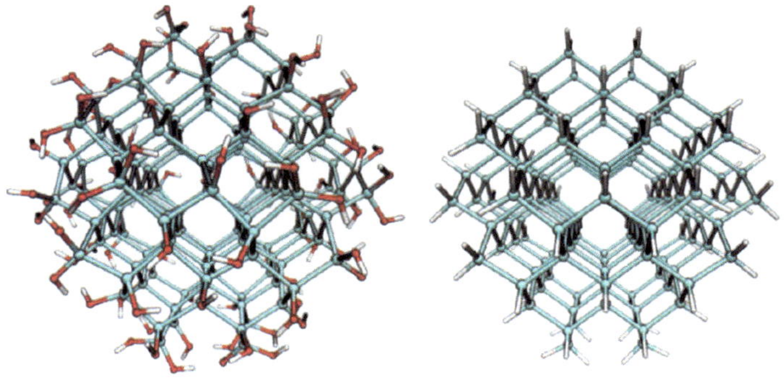

Figure 3.2 The freestanding Si_{147} passivated by (left) OH or (right) H groups. Dark spheres represent the O atoms, gray spheres represent the Si of the matrix, and white sticks represent bonds with H atoms.

3.2.2 *Freestanding Quantum Dots*

The embedded QDs results in a strained interface due to the mismatch between the lattice constants of Si and the embedding matrix. The QDs present Si–Si bonds that are strained with respect to the Si bulk case and such strain is removed when the QD is relaxed in vacuum. Therefore, we distinguish between strained QDs (i.e., relaxed in the embedding matrix), and unstrained QDs (relaxed in vacuum).[a] As reference for the discussion we will consider in the following the freestanding counterparts of the SiO_2-embedded QDs. The hydroxidized-strained QDs (s-Si_{xx}-OH) can be obtained by extracting the QDs together with the first interface oxygens from the relaxed QD–silica complexes, and then passivating the surface with hydrogen atoms (Fig. 3.2, left QD). It is interesting to consider also the case of hydrogenated-strained QDs (s-Si_{xx}-H), obtained by replacing the OH groups with hydrogens (Fig. 3.2, right QD). In the last two cases, to preserve the original strain, only the hydrogen atoms are relaxed.

---

[a] Actually, also the QDs relaxed in vacuum are known to present some strain (of much smaller proportions than that induced by the surrounding matrix) with respect to the ideal (bulk) configuration [109]. Nevertheless, in the following we always refer to the strain as the difference between ground-state configurations, that is, fully relaxed conditions of embedded or freestanding QDs.

The equivalent hydroxidized-relaxed (r-Si_{xx}-OH) and hydrogenated-relaxed (r-Si_{xx}-H) QDs can be obtained by a full relaxation in vacuum of the systems. In this case the strain is totally removed and the Si–Si bond lengths in the QD core match the bulk value.

3.3 Results

It is known that silicon forms a type I band offset when interfaced to wide-band-gap materials like SiO_2, Si_3N_4, or SiC [18]). In the case of nanostructured silicon, the conduction and valence band offsets are reduced proportionally to the confinement energy (see Fig. 3.3).

From a theoretical point of view, the simplest model for the QC is provided by the particle-in-a-box scheme, in which the box size is given by the QD diameter and the potential barrier represents the host insulating matrix. When QC effect dominates over other quantum phenomena the band-gap value can be described by an inverse power law, $E_G(R) = E_0 + A/R^{\alpha}$, where E_0 is the bulk Si band gap, R is the radius of the QD, A is a positive constant and $\alpha \leq 2$

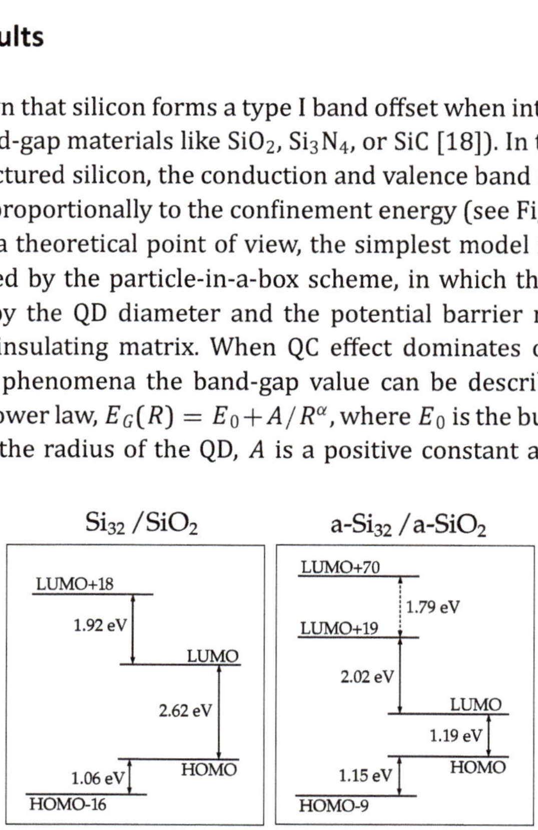

Figure 3.3 Band offset of (left) crystalline and (right) amorphous Si_{32} QD embedded in SiO_2. The values are derived from the localization of the orbitals on the QD (internal levels) or on the SiO_2 (external levels). For the amorphous case the states from LUMO+19 to LUMO+70 extend progressively from the QD to the matrix. Note that the reported values suffer the limitations of DFT–LDA and therefore cannot be directly compared to experimental measurements. Adapted from [43]. Copyright © 2010 WILEY-VCH Verlag GmbH & Co. KGaA, Weinheim.

[5].[a] This model, however, is often too simple to describe correctly systems in which the interfacial effects dominate: since in the QC picture the opening of the gap depends on the confining potential, one could expect a proportionality of the confinement energy with the gap of the hosting material, but this is not the case. Instead, in small QDs ($d \lesssim 3$ nm) the interface governs the electronic structure, with the band gap depending mainly on the polar nature of the Si bonds at the interface [41, 62]. In particular, for a given QD size, the gap of the embedded system results larger for less insulating hosting materials (i.e., with smaller gaps) [18].

For this reason, the electronic transport is particularly unfavored for Si/SiO_2 QDs, because the electrons that move across the supercell need to pass a high barrier, resulting in slow drift velocities, which significantly decrease for larger oxide quantities [91]. Such very high conduction threshold limits the utilization of pure Si/SiO_2 systems on applications that rely on current pass-through. Despite this limitations, some studies demonstrated electroluminescence (EL) emission of light-emitting devices, such as Si/SiO_2 QDs [56, 67], Si QDs embedded in Si nitride [102], or hydrogenated amorphous Si [105].

Other works have instead explored the employment of reduced-gap matrices (e.g., Si_3N_4 or SiC), in the view of a double advantage from large confinement energies and small conduction barriers [45, 69].

3.3.1 *Amorphization Effects*

The QD, when formed in the glass, completely loses memory of the starting tetrahedral symmetry. Despite the fact that dramatic structural changes occurs with respect to the crystalline case, similar considerations can be done concerning the optoelectronic properties. An important point is that, when the glass is well produced [113], it presents a band-gap value that is very similar (sometimes even greater) to that of the crystalline phase. On the other side, the amorphization process strongly reduces the energy gap of the freestanding QDs, and as a consequence that of the

[a] The upper limit holds for an infinite potential barrier. For finite barriers is $\alpha < 2$.

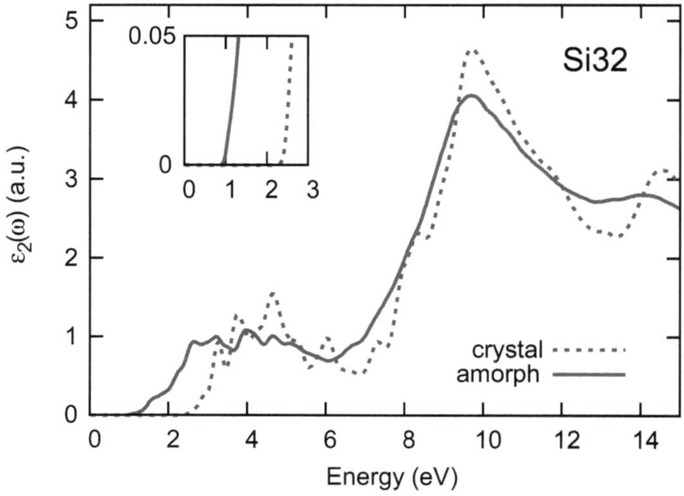

Figure 3.4 Imaginary part of the dielectric function (without local fields) for the crystalline (dotted) and amorphous (solid) Si_{32} embedded QDs. The spectra are per unit volume [43]. Copyright © 2010 WILEY-VCH Verlag GmbH & Co. KGaA, Weinheim.

embedded system. This result suggests that the phase (crystalline or amorphous) of the QD is what ultimately determines the band gap of the composite system. Besides, the phase of the embedding matrix is important in determining the interface geometry, eventually producing different types of bonds at the interface. For example, when the QD is embedded in a glass, some bridge bonds appear (not present in the crystalline case), especially for larger QDs (see Section 3.2.1). The presence of these alternative bond types plays an important role on the system stability and on the localization of the band-edge states near the interface [57, 78, 85, 94].

In the calculated optical absorption (see Fig. 3.4), the features due to the hosting matrix and the Si QD are clearly distinguishable.

We note that the amorphization induces a red shift of the spectrum, related to the reduced gap of the QD. The calculated dielectric function of Si_{32}/SiO_2 shows a nice agreement with experimental measurements on small ($d \simeq 1$ nm) QDs, especially in the 2–6 eV range [34].

The most important differences between the calculated optical absorption spectra for both the amorphous and the crystalline embedded QDs are found in the 1–3 eV range. In this region the amorphous system shows more intense peaks with respect to the crystalline case, suggesting a possible higher absorption (and emission) in the visible range.

Despite the possibility to form amorphous QD in silica has been recently explored [25], experimental measurements on single Si QDs with diameters of the order of 1 nm could not yet be achieved. Thus, a straightforward comparison of our results with experimental data is not possible. Besides, the comparison with other works sustains the idea that the strong deformation of the QD influences the optical absorption at low energies [12]. This idea is also supported by the fact that, for larger systems, when the shape of the QDs tends to be spherical and the distortion is usually less pronounced, crystalline and amorphous systems produce more similar absorption spectra [25, 48].

3.3.2 *Size and Passivation*

It is worth to stress that in the small QD size limit, most of the QD atoms are positioned at the interface, where the effects of stress and passivation type are stronger. For bigger QDs, when Si bulk states emerge and the surface-to-volume ratio decreases, these effects are limited; in this case we expect a response less sensitive to local variations of the interface configuration. Thus, to understand all the interface-related phenomena, an investigation of the small size limit becomes necessary.

In particular, DFT calculations have revealed that the passivation type and regime at the QD interface can overcome the QC in determining the band gap of the embedded system [41, 62]. By defining Ω as the number of passivants per QD silicon at the interface, it is possible to observe in Fig. 3.5 the effect of Ω on the highest occupied molecular orbital (HOMO)–lowest unoccupied molecular orbital (LUMO) gap of H- and OH-terminated QDs. Clearly, the gap of the former does not depend on Ω but simply follows the QC [107], while in the latter case the gap becomes strongly correlated with Ω, producing large variations of the gap value also

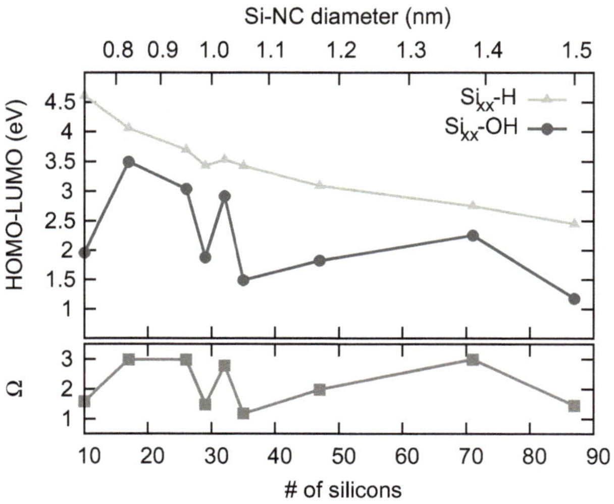

Figure 3.5 HOMO–LUMO gaps for the hydroxidized (circles) and hydrogenated (triangles) QDs, together with the oxidation/hydrogenation degree Ω (squares) [41]. Copyright (2009) by the American Physical Society.

for small variations of the QD size. In this case, the effect of QC is still visible, but strongly modulated by that of the oxidation degree at the interface.

The importance of this result is enforced by a recent size-dependent experimental study showing that the shell region around the Si QD bordered by SiO_2 consists of the three Si suboxide states, Si^{1+}, Si^{2+}, and Si^{3+}, whose densities strongly vary in the small QD size limit [61].

The small QDs, presenting such strong fluctuations of the gap with the size and stronger emission rates with respect to the larger QDs [42, 75], could be responsible for the large broadening of the PL spectra, observed also at low temperatures, [3, 4, 7, 23, 27, 35, 51, 59, 68, 88, 90, 98] even for apparently monodispersed multilayered samples [15]. Also the multiexponential PL decay of Si QDs has been associated with QDs that emits at the same energy but presenting very different decay rates [23].

For large clusters ($d > 3$ nm [103]) we expect the gap recovering a smooth (QC-driven) trend, becoming independent on the passivation regime.

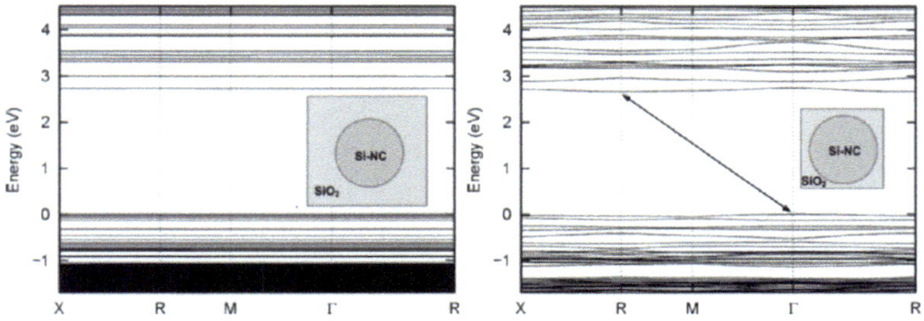

Figure 3.6 Band structure along high symmetry points of the BZ for the crystalline Si_{32} QD embedded (left) in a 3×3×3 and (right) in a 2×2×2 SiO_2 matrix. The insets depict the proportion between the QD and the periodic simulation box. The arrow indicates the HOMO–LUMO band gap.

3.3.3 *Embedding Insulating Materials*

As anticipated in Section 3.2.1, for a given embedding material the insulation of the QD depends on the distance between its replica in the periodic simulation box. From Fig. 3.6 it is possible to observe the variation of the band structure of a 1 nm sized QD as a function of the thickness of the surrounding SiO_2. Clearly, for well insulated QDs the strong confinement generates flat bands corresponding to states localized on the QD. Conversely, for an insufficiently insulated QD the wavefunctions extend over the whole cell, and minibands start to form. We note that in the case of SiO_2, a very thin insulating layer is sufficient to keep the QC active (in the right panel of Fig. 3.6 the QD was separated by its periodic replica by only one oxygen in the x, y, and z directions).

3.3.4 *Optical Absorption*

The extreme efficiency of SiO_2 in passivating the QD (see Chapter 4) can be revealed also by the analysis of the absorption spectrum of Fig. 3.7. For the embedded system we can identify two absorption features, below and above 6 eV,[a] associated to the QD and to the SiO_2, respectively. By removing the SiO_2 matrix while keeping the strain

[a]This value is related to the gap of the SiO_2, underestimated in DFT, and should be larger in experimental samples. However, the concept discussed here holds in general, and does not depend on the threshold value.

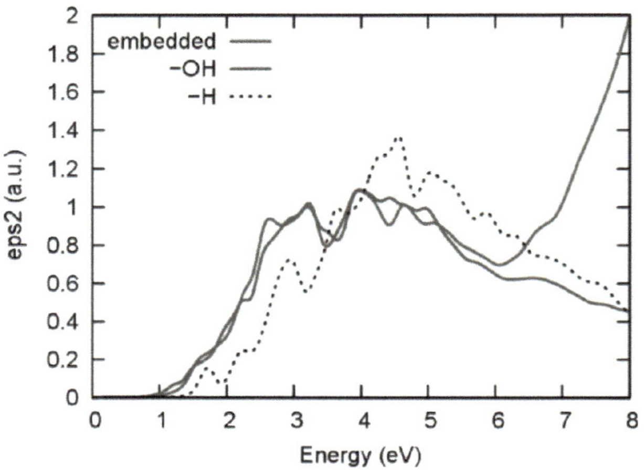

Figure 3.7 Imaginary part of the dielectric function (without local-field effects) for the Si_{32}/SiO_2 QD, and for the freestanding s-Si_{10}-OH and s-Si_{10}-H QDs.

induced by it (see Section 3.2.2), it is possible to observe how a single shell of interfacial oxygens is able to mimick the presence of the SiO_2, well reproducing the absorption of the embedding QD in the low-energy range. Conversely, the absorption of the H-terminated QD completely differs from the others, as a consequence of different confinement potential and bond type at the interface.

3.3.5 *Applicability of Effective Medium Approximation*

The effective medium approximation (EMA) is an approach ideally suited for treating heterogeneous materials, which are mixtures of constituents of different polarizabilities, α_a and α_b. By the mean of this theory it is possible to calculate the dielectric function of the composed material, ε, starting from the dielectric functions and from the volume fractions of the two constituents. The expression for evaluating the total dielectric function depends strongly on the geometrical configuration of the mixtures. A general expression valid in the case that a spherical inclusion of dielectric function ε_a is embedded (but noninteracting) in a medium of dielectric function ε_b has been proposed by Bruggeman [8], valid for every proportion of the volume fractions f_a and f_b:

$$0 = f_a \frac{\varepsilon_a - \varepsilon}{\varepsilon_a + 2\varepsilon} + f_b \frac{\varepsilon_b - \varepsilon}{\varepsilon_b + 2\varepsilon}. \tag{3.1}$$

This is the Bruggeman effective medium expression (BEMA), or in conventional terminology, the EMA.

The case of QDs freestanding in vacuum is a particular case of two systems with different dielectric functions. The dielectric function of a such system calculated with DFT-supercell methods will therefore depends on the quantity of vacuum in the supercell.

Our effective medium theory for the dielectric function of a QD from the dielectric function of the supercell (c=crystallite, s=supercell, v=vacuum) assumes that

$$0 = (1 - f)\frac{\varepsilon_v - \varepsilon_s}{1 + 2\varepsilon_s} + f\frac{\varepsilon_c - \varepsilon_s}{\varepsilon_c + 2\varepsilon_s}. \tag{3.2}$$

where $f = V_c/V_s$ is the crystallite-supercell volume ratio.

Therefore, using $\varepsilon_v = 1$ and solving for ε_s, ε_c we get

$$\varepsilon_s = \frac{1}{4}\left(2 - 3f - \varepsilon_c + 3f\varepsilon_c + \sqrt{8\varepsilon_c + (-2 - 3f(-1 + \varepsilon_c) + \varepsilon_c)^2}\right), \tag{3.3}$$

and

$$\varepsilon_c = \frac{\varepsilon_s(-2 + 3f + 2\varepsilon_s)}{1 + (-1 + 3f)\varepsilon_s}. \tag{3.4}$$

The application of the EMA on Si QDs is the natural choice to investigate the interplay between the dielectric function of the QD, the embedding matrix, and the composite system [60]. It can be applied to calculate the dielectric function of the composite system, ε, by combining that of the host matrix, ε_h, and that of the QD, ε_c. The real parts of the dielectric functions can be derived from the corresponding DFT-RPA imaginary parts through the Kramers–Kronig relations. Then, Eq. 3.4 can be applied on the ε_s calculated by the supercell method in order to obtain ε_c.

The expression for the total dielectric function is obtained by substituting ε_a and ε_b in Eq. 3.1, with ε_c and ε_h, and solving for ε:

$$\varepsilon = \frac{1}{4}\Big(-\varepsilon_c + 3f\varepsilon_c + 2\varepsilon_h - 3f\varepsilon_h + \sqrt{8\varepsilon_c\varepsilon_h + ((-1+3f)\varepsilon_c + (2-3f)\varepsilon_h)^2}\Big)\,, \tag{3.5}$$

where $f = V_c/V_s$ is the ratio between the QD volume, V_c, and the supercell volume, V_s. The QD volume is estimated from the average Si–Si (l_{SS}) and Si–O (l_{SO}) bond lengths of the considered system, through the coupled equations

$$(N-n)l_{SO}^3 + nl_{SS}^3 = V_s/\rho\,, \tag{3.6}$$

$$V_c = n\rho l_{SS}^3\,, \tag{3.7}$$

where N is the total number of atoms, n is the number of atoms composing the QD, and ρ is a free parameter representing the average atomic density of the QD.

Note that, since a crystalline embedding matrix looses its crystallinity after the creation of the QD, the EMA for crystalline embedded QDs ought to be performed using the ε_h of an amorphous matrix instead of that of a crystalline one.

As discussed above, the EMA is based on the assumptions that the embedded QD is spheric and not interacting with the SiO_2. We expect that the first assumption is satisfied by construction, because we built the QD using a spherical cutoff. Then, to find the dielectric function of the noninteracting system, we adopt the alternative approach of the BEMA, in which we obtain the ε_c of the QD starting from the DFT-calculated ε (composite system) and ε_h (embedding matrix). By solving the Bruggeman equation 3.5 for ε_c, we get

$$\varepsilon_c = \frac{\varepsilon((-2+3f)\varepsilon_h + 2\varepsilon)}{\varepsilon_h + (3f-1)\varepsilon}\,. \tag{3.8}$$

We apply Eqs. 3.8 and 3.3 using the ε_h of the pure silica glass and the ε of the a-Si_{32}/a-SiO_2 system. The result is shown in Fig. 3.8 (left panel). We note that the dielectric function of the noninteracting system matches that of the a-s-Si_{32}-OH, especially in the 0–6 eV energy range. This strongly supports the idea that a true separation is established between the QD+interface system and the remaining silica matrix. These separation is allowed due to the noninteracting behavior of the parts, allowing an excellent description of the absorption at low energies.

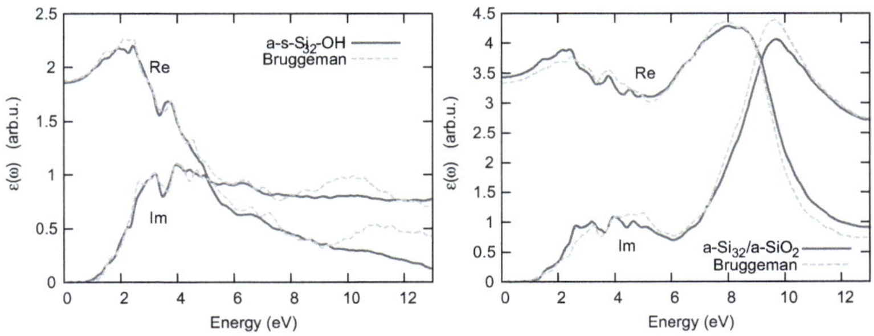

Figure 3.8 Spectrum decomposition (left) and composition (right) through BEMA using the a-Si_{32}/a-SiO_2 QD and its freestanding OH-terminated counterpart [43]. Copyright © 2010 WILEY-VCH Verlag GmbH & Co. KGaA, Weinheim.

We finally report the conventional application of the BEMA by compositing the dielectric function of the a-s-Si_{32}-OH QDs and that of the a-SiO_2 through Eqs. 3.4, 3.5. The result is shown in Fig. 3.8 (right panel) where, as expected, a nice agreement with the DFT-calculated ε of the embedded system is obtained.

As discussed in the following, the application of the EMA to calculate the absorption of the embedded system by combining that of the freestanding QD with that of the embedding matrix is limited by the fact that the strain forming at the inferface of the embedded system holds an important role in determining the final properties of the complex. It is therefore not possible to approximate the embedded Si QD with the corresponding free-standing counterpart without considering the influence of the SiO_2 matrix on the QD+interface geometry. Despite the validity of the EMA has already been demonstrated for several embedded Si nanostructures, the importance of strain for its applicability has been sometimes underestimated [107, 108]. In the next we will briefly discuss the role of strain in embedded systems.

3.3.6 *Strain*

Being the bonds near the interface region much more strained with respect to those in the QD core [12, 25, 65], the average strain tends to zero for very large clusters. Therefore, the effect of strain are

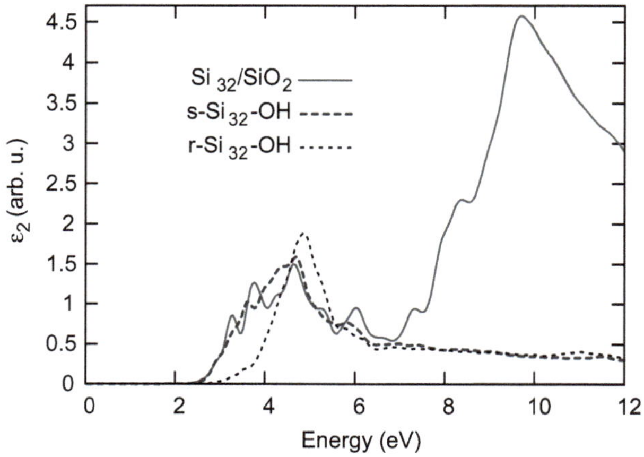

Figure 3.9 DFT–RPA imaginary part of the dielectric function for the Si_{32}/SiO_2 (solid), s-Si_{32}-OH (dashed), r-Si_{32}-OH (dotted) [41]. Copyright (2009) by the American Physical Society.

significant when also interface passivation effects are relevant, that is, at small QD size (See Section 3.3.2).[a]

First of all, we note that the removal of the strain generally increase the gap of an amount that depends on the considered system [41]. This result reveals that, at small size, the QDs can be subjected to amounts of strain that differs very much one from another. Few works have made attempts to correlate the strain with the passivation regime [41], but no simple model to predict the strain level as a function of other parameters is to date yet available. However, it is possible to investigate the role of the embedding matrix, as for the contribution to the absorption spectrum, like as for the induced strain, by considering the reference case of the Si_{32} QD. The optical absorption of Si_{32}/SiO_2, s-Si_{32}-OH, and r-Si_{32}-OH are presented in Fig. 3.9.

We observe that, while the strained QD is able to reproduce very well the spectrum of the full Si/SiO_2 system in the 0–6 eV

[a]Other mechanisms of strain, not considered here, can arise in real samples as a consequence of the "caging effect" of the embedding matrix on the growing QD during annealing. In this case the (compressive) strain could increase with the QD size [112]. Another source of compressive strain may lie in the degree of matrix structural order [114].

region, the removal of the strain produces an enlargement of the gap [32, 66, 83], and a strong blue shift of the absorption spectrum in this region.[a] These are very general results that can be verified in all the cases of embedded QDs (at least for diameters below 2 nm), and demonstrate the importance of considering the strain for a correct modelling of the nanostructures. For example, a recent work has evidenced a crucial impact of the strain on the collective properties of closely packed QD ensembles [46].

3.3.7 *Local-Field Effects*

The abrupt change in the dielectric constant between different chemical species gives rise to the charge polarization at the surfaces when the material is immersed in an external field. Such "surface polarization effects" are of great importance for the optical absorption, since they tend to screen the incoming field, producing a dramatic reduction of the absorption at low-energy (0–5 eV), beside a blue-shifting of the resonant peak. This effect is theoretically described by taking into account the fluctuations of the density charge at the microscopic level, the LFs. It is well known that local-field effects (LFEs) play a crucial role for systems characterized by strong charge inhomegeneities. Instead, for ordered systems like bulk Si and SiO_2-betacristobalite, LFE tend to vanish out [70]. The latter rule applies also for completely amorphized systems, like silica-glasses, that at last behave as homogeneous materials. In the case of QDs, the inhomogeneity is represented by the interface that the they form with the surrounding matrix, and it is therefore important to investigate the role of LFE for systems with different interface conditions.

The importance of LFE has been recently demonstrated by both experiments and calculations, showing dramatic corrections to the absorption spectra also for very large structures [9, 12, 37, 101]. LFE can be included at the DFT–RPA level by connecting the macroscopic dielectric function, $\epsilon_M(\omega)$, to the inverse of the microscopic dielectric

[a] For energies above 6 eV the spectra of the freestanding clusters do not match that of the composite system due to the absence of the silica matrix.

function $\epsilon_{GG'}^{-1}(q, \omega)$ through the so called 'macroscopic average' [79]:

$$\epsilon_M(\omega) = \lim_{q \to 0} \frac{1}{\epsilon_{00}^{-1}(q, \omega)}. \tag{3.9}$$

When LFE are neglected at the RPA level, $\epsilon_M(\omega) = \lim_{q \to 0} \epsilon_{00}(q, \omega) = 1 - \lim_{q \to 0} v(q) P_{00}^0$, where $v(q)$ is the Coulomb interaction and P^0 is the irreducible RPA polarizability. This procedure is in fact exact in the case of an homogeneous system for which the off-diagonal terms of $\epsilon_{GG'}^{-1}(q, \omega)$ are null. On the other hand, when LFE are included the quantity $\epsilon_{00}^{-1}(q, \omega)$ must be accessed. Very briefly, ϵ^{-1} is linked to the reducible polarizability χ by the relation $\epsilon^{-1} = 1 + v\chi$. At the RPA level we have that $\chi = P^0 + P^0 v \chi$. Hence by calculating $P^0 = -iG^0G^0$ with G^0 single particle Green function, we can obtain χ and ϵ^{-1}.

Unfortunately, the ab initio calculation of the full dielectric response requires a computational effort that increases dramatically with the system size, setting a strong limit on the maximum processable QD size. To circumvent such limitation it is possible to to include the LF correction to the dielectric function making use of the EMA. Since in QDs the LF are mostly given by surface polarization effects, the *Clausius–Mossotti* equation can be employed to describe the polarizability α of a dielectric sphere with dielectric constant ε and volume V, embedded into a background with dielectric constant ε_0:

$$\alpha = 3\, V\, \varepsilon_0\, (\varepsilon - \varepsilon_0)/[4\pi(\varepsilon + 2\varepsilon_0)]. \tag{3.10}$$

Then, the LF-corrected ε is given by

$$\varepsilon_{LF} = \varepsilon_0 + \frac{4\pi\alpha}{V} = \varepsilon_0 \frac{4\,\varepsilon_{nc} - \varepsilon_0}{\varepsilon_{nc} + 2\,\varepsilon_0}. \tag{3.11}$$

For freestanding QDs (i.e., embedded in vacuum), by posing $\varepsilon_0 = 1$ we get

$$\varepsilon_{LF} = \frac{4\,\varepsilon_{nc} - 1}{\varepsilon_{nc} + 2} \tag{3.12}$$

The validation of the Eq. 3.12 is performed by a comparison with ε calculated by full-response ab initio techniques [44]. In Fig. 3.10 such comparison is reported for the $Si_{32}(OH)_{56}$ case. We observe that Eq. 3.12 is able to produce a LF correction nicely matching the ab initio one.

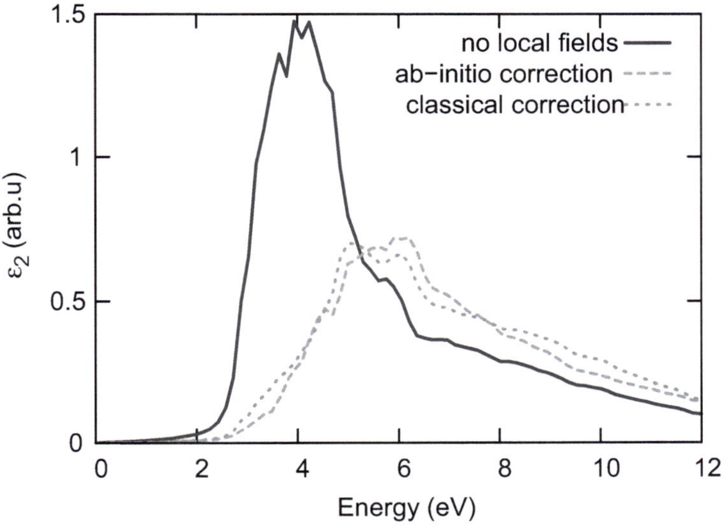

Figure 3.10 Imaginary part of dielectric function of $Si_{32}(OH)_{56}$ QD without LF (solid curve), with LF correction calculated by ab initio (dashed curve), and with LF correction calculated by Eq. 3.12 (dotted curve) [47]. Reprinted with permission from [Guerra, R., Cigarini, F., Ossicini, S. (2013). *J. Appl. Phys.* **113**, 143505]. Copyright [2013], AIP Publishing LLC.

We note that the microscopic field fluctuations produce important screening effects on the spectrum, with a damping of the absorption that is evident at low energies. In fact, at low energies the induced polarizations oscillates in phase with the incoming field, which gets strongly screened. Besides, in the limit of very high frequencies the polarizations can not follow the external field, leading to a vanishing screening. Finally, near to the resonance frequency the absorption is maximized, forming a peak in the absorption, blue-shifted with respect to the original one.

It is worth to stress that interface polarization effects have a strong influence at every QD size. This result is in agreement with previous works reporting strong effects on very large nanostructures, supporting the idea that they arise almost entirely by classical effects [9, 101]. In addition, we observe strong similarities on the spectra of Si_{32} and a-Si_{32} QDs [44], suggesting that for large QDs the response depends mainly on the interface polarization effects and not on the amorphization degree, nor on the particular geometry of the interface.

3.3.8 *Ensembles of Quantum Dots*

As discussed above, one of the most challenging aspect of Si QDs concerns the high sensitivity of the measured response to the precise structural configuration of the QD and of its surrounding environment. In fact, size, shape, interface, defects, impurities, embedding medium, and cristallinity level, among others, constitute a set of mutually dependent parameters that drastically change the optoelectronic properties of the QDs. Many theoretical and experimental works have contributed to characterize the connection between the above parameters and the observed QD response.

While the theoretical approach is more suitable to deal with single QDs, especially when making use of simulations at the atomistic level, experiments usually make use of samples containing a large number of different QDs. Therefore, despite the tremendous advances of the latest years, a direct comparison between theoretical simulations and experimental observations is still a complicated task.

The simplest way to provide a connection with the experiments is by describing the optical response of an ensemble of QDs as the sum of the responses of the individual QDs [47]. The main approximation regards the absence of QD–QD interaction mechanisms, which implies QD–QD surface distances larger than about 0.5 nm, 2 nm, and 4 nm for SiO_2, Si_3N_4, and SiC embedding matrices, respectively [46, 95, 100]. The latter conditions can be satisfied in real embedded or freestanding QD samples by varying the silicon excess or the QD concentration, respectively.

In Fig. 3.11 we report the calculated optical emission (PL) and absorption of an ensemble of 10^6 freestanding OH-terminated QDs. The depicted results are in good agreement with the experimental outcomes showing that a very weak absorption exists in the region where luminescence peaks [36, 82]. The origin of the large Stokes shift between absorption and emission peaks has been subject of intense debate from twenty years to date. While contributions from tunneling between QDs and from structural deformation of the excited QDs have been proposed, the most acknowledged

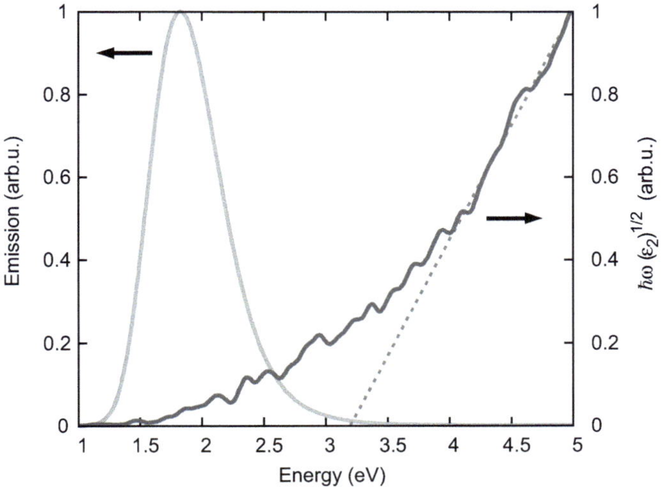

Figure 3.11 Optical emission (left curve) and absorption (right curve) spectra of an OH-terminated QD ensemble with log-normal distribution of the QD radius parametrized by $\mu = 0.84$ nm and $\sigma = 0.01$ nm. The Tauc fitting is reported by the dotted curve [47]. Reprinted with permission from [Guerra, R., Cigarini, F., Ossicini, S. (2013). *J. Appl. Phys.* **113**, 143505]. Copyright [2013], AIP Publishing LLC.

contribution to the Stokes shift comes from associating emission and absorption to interface states and to quantum-confined states in the QD, respectively. Within the latter picture, the Tauc gap helps in distinguishing the absorption due to interface (surface) states ($E < E_{Tauc}$) and due to QD states ($E > E_{Tauc}$). Experimental measurements on samples made by SiO_2-embedded QDs with average diameter smaller than 2 nm report a Tauc gap of about 2.5 eV and a PL peak centered at about 1.7 eV [36]. As discussed above, the difference between the experimental and computed values should be attributed, in part, to the SiO_2-induced strain on the QDs, and to the lack of phonon-assisted transitions, not included in the calculations.

3.3.9 *Beyond DFT*

Self-energy and excitonic effects are known to play a very important role both in low-dimensional systems (as QDs) and in 3D systems as the SiO_2. In this last case it is known that DFT underestimates the electronic gap to about 5.5–7.0 eV, and important excitonic effects are responsible of the strong absorption peak at about 10 eV. The inclusion of the many-body effects is thus of fundamental importance in order to obtain a better description of the optical properties of embedded QDs.

A correction to the fundamental band gap is usually obtained by calculating the separate electron and hole quasiparticle energies via the GW method [79]. The knowledge of the quasiparticle energies, however, is still not sufficient to correctly describe a process in which electron–hole pairs are created. In the optical absorption, the interaction between the positively and negatively charged quasiparticles can lead to a strong shift of the peak positions as well as to distortions of the spectral line shape. Within many-body perturbation theory (MBPT) framework such interaction is taken into account by the solution of the Bethe–Salpeter equation (BSE) for the polarizability [29].

An alternative approach to MBPT for the computation of neutral excitations is represented by time-dependent density functional theory (TDDFT) [92]. TDDFT is expected to be more efficient than the MBPT-based approach; however many conceptual and computational problems remains still unsolved preventing its application to complex systems [79]. Moreover, a recent comparison of the two techniques applied to Si QDs revealed that TDDFT does not take into account correctly the screened Coulomb interaction, also for small QDs [87].

Table 3.1 shows the optical gap (i.e., absorption threshold) calculated within DFT, GW, GW+BSE approximations, for the Si_{10} and a-Si_{10} QDs embedded in SiO_2. It is possible to observe that, in the crystalline (amorphous) case, the inclusion of the GW corrections

Table 3.1 Many-body effects on the optical gap values, in eV, for the crystalline and amorphous Si_{10} QDs embedded in SiO_2. [40]

	DFT	GW	GW+BSE
Crystalline	1.77	3.67	1.86
Amorphous	1.41	3.11	1.41

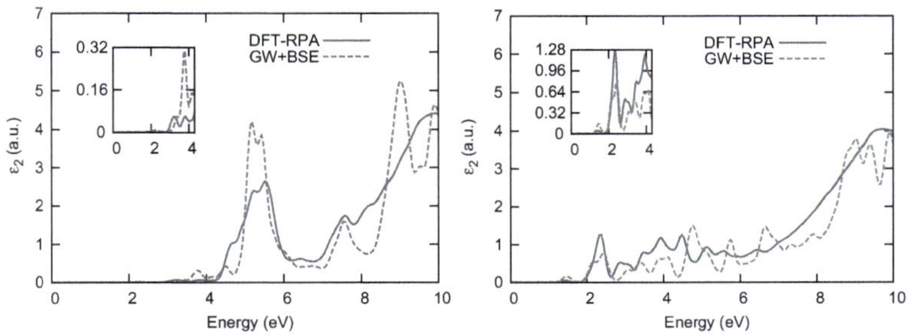

Figure 3.12 DFT–RPA and GW+BSE calculated imaginary part of the dielectric function for the crystalline (left) and the amorphous (right) embedded Si_{10} QDs [40]. Copyright (2009) by the American Physical Society.

opens up the gap by about 1.9 (1.7) eV, while the excitonic correction reduces it by about 1.8 (1.7) eV. Thus, the total correction to the gap results very small, around 0.1 (0.0) eV. The difference between the GW electronic gap and the GW+BSE optical excitonic gap gives the exciton binding energy E_b. Our calculated exciton binding energies are quite large: 1.9 eV (crystalline) and 1.7 eV (amorphous). They are very large if compared with that of bulk SiO_2 (almost 0 eV) [16, 73, 86], bulk Si ($\sim$ 15 meV) or with carbon nanotubes [17, 96] where $E_b \sim 1$ eV, but they are similar to those calculated for undoped and doped Si QD [28, 53] of similar size and for Si and Ge small nanowires [10, 11].

Figure 3.12 shows the calculated DFT–RPA and GW+BSE absorption spectra for the embedded Si_{10} and a-Si_{10} QDs. The LFE have been intentionally neglected, since we have already treated them separately. The results show that the inclusion of the many-body effects does not substantially modify the absorption spectra. In both cases the energy position of the absorption onset is practically not modified (see insets). Delerue et al. [22] found that, for Si QD larger than 1.2 nm, the self-energy and Coulomb corrections almost cancel each other. Also for the Si_{10} case (diameter $\simeq$ 0.7 nm), this cancellation seems to occur. This is a very convenient result that demonstrates the reliability of the DFT–LDA scheme for the description of the optoelectronical properties of Si/SiO_2 QDs.

In summary, even if complex treatments should be invoked to include self-energy and excitonic effects, some many-body calculations on Si QDs reported fundamental gaps [22] and absorption spectra [37, 40, 87] very close to the independent-particle calculated ones when LFE are neglected [40]. As reported above, such simplification is due to the cancellation between self-energy corrections (calculated through the GW method) and electron–hole Coulomb corrections (calculated through the Bethe–Salpeter equation). These considerations justify the choice of DFT–LDA for evaluating the optical properties of small QDs, allowing a good compromise between results accuracy and computational effort.

3.4 Conclusions

In the above sections we have shown how ab initio simulations can reveal some of the fundamental properties of nanostructured materials.

In the case of Si QDs we have observed a type I band offset between the embedding and the embedded materials, and an increase of the band gap with smaller QD diameters. For small QDs such increase is interfered with by interface effects, which become dominant below a threshold diameter of about 2 nm. In particular, interface effects are relevant in the presence of strongly polar atoms at the interface (e.g., oxygen), while for hydrogenic bonds a QC picture is recovered.

Beside interface effects, small QDs are always present in real samples, and since they are the most optically active (due to QC the optical emission rate increases for smaller QD size), their contribution to the observed response can be very important.

Since many works have observed a residual amorphization in small QDs, it is important to understand the response of amorphous QDs beside crystalline ones. We have shown that amorphized QDs have a reduced band gap and a red-shifted absorption onset. The amount of amorphization shall depend on the QD size and on the embedding matrix.

The embedding matrix determines also the insulation level of the QD. For wide-band-gap matrices like SiO_2 we have observed no hybridization between QD and matrix states, making possible to describe the characteristic of the composite material as a function of its components (e.g., using the effective-medium approximation). Also, we have shown how the response of an ensamble of strongly insulated QDs can be conveniently described as a superposition of the responses of the single QDs, comparable with the experimental observations on real samples.

Conversely, for poorly insulating matrices (e.g., SiC), the phase separation between embedding and embedded materials is reduced and strongly hybridized states appear, while QD–QD interaction is enhanced favoring transport to the detriment of QC.

The type of embedding matrix has also an important role on the amount (and type) of strain forming at the interface of the nanostructure. We have observed that such strain has a fundamental impact on the optoelectronic properties of the embedded system. Similarly to the amorphization effects, it reduces the band gap and determines a strong red shift of the absorption onset.

An important contribution to the absorption (and emission) properties is dictated by the LFs that arise due to the polarization at the interface/surface of the QD, a process describable by classical effects. We have shown that the surface polarization effects strongly screen the optical absorption and produce a blue shift of the main absorption peak, and should be included whenever an interface is present, also for very large nanostructures.

Finally, we have discussed the possible advancements obtainable by many-body methods. We have showed that quasiparticle and excitonic effects play a fundamental role in nanostructures. Coincidentally, for small-sized Si QDs their correction to the band gap almost exactly cancel out, making the DFT scheme, in a first approximation, a surprisingly reliable and convenient tool for exploring the properties of embedded and freestanding Si QDs.

References

1. Allan, G., Delerue, C., Lannoo, M. (1996). *Phys. Rev. Lett.* **76**, 2961.
2. Antonova, I. V., Gulyaev, M., Savir, E., Jedrzejewsky, J., Balberg, I. (2009). *Phys. Rev. B* **77**, 125318.
3. Averboukh, B., Huber, R., Cheah, K. W., Shen, Y. R., Qin, G. G., Ma, Z. C., Zong, W. H. (2002). *J. Appl. Phys.* **92**, 3564–3568.

4. Belomoin, G., Therrien, J., Nayfeh, M. (2000). *Appl. Phys. Lett.* **77**, 779–781.
5. Belyakov, V. A., Burdov, V. A., Lockwood, R., Meldrum, A. (2008). *Adv. in Opt. Tech.* **2008**, 279502.
6. Bisi, O., Ossicini, S., Pavesi, S. (2000). *Surface Science Reports* **38**, 1.
7. Brongersma, M. L., Kik, P. G., Polman, A. (2000). *Appl. Phys. Lett.* **76**, 351–353.
8. Bruggeman, D. A. G. (1935). *Ann. Phys.* (Leipzig) **24**, 636.
9. Bruneval, F., Botti, S., Reining, L. (2005). *Phys. Rev. Lett.* **94**, 219701.
10. Bruno, M., Palummo, M., Marini, A., Del Sole, R., Olevano, V., et al. (2005). *Phys. Rev. B* **72**, 153310.
11. Bruno, M., Palummo, M., Marini, A., Del Sole, R., Ossicini, S. (2007). *Phys. Rev. Lett.* **98**, 036807.
12. Bulutay, C. (2007). *Phys. Rev. B* **76**, 205321.
13. Calcott, P. D. J., Nash, K. J., Canham, L. T., Kane, M. J., Brumhead, D. (1993). *J. Phys. Cond. Mat.* **5**, L91.
14. Canham, L. T. (1990). *Appl. Phys. Lett.* **57**, 1046.
15. Cazzanelli, M., Navarro-Urriós, D., Riboli, F., et al. (2004). *J. Appl. Phys.* **96**, 3164–3171.
16. Chang, E. K., Rohlfing, M., Louie, S. G. (2000). *Phys. Rev. Lett.* **85**, 2613.
17. Chang, E., Bussi, G., Ruini, A., Molinari, E. (2004). *Phys. Rev. Lett.* **92**, 196401.
18. Conibeer, G., Green, M., Cho, E.-C., et al. (2008). *Thin Solid Films* **516**, 6748–6756.
19. Credo, G. M., Mason, M. D., Buratto, S. K. (1999). *Appl. Phys. Lett.* **74**, 1978.
20. Daldosso, N., Luppi, M., Ossicini, S., Degoli, E., Magri, R., et al. (2003). *Phys. Rev. B* **68**, 085327.
21. Delerue, C., Lannoo, M. (2004). *Nanostructures. Theory and Modelling* (Springer Verlag, Berlin).
22. Delerue, C., Lannoo, M., Allan, G. (2000). *Phys. Rev. Lett.* **84**, 2457.
23. Delerue, C., Allan, G., Reynaud, C., Guillois, O., Ledoux, G., Huisken, F. (2006). *Phys. Rev. B* **73**, 235318.
24. Derr, J., Dunn, K., Riabinina, D., Martin, F., Chacker, M., Rosei, F. (2009). *Physica E* **41** 668–670.
25. Djurabekova, F., Nordlund, K. (2008). *Phys. Rev. B* **77**, 115325.

26. Dorigoni, L., Bisi, O., Bernardini, F., Ossicini, S. (1996). *Phys. Rev. B* **53**, 4557.
27. Dorigoni, L., Bisi, O., Bernardini, F., Ossicini, S. (2009). *Phys. Rev. B* **79**, 125306.
28. Luppi, E., Iori, F., Magri, R., Pulci, O., Ossicini, S., et al. (2007). *Phys. Rev. B* **75**, 033303.
29. Fetter, A. L., Walecka, J. D. (1971). *Quantum Theory of Many-Particle Systems* (McGraw-Hill - New York).
30. Filonov, A., Ossicini, S., Bassani, F., d'Avitaya, F. A. (2002). *Phys. Rev. B* **65**, 195317.
31. Fiolhais, C., Nogueira, F., Marques, M. (2003). *A Primer in Density Functional Theory*, Lecture Notes in Physics - Vol. 620 (Springer Series); ISBN 978-3-540-03083-6.
32. Franceschetti, A. (2007) *Phys. Rev. B* **76**, 161301(R).
33. Franzò, G., Miritello, M., Boninelli, S., Lo Savio, R., Grimaldi, M. G., et al. (2008). *J. Appl. Phys.* **104**, 094306.
34. Gallas, B., Stenger, I., Kao, C.-C., Fisson, S., Vuye, G., Rivory, J. (2005). *Phys. Rev. B* **72**, 155319.
35. Gardelis, S., Nassiopoulou, A. G., Vouroutzis, N., Frangis, N. (2009). *J. Appl. Phys.* **105**, 113509.
36. Gardelis, S., Manousiadis, P., Nassiopoulou, A. G. (2011). *Nanoscale Research Letters* **6**, 227.
37. Gatti, M., Onida, G. (2005). *Phys. Rev. B* **72**, 045442.
38. Godefroo, S., Hayne, M., Jivanescu, M., Stesmans, A., Zacharias, M., Lebedev, O. I., Tendeloo, G. V., Moshchalkov, V. V. (2008). *Nature Nanotech.* **3**, 174.
39. Gourbilleau, F., Ternon, C., Maestre, D., Palais, O., Dofour, C. (2009). *J. Appl. Phys.* **106**, 013501.
40. Guerra, R., Marri, I., Magri, R., Martin-Samos, L., Pulci, O., et al. (2009). *Phys. Rev. B* **79**, 155320.
41. Guerra, R., Degoli, E., Ossicini, S. (2009). *Phys. Rev. B* **80**, 155332.
42. Guerra, R., Ossicini, S. (2010). *Phys. Rev. B* **81**, 245307.
43. Guerra, R., Degoli, E., Marsili, M., Pulci, O., Ossicini, S. (2010). *Phys. Stat. Solidi B* **247**, 2113.
44. Guerra, R., Marsili, M., Pulci, O., Ossicini, S. (2011). *Phys. Rev. B* **84**, 075342.

45. Guerra, R., Ippolito, M., Meloni, S., Ossicini, S. (2012). *Appl. Phys. Lett.* **100**, 81905.
46. Guerra. R., Ossicini, S. (2013). *Phys. Rev. B* **87**, 165441.
47. Guerra, R., Cigarini, F., Ossicini, S. (2013). *J. Appl. Phys.* **113**, 143505.
48. Hadjisavvas, G., Kelires, P. C. (2007). *Physica E* **38**, 99.
49. Hao, X. J., Podhorodecki, A. P., Shen, Y. S., Zatryb, G., Misiewicz, J., Green, M. A. (2009). *Nanotechnology* **20**, 485703.
50. Hapala, P., Køusová, K., Pelant, I., Jelńek, P. (2013). *Phys. Rev. B* **87**, 195420.
51. Hernandez, A. V., Torchynska, T. V., Matsumoto, Y., Sandoval, S. J., Dybiec, M., Ostapenko, S., Shcherbina, L. V. (2005). *Microelectron. J.* **36**, 510–513.
52. Hill, N. A., Whaley, K. B. (1995). *Phys. Rev. Lett.* **75**, 1130.
53. Iori, F., Degoli, E., Magri, R., Marri, I., Cantele, G., et al. (2007). *Phys. Rev. B* **76**, 085302.
54. Ippolito, M., Meloni, S., Colombo, L. (2008). *Appl. Phys. Lett.* **93**, 153109.
55. Iwayama, T. S., Hama, T., Hole, D. E., Boyd, I. W. (2001). *Solid-State Electron.* **45**, 1487–1494.
56. Jambois, O., Rinnert, H., Devaux, X., Vergnat, M. (2005). *J. Appl. Phys.* **98**, 046105.
57. Jurbergs, D., Rogojina, E., Mangolini, L., Kortshagen, U. (2006). *Appl. Phys. Lett.* **88**, 233116.
58. Kanemitsu, Y. (1996). *Thin Solid Films* **276**, 44–46.
59. Kanemitsu, Y., Fukunishi, Y., Iiboshi, M., Okamoto, S., Kushida, T. (2000). *Physica E* **7**, 456–460.
60. Khriachtchev, L., Nikitin, T., Velagapudi, R., Lahtinen, J., Novikov, S. (2009). *Appl. Phys. Lett.* **94**, 043115.
61. Kim, S., Kim, M. C., Choi, S.-H., Kim, K. J., Hwang, H. N., Hwang, C. C. (2007). *Appl. Phys. Lett.* **91**, 103113.
62. König, D., Rudd, J., Green, M., Conibeer, G. (2008). *Phys. Rev. B* **78**, 035339.
63. Koponen, L., Tunturivuori, L. O., Puska, M. J., Nieminen, R. M. (2009). *Phys. Rev. B* **79**, 235332.
64. Kroll, P., Schulte, H. J. (2006). *Phys. Stat. Sol. B* **632**, 2089.
65. Kroll, P., Schulte, H. J. (2006). *Phys. Stat. Sol. B* **243**, R47–R49.
66. Kůsová, K., Ondič, L., Klimešová, E., Herynková, K., Pelant, I., et al. (2012). *Appl. Phys. Lett.* **101**, 143101.

67. Lacombe, A., Kadari, B., Beaudoin, F., Barba, D., Martin, F., Ross, G. G. (2008). *Nanotechnology* **19**, 465702.
68. Lin, S.-W., Chen, D.-H. (2009). *Small* **5**, 72–76.
69. Löper, P., Canino, M., López-Vidrier, J., Schnabel, M., Schindler, F., et al. (2013). *Phys. Stat. Solidi A* **210**, 669–675.
70. Louie, S. G., Chelikowsky, J. R., Cohen, M. L. (1975). *Phys. Rev. Lett.* **34**, 155.
71. Luppi, M., Ossicini, S. (2003). *J. Appl. Phys.* **94**, 2130.
72. Luppi, M., Ossicini, S. (2005). *Phys. Rev. B* **71**, 035340.
73. Marini, A., Del Sole, R., Rubio, A. (2003). *Phys. Rev. Lett.* **91**, 256402.
74. Martin, J., Cichos, F., Huisken, F., von Borczyskowski C. (2008). *Nano Letters* **8**, 656.
75. Meier, C., Gondorf, A., Lüttjohann, S., Lorke, A., Wiggers, H. (2007). *J. Appl. Phys.* **101**, 103112.
76. Mirabella, S., Agosta, R., Franzò, G., Crupi, I., Miritello, M., et al. (2009). *J. Appl. Phys.* **106**, 103505.
77. Moskalenko, A. S., Berakdar, J., Prokofiev, A. A., Yassievich, I. N. (2007). *Phys. Rev. B* **76**, 085427.
78. Nguyen, N. B., Dufour, C., Petit, S. (2008). *J. Phys. Cond. Matt.* **20**, 455209.
79. Onida, G., Reining, L., Rubio, A. (2002). *Rev. Mod. Phys.* **74**, 601.
80. Ossicini, S., Pavesi, L., Priolo, F. (2003). *Light Emitting Silicon for Microphotonics*, Springer Tracts on Modern Physics, **194** (Springer-Verlag, Berlin).
81. Palummo, M., Del Sole, R., Ossicini, S. (2010). *Pysica Status Solidi B* **247**, 2089.
82. Pavesi, L., Dal Negro, L., Mazzoleni, C., Franzò, G., Priolo, F. (2000). *Nature* **408**, 440.
83. Peng, X.-H., Ganti, S., Alizadeh, A., Sharma, P., Kumar, S. K., Nayak, S. K. (2006). *Phys. Rev. B* **74**, 035339.
84. Prezioso, S., Hossain, S. M., Anopchenko, A., Pavesi, L., Wang, M., et al. (2009). *Appl. Phys. Lett.* **94**, 062108.
85. Puzder, A., Williamson, A. J., Grossman, J. C., Galli, G. (2003). *J. Am. Chem. Soc.* **125**, 2786–2791.
86. Ramos, L. E., Furthmüller, J., Bechstedt, F. (2004). *Phys. Rev. B* **69**, 085102.
87. Ramos, L. E., Paier, J., Kresse, G., Bechstedt, F. (2008). *Phys. Rev. B* **78**, 195423.

88. Ray, M., Jana, K., Bandyopadhyay, N. R., Hossain, S. M., Navarro-Urriós, D., Chattyopadhyay, P. P., Green, M. A. (2009). *Solid State Commun.* **149**, 352–356.
89. Rinnert, H., Vergnat, M., Burneau, A. (2001). *J. Appl. Phys.* **89**, 237.
90. Rolver, R., Först, M., Winkler, O., Spangenberg, B., Kurz, H. (2006). *J. Vac. Sci. Technol. A* **24**, 141–145.
91. Rosini, M., Jacoboni, C., Ossicini, S. (2002). *Phys. Rev. B* **66**, 155332.
92. Runge, E., Gross, E. K. U. (1984). *Phys. Rev. Lett.* **52**, 997.
93. Rurali, R. (2010). *Rev. Mod. Phys.* **82**, 427.
94. Seino, K., Wagner, J.-M., Bechstedt, F. (2007). *Appl. Phys. Lett.* **90**, 253109.
95. Seino, K., Bechstedt, F., Kroll, P. (2012). *Phys. Rev. B* **86**, 075312.
96. Spataru, C. D., Ismail-Beigi, S., Benedict, L. X., Louie, S. G. (2004). *Phys. Rev. Lett.* **92**, 077402.
97. Sychugov, I., Juhasz, R., Valenta, J., Linnros, J. (2005). *Phys. Rev. Lett.* **94**, 087405.
98. Sykora, M., Mangolini, L., Schaller, R. D., Kortshagen, U., Jurbergs, D., Klimov, V. I. (2008). *Phys. Rev. Lett.* **100**, 067401.
99. Takagi, H., Ogawa, H., Yamazaki, Y., Ishikazi, A., Nakagiri, T. (1990). *Appl. Phys. Lett.* **56**, 2379.
100. Timmerman, D., Izeddin, I., Stallinga, P., Yassievich, I. N., Gregorkiewicz, T. (2008). *Nat. Photon.* **2**, 105.
101. Trani, F., Ninno, D., Iadonisi, G. (2007). *Phys. Rev. B* **75**, 033312.
102. Tsai, T.-C., Yu, Li-Z., Lee, C.-T. (2007). *Nanotechnology* **18**, 275707.
103. Vasiliev, I., Chelikowsky, J. R., Martin, R. M. (2002). *Phys. Rev. B* **65**, 121302(R).
104. Veprek, S., Iqbal, Z., Sarott, F. A. (1982). *Philos. Mag. B* **45**, 137.
105. Wang, K., Han, D., Kemp, M., Silver, M. (1993). *Appl. Phys. Lett.* **62**, 157.
106. Watanabe, T., Tatsamura, K., Ohdomari, I. (2004). *Appl. Surf. Sci.* **237**, 125.
107. Weissker, H.-Ch., Furthmüller, J., Bechstedt, F. (2002) *Phys. Rev. B* **65**, 155328.
108. Weissker, H.-Ch., Furthmüller, J., Bechstedt, F. (2003) *Phys. Rev. B* **67**, 165322.
109. Weissker, H.-Ch., Furthmüller, J., Bechstedt, F. (2003). *Phys. Rev. B* **67**, 245304.

110. Wolkin, M. V., Jorne, J., Fauchet, P. M., Allan, G., Delerue, C. (1999). *Phys. Rev. Lett.* **82**, 000197.

111. Yilmaz, D. E., Bulutay, C., Çağin, T. (2008). *Phys. Rev. B* **77**, 155306.

112. Yilmaz, D. E., Bulutay, C., Çağin, T. (2009). *Appl. Phys. Lett.* **94**, 191914.

113. Zachariasen, W. H. (1932). *J. Am. Chem. Soc.* **54**, 3841.

114. Zatryb, G., Podhorodecki, A., Misiewicz, J., Cardin, J., Gourbilleau, F. (2013). *Nanoscale Res. Lett.* **8**, 40.

115. Zhou, Z., Brus, L., Friesner, R. (2003). *Nano Lett.* **3**, 163.

Chapter 4

Silicon Nanoclusters Embedded in Dielectric Matrices: Nucleation, Growth, Crystallization, and Defects

Daniel Hiller
Laboratory for Nanotechnology, Department of Microsystems Engineering (IMTEK), University of Freiburg, Georges-Köhler-Allee 103, 79110 Freiburg, Germany
daniel.hiller@imtek.uni-freiburg.de

4.1 Introduction

Silicon quantum dots (Si QDs) or, if crystalline also known as, silicon nanocrystals (Si NCs) are zero-dimensional nanostructures which are strongly influenced by quantum confinement (QC) effects. To estimate the significance of the influences of QC, the dimension of the QD has to be put into relation with the exciton Bohr radius (a_B) of the respective material. For silicon a_B is typically reported to be 4.3 to 5.3 nm [1–3], that is, in the bulk material the electron and the hole have a distance of around 10 nm. Different classification schemes exist to distinguish between weak, moderate/intermediate, and strong QC [1, 4], however, strong QC is generally assumed for $r_{QD} < a_B$. In other words, the spatial extension of the Wannier–Mott

Nanotechnology and Photovoltaic Devices: Light Energy Harvesting with Group IV Nanostructures
Edited by Jan Valenta and Salvo Mirabella

ISBN 978-981-4463-63-8 (Hardcover), 978-981-4463-64-5 (eBook)
www.panstanford.com

exciton is restricted to the dimensions of the QD. Please note, in the case of distinctly varying effective masses of electron and hole, the respective Bohr radii of both have to fulfill the relation $r_{QD} < a_{B,e}$, $a_{B,h}$ to allow for the strong QC regime. As a consequence the ground state energy of the exciton (or respectively the band gap of the QD) increases with decreasing QD size. The same applies for the exciton-binding energy which increases from the bulk Si value of $\sim$15 meV [5] to $>$ 100 meV for small Si NCs [6]. According to Heisenberg's uncertainty principle the confinement of the carriers in real space causes a spreading of their wavefunctions in momentum space which is accompanied by a high oscillator strength of the transition. For an indirect semiconductor like Si this leads to an increased probability for radiative exciton recombination. In numbers, the radiative quantum efficiency of bulk Si of $\sim 10^{-6}$ [7] can be increased by 5 orders of magnitude [8–10]. Despite the experimental observation of a substantial no-phonon transition probability at LHe temperatures [11, 12], it has to be noted that even the smallest Si NCs ($\leq$2 nm) remain an indirect semiconductor material.

One of the most widely used measurement techniques for the characterization of Si NCs is photoluminescence (PL) spectroscopy. Due to the influence of QC a substantial amount of the excitons generated by the excitation light will recombine radiatively and the emitted light is blue-shifted to higher energies than the bulk Si band-gap energy (1.1 eV). However, the excitation light can also cause radiative emission via point defects in the sample and hence any PL spectrum of a Si NC sample has to be carefully analyzed to exclude defect luminescence. As mentioned above, the band-gap configuration of Si NCs remains indirect despite the breakdown of the momentum conservation rule. Therefore, the lifetime of an exciton in a Si NC is rather long (μs to ms timescale) compared to direct semiconductors or radiative defects (ns to ps timescale). Time-resolved photoluminescence (TRPL) measures the decay of the PL signal after an excitation pulse and can thereby deliver information about the lifetime of excited states in the sample.

The PL peak width also contains important information about the Si NCs. Two factors make the PL peak of Si NCs rather broad compared to direct gap semiconductor NCs. At room temperature most of the radiative transitions require the contribution of phonons which do not only supply momentum but also energy. Different phononic contributions lead to PL peak full-width at half-maximum (FWHM) values of 120–150 meV even for single Si NCs [13]. Another factor is the NC size distribution (different NC size = different band gap) so that porous Si exhibits an FWHM of up to 500 meV [11] whereas size-controlled Si NC ensembles in a superlattice (SL) configuration have FWHM values of only 250–280 meV. Hence, emission lines sharper than ~100 meV are unlikely to have an origin in the exciton recombination in Si NCs.

Another criterion often used to distinguish between a PL origin in the Si NCs or defects is the blue shift of the PL peak with decreasing NC size. Though, this observation might be consistent with the QC model, optical interference artifacts also have to be taken into account. Using the transfer matrix method (TMM) it was demonstrated that different SiO_2 buffer and capping layer thicknesses that sandwich a PL emitting Si NC layer changes both the PL peak energy and the intensity [14].

Finally, the surrounding matrix of Si QDs plays a major role for the optical and electrical properties. In Ref. [15] it was demonstrated that the PL of Si–H terminated Si NCs changes from up to 3 eV to $\leq$2 eV when the samples are exposed to ambient air which causes an encapsulation in SiO_2 via oxidation. It turns out that the polarity of the surface terminating groups influences the highest occupied molecular orbital-lowest unoccupied molecular orbital (HOMO–LUMO) gap energy of the Si QD [16]. For instance, the polarity of the Si–N bond is smaller than that of the Si–O bond and hence the HOMO–LUMO gap energy of a nitride terminated Si NC is supposed to be larger than that of an oxide terminated Si NC [17]. Also, the exciton itself is influenced by the surface termination: The higher its electronegativity (EN) the more the electron is attracted toward the interface near region of the QD which increases the probability of the hole to be located in the center of the QD. This additional localization of the exciton increases the wavefunction overlap in momentum space and thereby the radiative recombination probability further. Interestingly, the impact of the polarity of the surface terminating groups even dominates the electronic structure for small Si NCs irrespective of QC effects [16].

It has been a long (and still ongoing) debate whether excitons in oxide embedded Si NCs recombine intrinsically or if the emission of luminescence is mediated by localized radiative defect states at the Si NC/SiO_2 interface (see also Chapter 5). Initially, the discovery of PL from porous Si was dedicated to QC [18]. The above-mentioned strong PL red shift upon oxidation led to a model in which the recombination proceeds via O-related states (Si=O double bonds) that trap either the electron (for midsize NCs and thereby weaker NC size dependence of the PL energy) or the whole exciton (for very small NCs with the consequence of size independent PL energy) [15]. Alternatively, it was suggested that after inter band excitation the electron relaxes into oxide inherent interfacial defect states that are in the energetic vicinity of the NC conduction band (CB) edge and the hole is trapped by midgap SiO_2 defects in the energetic vicinity of the NC valence band (VB) edge [19]. Using high-field magneto-PL the extent of the wavefunction that causes the luminescence can be derived [20]. Applied to Si NCs an unexpected explanation has been suggested: For oxide embedded Si NCs prepared by the normal inert gas annealing method the extent of the wavefunction is infinitesimally small, indicating a defect mediated PL origin [21]. If, however, the sample is subjected to a postannealing in H_2 (to passivate defects), the detected extent of the wavefunction is of the dimension of the Si NC, indicating QC as PL origin [21]. It remains a mystery why the PL spectra as well as their dynamics do apparently not change significantly by H_2 passivation although this model suggests a fundamental switching of the PL emission mechanism.

Anyhow, a comprehensive understanding of the properties of Si QDs cannot be obtained by solely focusing on the intrinsic properties of quantum-confined Si. The (partially dominating) impact of the interaction with the matrix material has also to be taken into account.

4.2 Silicon Quantum Dot Formation

4.2.1 *Preparation Methods*

Historically, the first QC effects were observed in the early 1990s on porous silicon prepared by electrochemical etching in light emission [22] and optical absorption [23]. Due to a lack of long-term stability of the nanoporous structures at ambient conditions and the problematic integration into device technology, Si NCs were later fabricated by implantation of Si ions into SiO_2 and subsequent high-temperature annealing [24, 25]. An alternative fabrication procedure employs the deposition of Si-rich dielectric films and subsequent high-temperature annealing to from the Si NCs. Widely used deposition techniques are sputtering [26, 27], chemical vapor deposition (CVD) or plasma-enhanced chemical vapor deposition (PECVD) [28, 29], and thermal evaporation [30, 31].

Alternatively, Si NCs can also be fabricated by wet chemical synthesis routines [32, 33] or by nonthermal plasma synthesis [34–36]. In both cases free-standing (in contrast to matrix-embedded) Si NCs are produced which has a significant impact on the Si NC properties [37]. In the case of synthesis methods that fabricate free-standing and H-terminated Si NCs, wet-chemical postprocessing is important to obtain an organic capping that prevents the Si from oxidation. Suitable cappings are xylene-based suspensions [32], decyl [33], or various alkenes such as dodecene [38]. Under suitable conditions free-standing Si NCs can also be oxidized and subsequently dissolved in a liquid [35]. However, the extreme surface-to-volume ratio of the Si QD powder can also cause a vigorous reaction with ambient air and might even cause an explosion under special conditions [39, 40]. In terms of semiconductor technology free-standing Si NCs have the great advantage that no annealing is required. In contrast, a high-temperature annealing is mandatory for the synthesis via excess Si in dielectric materials. On the other hand, Si NCs embedded in, for example, SiO_2 have unprecedented long-term stability against ambient influences. Moreover, the required depositions techniques for SiO_x (e.g., PECVD or sputtering) are already used in standard microelectronic or photovoltaic fabrication lines and such films can be easily treated with the standard process technology (photolithography, wet or dry etching, chemical-mechanical planarization, etc.). Apparently, the choice of the Si NC fabrication method is a trade-off and both classes (free standing and matrix embedded) have their advantages and disadvantages depending on the intended application. In the following of this chapter only matrix-embedded Si QDs will be discussed.

4.2.2 *Phase Separation for Matrix-Embedded Si QDs*

Besides small dimensions in the range of the exciton Bohr radius, significant band offsets are another requirement to observe significant QC effects in QDs. The surrounding matrix of the Si QD must therefore not only provide a higher band gap but also the band offsets between the respective CBs and VBs have to be large enough to confine the exciton. If one band offset is too small the wavefunction of either the electron or the hole will leak into the matrix and the confinement is attenuated. Three important Si-based dielectric compounds with group IV, V, and VI elements exist that can serve as matrix material: SiC, Si_3N_4, and SiO_2. The assumed band alignments of these dielectrics with a Si NC are shown schematically in Fig. 4.1. Assuming a band gap of 1.7 eV for very small Si NCs, the band offsets to SiC become rather small. On the other hand, the band gap of SiC varies between 2.4–3.3 eV depending on its polytype [42]. For Si_3N_4 and SiO_2 the fundamental band offsets are large enough even for the case of a substantially widened band gap of the Si NC. It has to be pointed out that QC does generally not affect VB and CB symmetrically, that is, a total confinement energy of 200 meV is not equally distributed onto both band edges with 100 meV each. Theoretically [43] and experimentally [44] the shift of the VB edge has been identified to be approximately twice the shift of the CB edge. In contrast, other experimental data suggested that the CB edge is shifted by more than twice the energy of the VB edge [45].

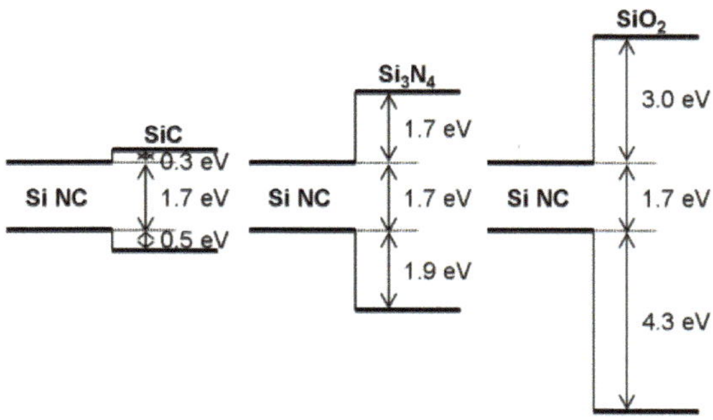

Figure 4.1 Band alignments of a Si NC with a VB:CB shift of 2:1 in SiC, Si_3N_4, and SiO_2; schematic in analogy to Ref. [41].

The issue of energy band alignments of Si NCs is apparently still a matter of debate.

The self-organized growth of Si QDs requires a Si excess concentration which is either introduced by implantation or the Si-rich dielectrics are directly deposited. The respective materials are often referred to as Si-rich carbide (SRC), Si-rich nitride (SRN), and Si-rich oxide (SRO). Equations 4.1, 4.2, and 4.3 illustrate the thermally activated phase separation reactions into the stoichiometric matrix and pure silicon:

$$\mathrm{SiC}_x \rightarrow x\ \mathrm{SiC} + (1 - x)\ \mathrm{Si} \tag{4.1}$$

$$\mathrm{SiN}_x \rightarrow \frac{3x}{4}\mathrm{Si_3N_4} + \left(1 - \frac{3x}{4}\right)\ \mathrm{Si} \tag{4.2}$$

$$\mathrm{SiO}_x \rightarrow \frac{x}{2}\mathrm{SiO_2} + \left(1 - \frac{x}{2}\right)\ \mathrm{Si} \tag{4.3}$$

Please note, instead of SiC_x the form $\mathrm{Si}_{1-y}\mathrm{C}_y$ is also often used for SRC but both notations can be easily converted into each other by $x = y/(1 - y)$. The driving force of the phase separation reaction is the ambition of the compound elements to achieve inert-gas configuration—the higher the EN, the stronger the phase separation. Typically, the phase separation is studied by Fourier transform infrared spectroscopy (FTIR) as function of annealing temperature mainly by the observation of a shift of the asymmetric Si–O stretching mode from ~1000 to 1080 cm^{-1} (see Fig. 4.2). It turns out that onset of phase separation is at temperatures as low as ~400°C for SiO_x [46] and increases to ~650°C–800°C for SiN_x [47, 48] or ~700°C for SiC_x, respectively [49, 50]. Another factor that influences the phase separation is the atomic density of the material, the higher the density the more structural rearrangements are hampered. The rather low onset temperature of the phase separation for SiO_x is therefore not only a result of the higher EN difference between Si (EN = 1.8) and O (EN = 3.5) compared to N (EN = 3.0) and C (EN = 2.5). The atomic density of $\mathrm{SiO_2}$ is also only $\sim 7 \times 10^{22}$ cm^{-3} compared to $\sim 1 \times 10^{23}$ cm^{-3} for $\mathrm{Si_3N_4}$ and 3C–SiC. The difficulties to efficiently form carbide embedded Si NCs is also reflected in the fact that often very high excess Si concentrations with $\mathrm{SiC}_{x\leq 0.1}$ are chosen [49, 50], which correspond

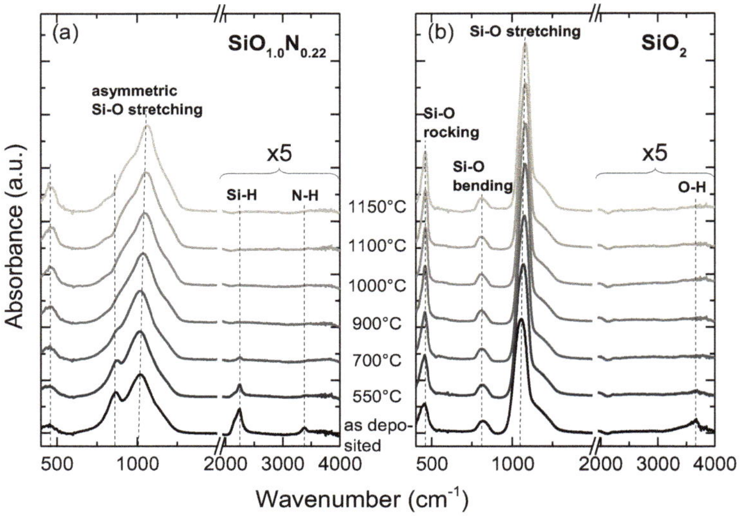

Figure 4.2 FTIR absorption spectra of 200 nm thick (a) SRON and (b) SiO_2 films subjected to different annealing temperatures. The spectral region from 2000 to 4000 cm^{-1} is magnified by a factor of 5 to improve the visibility of the N–H, Si–H, and O–H bands.

more to a C-containing a-Si material than to a real SRC. Very recently, this situation changed and SRC materials with $SiC_{x=0.66-0.33}$ were successfully fabricated to investigate fundamental properties [51] and its applicability on the device level [52].

A PECVD inherent issue for SRO deposition arises from the generally used precursor gasses SiH_4 and N_2O. Whereas for the deposition of SiO_2 very high $[N_2O]/[SiH_4]$ gas flow ratios are used to provide sufficient oxygen for the stoichiometric reaction, the deposition of SRO requires rather low $[N_2O]/[SiH_4]$ gas flow ratios. Although the reactivity of the O radicals is much higher than that of the N≡N molecules left behind from the dissociation of N≡N–O/N=N=O molecules in the plasma, some of the N_2 molecules also dissociate and react with Si. As a consequence pure SRO cannot be deposited from a SiH_4-/N_2O-based plasma chemistry. The deposited material is better described as a Si-rich oxinitride (SRON) and typical (and inevitable) N concentrations range between 5 and 15 at.% [53–55]. It has been shown that the nitrogen slightly hampers the phase separation but the overall impact, for instance, on the luminescence properties of the Si NCs embedded in SiON is negligible. An important feature of the oxinitride matrix is the finding that it does not represent a microscopic mixture of SiO_2 and Si_3N_4 phases but a homogenous composition in which N atoms are just substituting O atoms [54, 56]. The use of SiH_4 and O_2 in a PECVD system is potentially dangerous (explosive) and therefore interlocked in most commercial systems. Especially at increased pressure and high precursor gas concentrations (low inert gas dilution) the formation of dust can occur even without plasma. However, using suitable conditions a clean PECVD process is possible and successful Si NC fabrication in N-free SiO_2 has been demonstrated [57].

For all PECVD deposition techniques the films contain often quite high H_2 concentration (in the range 10 at.% H) due to the use of hydrogen based precursor gasses (e.g., SiH_4, NH_3, CH_4, etc.). As shown in Fig. 4.2 the hydrogen effuses starting from $\sim$500°C. FTIR demonstrates that in the SRON material the majority of the hydrogen is configured as Si–H but also N–H and O–H bonds are present. It is mandatory that the heating ramps for the annealing of H-containing films are reduced to values low enough (sometimes down to single K/min ramps) to allow for effusion until all the hydrogen in the sample is depleted. Otherwise, a blistering of the film or the occurrence of micro bubbles deteriorates the structure of the sample.

The dynamics of the phase separation provides fundamental insight in the formation of Si QDs and is therefore discussed in the following for SiO_x as exemplary case. Isochronal FTIR measurements at different annealing temperatures reveal the evolution of the phase separation [46], however, isothermal FTIR measurement series at different annealing times show that the phase separation is a very fast process: The shift of the Si–O stretching mode reaches its maximum already after a couple of seconds [58]. Considering the low diffusion coefficient of Si in SiO_2 of 5×10^{-17} cm^{2}s^{-1} at 1000°C [59] the fast phase separation can hardly be explained by solely considering Si diffusion. As shown by a simulation in Ref. [60] annealing times beyond 10^3 hours would be required to achieve complete phase separation at 1000°C. The authors therefore suggested a different mechanism which is based on the local out-diffusion of oxygen atoms from a SiO_x region, which leaves a Si nanocluster surrounded by SiO_2 behind [60].

Typically, the phase separation is accomplished in the range of 900°C for all matrices and amorphous Si QDs are formed. In principle, QC affects amorphous QDs in the same way as crystalline QDs. However, in the case of Si nanocluster PL and electroluminescence quantum yields (QYs) are at least 1 order of magnitude higher for c-Si QDs [10, 61]. The absorption edge of oxide embedded a-Si QDs is masked by the absorption of defects which most likely also act as luminescence quenching centers [10]. Also, the current densities through a-Si QDs ensembles are much higher than through c-Si QDs which can be dedicated to a defect assisted current transport [61]. However, the trends of low PL efficiencies hold true even for Si QDs not embedded in SiO_2 [62, 63]. Anthony et al. reported that even perfectly surface functionalized a-Si QDs do not emit PL with more than 2% QY, whereas equally treated c-Si QDs exceed 40% [62]. Therefore, defect centers inherent to the disordered amorphous Si network like, for example, D-centers (cf. Section 4.5) represent a fundamental limit to the photovoltaic or optoelectronic applicability of a-Si QDs.

4.3 Silicon Quantum Dot Crystallization

The crystallization temperature of bulk Si is ~700°C. Due to the Gibbs–Thomson effect and the huge surface-to-volume ratio at the nanoscale the melting temperature of, for example, Au nanoparticles is decreased by several hundred degrees Celsius [64]. In contrast, the crystallization temperature of a-Si is increasing with decreasing nanocluster size. This can be explained by the classical nucleation theory for a spherical nucleus. The crystallization of a-Si liberates a Gibbs free energy $G_{\text{V,ac}}$ that scales with the volume of the nanocluster. On the other hand, the energy $\Delta\sigma_{\text{oc-oa}}$ is consumed because a new interface has to be formed (a-Si/SiO_2 → c-Si/SiO_2). The resulting Gibbs free energy ΔG is therefore:

$$\Delta G = -\frac{4}{3}\pi r^3 \cdot G_{\text{V,ac}} + 4\pi r^2 \Delta\sigma_{\text{oc-oa}} \tag{4.4}$$

Apparently, the stable crystallization of a Si nanocluster requires to overcome the energy expense of the interface formation. The critical crystallization radius r^* is defined as the point where the further growth of the nucleus liberates Gibbs free energy. On the contrary a cluster cannot crystallize if $r < r^*$. The critical crystallization radius r^* can be derived by solving Eq. 4.4 for:

$$\frac{d\Delta G}{dr} = 0 \Rightarrow r^* = \frac{2\Delta\sigma_{\text{oc-oa}}}{G_{\text{V,ac}}} \tag{4.5}$$

Using the values for $G_{\text{V,ac}}$ and $\Delta\sigma_{\text{oc-oa}}$ given in Refs. [65–67], a critical crystallization size of $r^* \approx 1.5$ nm can be derived. Significantly smaller a-Si QDs do not crystallize irrespective temperature. Furthermore, the crystallization temperature of small Si clusters increases to around 1000°C [67].

Experimentally, annealing temperatures in the range of 1100°C to 1250°C are widely used to form Si NCs, so that the crystallization of the matrix has to be considered. In the case of SiO_2, matrix crystallization does not take place for the above-mentioned annealing temperature range. In contrast, the crystallization of the Si_3N_4 matrix has been reported for annealings at 1150°C [68], whereas our own results of SRN/Si_3N_4 samples did not show crystallization even at 1200°C [69]. Hence, the unintended crystallization of Si_3N_4 during annealing might also be influenced by the deposition technique and other parameters. The crystallization of the SiC matrix surrounding Si NCs, however, is inevitable in any case since its crystallization in the cubic polytype (3C or β-SiC) was shown to start already at 900°C [50, 70]. In this case the average size of the SiC crystals was determined to 2–3 nm. In general, the persistence of the matrix in its amorphous state during high-temperature annealing is beneficial for the Si NCs. An amorphous matrix can easily form a smooth interface around the nanocrystalline Si core. In contrast, a crystalline or nanocrystalline matrix implies a lattice mismatch with the c-Si and is thereby also prone to create dislocations, grain boundaries, stacking faults, and other structural defects that are characteristic for crystalline materials and their interfaces. All of them have potentially deleterious effects on the optical and electrical properties of the Si NCs.

Three major methods are used to investigate the crystallinity of a sample: Raman spectroscopy, X-ray diffraction (XRD), and certain modes of transmission electron microscopy (TEM) such as high-resolution transmission electron microscopy (HRTEM) and dark-field transmission electron microscopy (DFTEM). From the TEM investigations the Si NC size can be measured directly. Raman and XRD can also determine the Si NC size via the phonon confinement model [71, 72] or the Scherrer equation [73], respectively. However, in the limit of 2 nm Si NCs essentially only six lattice planes with (111) orientation persist (lattice plane spacing 3.134 Å) and the outer two planes might be disturbed due to the adjacent matrix interface. The consequence is an insufficient X-ray scattering volume or a limited phonon propagation, respectively, which leads to rather large error bars in the Si NC size determination.

With Raman spectroscopy also the ratio between the a-Si and the c-Si fraction can be determined via the two peaks at 480 cm^{-1} and 520 cm^{-1}, respectively [74]. This method requires very careful measurements and analysis since (i) also the surrounding SiO_2 matrix contributes a broad background signal which might be misinterpreted as a-Si signal and (ii) strain causes a broadening and shift of the c-Si Raman line. If the laser light is efficiently absorbed in the sample, Raman peak shifts induced by heating have also to be taken into account. Together with other experimental evidence from TEM a core–shell model was suggested for Si NCs [75, 76]. Thermodynamically, an extended a-Si shell around a c-Si core is unlikely since it would involve a second interface formation energy between c-Si and a-Si. It is well established that the bulk Si/SiO_2 interface is not absolutely abrupt but rather comprised of a few Å thick suboxide (SiO_x) transition layer [77]. Indeed, the same feature was found for oxide embedded Si NCs by soft X-ray spectroscopy [78]. This only Å-thick suboxide transition shell around the Si NC limits the maximum crystallinity values that can be derived from Raman spectroscopy to 70%–80% [79]. The Raman inherent probing of this transition layer can therefore be regarded as an inevitable artifact and should not be misinterpreted as a generally limited crystallization fraction of oxide embedded Si NCs—sufficient annealing temperatures provided, the core of the Si NC is fully crystalline.

Throughout the literature the most widely used annealing dwell time is 1 h. Is it possible to accomplish the whole Si NC formation also within a much short time by using, for example, rapid thermal annealing (RTA) or even flash-lamp annealing (FLA)? Considering the general Arrhenius relation, that is, the speed of a reaction is proportional to $\exp(-1/T)$, we would have to increase the temperature T by almost 1 order of magnitude to accomplish the same reaction in a time t of 1 s rather than 1 h (because T is proportional to $1/\ln t$). However, the phase separation has been identified as a process that is accomplished within seconds. Also the crystallization of the Si QDs is expected to be a spontaneous reaction. Hence, it is well justified to substantially decrease the annealing time while keeping the temperature constant. Indeed, the successful formation of Si NCs after rapid thermal annealings (RTA) [80–82] and even 20 ms FLAs [83, 84] with structural properties comparable to 1 h annealings were demonstrated in literature. It turns out that the major challenge is not posed by the formation of the bare Si NCs but by a good interface between Si and SiO_2 with a minimized defect density [80]. The process of interface reconfiguration that eliminates dangling bonds at Si NC or respectively in the SiO_x transition shell apparently requires a long time at high temperatures, so that superior PL intensity is found for 1 h rather than for much shorter annealings.

4.4 Silicon Nanocrystal Size Control and Shape

The growth of Si NCs from excess Si in a dielectric matrix during annealing is a self-organized process (bottom-up approach). On the basis of the phase separation dynamics and the low diffusion coefficient of Si in the matrices, the typical NC size in, for example, $SiO_{x\approx 1}$ films after conventional annealing is around 5 nm, irrespective of the fabrication method [27, 29, 85]. The mean Si NC size can be decreased by providing less excess Si either by a lower implantation dose [86] or by depositing SiO_x with a high x value [85]. When the stoichiometry of the film is decreased also the total amount of Si NCs or respectively the volume density (NCs per cm^3) is decreased. Hence, this method does not allow for an independent control of Si NC size and density. Depending on the amount of excess Si even a network of interconnected Si nanostructures can occur (percolation threshold). In this case the classical nucleation process is replaced by spinodal decomposition. The excess Si threshold for these two growth regimes is in the range of ~40 at.%, corresponding to $SiO_{x\approx 0.6}$ [87]. A general drawback of the self-organized NC growth process is the rather broad size distribution which is of log-normal shape due to the influence of Ostwald ripening. At the beginning of the annealing, when the critical temperature for the crystallization of the QDs is reached, also Si NCs as small as the critical diameter are present. In the course of the typical 1 h annealing the diffusion of Si atoms will let the larger clusters grow while the smaller clusters shrink until dissolution. This classical Ostwald ripening has its origin in the surface-to-volume ratio of smaller Si NCs: The bigger the NC, the more atoms can be bonded in the energetically favorable interior of the volume and not at the surface. For oxide embedded Si NCs a typical size distribution and FHWM are, for instance, (4.5 ± 1.5) nm [88]. Such a broad size distribution involves a broad distribution of band gaps and hence the investigation of NC-size-dependent optical or electrical properties is difficult.

From TEM or HRTEM, which is typically used to image Si NCs, it was always expected that the self-organization creates mainly NCs with spherical-like shapes. This assumption is well justified, considering natures ambition for energy minimization that should favor a sphere due to its minimized surface-to-volume ratio. In Ref. [89] Yurtsever et al. demonstrated with impressive clarity how wrong the assumption of predominantly spherically shaped Si NCs grown from a SRO thin film actually is. As shown in Fig. 4.3a (reproduced from this work) complex surface morphologies and NC agglomerations with high surface-to- volume ratios were found. The authors also demonstrate the limitations of the commonly used conventional TEM projection imaging modes. In this paper an extended feature that would have been undoubtedly identified as two separated and round NCs in a projection image is revealed to be a single "horseshoe"-shaped NC by TEM tomography. In projection TEM modes, the finite thickness of the TEM specimen (~10–30 nm) causes an integration of image information along the viewing direction. Since the Si NC sizes are typically between 2 and 6 nm projection artifacts are inevitable. In contrast, TEM tomography provides three-dimensional structural information and represents thereby a route to circumvent projection artifacts [90].

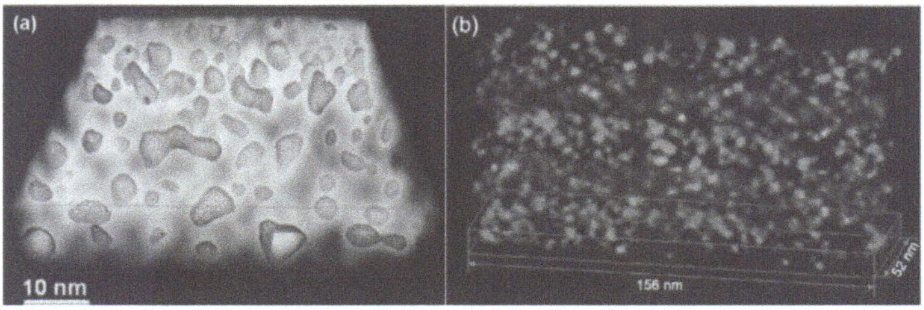

Figure 4.3 Energy-filtered transmission electron microscopy (EFTEM) tomographic reconstructions of (a) a bulk SiO_x film [89] and (b) a SiO_x/SiO_2 superlattice. Reprinted with permission from [Yurtsever A, Weyland M, Muller DA, Three-dimensional imaging of nonspherical silicon nanoparticles embedded in silicon oxide by plasmon tomography, *Appl. Phys. Lett.* **89**, 151920 (2006)]. Copyright [2006], AIP Publishing LLC.

The arbitrary Si NC shape is a major obstacle for both intended applications and fundamental science. Most simulations, theories, and models rely on a spherical shape of the QDs as well as reasonable inter-QD spacing [11, 16, 18, 91, 92]. Especially the coupling between QDs is very sensitive to size and symmetry fluctuations as has been simulated recently [6]. Collective effects in a QD array, such as the formation of energy minibands for electrical transport [91, 93, 94] would be impossible in an ensemble of significantly different Si NC sizes and spacings as well as arbitrary shapes (for more information about electrical transport see Chapter 7). Nonspherical shapes give also rise to anisotropic wavefunction distributions of the exciton within the NCs, so that it will favor the most extended dimension in a NC [94]. As a consequence, for instance, the assumption of isotropic light emission in PL spectroscopy is questionable. Finally, the Si NC agglomerates and other nonspherical shapes have a high surface-to-volume ratio and a larger surface area increases the probability for nonradiative interface defects. The spherical Si NC shape is therefore very desirable in all respects.

4.4.1 *The Superlattice Approach*

The SiO_x/SiO_2 SL approach was initially developed to disentangle NC size and density control [95] but allows also for the preservation

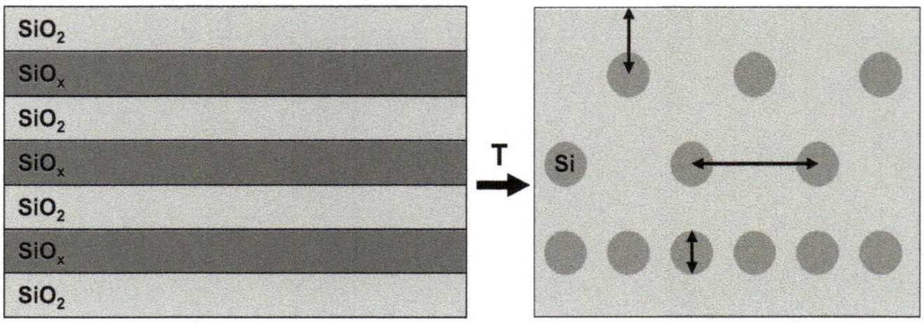

Figure 4.4 Schematic of the SiO_x/SiO_2 superlattice approach that allows for control over vertical and lateral spacing as well as the Si NC size.

of a predominantly spherical NC shape (as shown in Fig. 4.3b). The basic idea is that the Si NC size is controlled by the thickness of the Si rich layer, that is, via a preset deposition parameter. The phase separation and clustering is confined to a quasi-two-dimensional layer by the adjacent SiO_2 barriers, so the thickness of the Si rich layer determines the maximum diameter of the nanocluster. The excess Si concentration (another deposition parameter) provides control over the amount of available Si within the quasi-two-dimensional layer and thereby over the areal Si NC density. In other words, the spacing in all three dimensions can be controlled: The lateral QD spacing by the excess Si concentration and the vertical QD separation by the SiO_2 barrier thickness. This is shown schematically in Fig. 4.4 and in reality by TEM images in Fig. 4.5. Although, the SL approach was first demonstrated for oxide embedded Si NCs, it was later also adopted to oxinitrides [53], nitrides [68, 96], and carbides [50, 51, 70]. As evident from these papers the SL approach works likewise for various depositions methods like evaporation, sputtering, and PECVD. The size distribution of Si NCs in a SL is mostly of Gaussian shape and has a substantially smaller standard deviation of typically $\pm$ 0.5 nm [53, 95]. As explained in Section 4.1 the small size distribution causes comparatively narrow PL peaks. In general, the SL fabrication method allows for various optical, electrical, and structural investigations on size-selected and narrow-distributed Si NC ensembles with quite uniform confinement energy.

In addition to the homogeneous SLs that consist of a stoichiometric dielectric and its Si-rich counterpart, so-called hetero-SLs

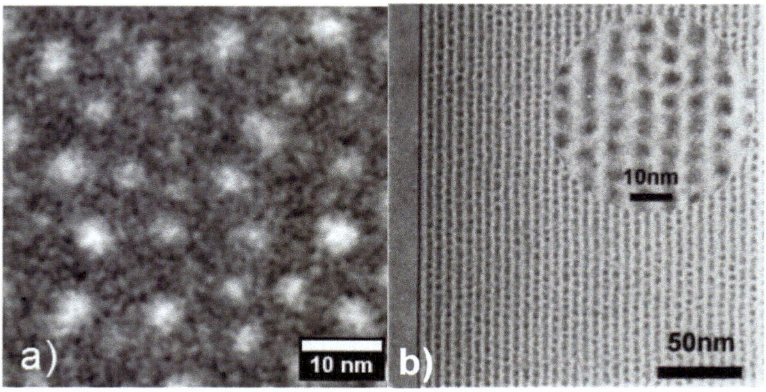

Figure 4.5 EFTEM plane-view image (a) and cross-sectional bright-field TEM image (b) of size-controlled Si NCs in a superlattice [95]. Reprinted with permission from [Zacharias M, Heitmann J, Scholz R, Kahler U, Schmidt M, Bläsing J, Size-controlled highly luminescent silicon nanocrystals: a SiO/SiO2 superlattice approach, *Appl. Phys. Lett.* **80**, 661 (2002)]. Copyright [2002], AIP Publishing LLC.

with two different matrix materials were introduced, sometimes also referred to as hybrid matrix. One example is the SiO_x/Si_3N_4 hetero-SL [97, 98] another one the SiO_x/SiC hetero-SL [99, 100]. The idea of the hetero-SLs is based on the opposing trends of carrier localization for QC (increases with higher matrix band gap) and transport of charge carriers through the QD ensemble for, for example, photovoltaic applications (increases with lower matrix band gap).

To convert higher energetic light more efficiently than a bulk Si solar cell, the QC-induced shift of the ground state to higher energies is required. On the other hand, the charge carriers created by sunlight have to be separated and transported over the CB and VB barriers to the contacts. Therefore the Si NCs are fabricated in a high-gap material (SiO_2) but for the stoichiometric barriers lower gap materials are chosen (Si_3N_4, SiC). Another reason to fabricate the Si NCs from SRO is the superior quality of the Si/SiO_2 interface which will be discussed in the following section. A different, non-oxide-based hetero-SL using SiC_x/Si_3N_4 multilayers was also suggested [101].

4.5 Silicon Nanocrystals: The Role of Point Defects

4.5.1 *Identification and Quantification of Defects*

The most important measurement technique to study paramagnetic point defects like dangling bonds (DBs) with utmost sensitivity is electron spin resonance (ESR). The measurement principle is based on the interaction of an unpaired electron with its atomic surrounding so that information about adjacent atoms and crystallographic characteristics is provided. The magnetic moment μ of the electron will orient itself parallel or antiparallel if a magnetic field **B** is applied. These two alignments have different energies, which is known as the Zeeman effect. By absorbing or emitting a photon of the energy $h\nu = g\mu_{\text{B}}\mathbf{B}$ (μ_{B}: Bohr magneton; g:g factor, that is, effective proportionality constant between observed magnetic moment and angular momentum) the electron can switch between these states. Since the electron has a spin quantum number $s = 1/2$ two states are present which are occupied following a Maxwell–Boltzmann distribution. Therefore ESR is often measured at temperatures as low as 4 K to allow for a maximum occupation of the lower energy state (parallel orientation of spin and **B**-vector). For typical laboratory B-amplitudes of $\sim 1\ T$ (equal to $\sim 10^4$ G) the electromagnetic radiation absorbed by the electron corresponds to microwave radiation in the GHz range. Spectral scans are carried out by varying the magnetic field and measuring the absorbed microwave power P_{μ} at a fixed frequency. Only when the resonance condition of the Zeeman transition is fulfilled significant microwave power is absorbed. For a free electron the g factor is 2.002319. However, the unpaired electron of a DB is influenced by local magnetic fields, including spin–orbit coupling effects mediated by its atomic surrounding. In other words, the g value is characteristic for a defect species. Please note, that the symmetry of the local magnetic fields is not necessarily spherical, so that the g value has to be replaced by a g tensor.

If planar interfaces are measured (e.g., a thermally oxidized Si wafer) another parameter is of concern: The orientation of the applied **B**-field to the sample normal n. In general, the defects at a planar interface can only exist in a certain geometric form so that they will experience significant differences in the **B**-field, depending on its orientation. In contrast, the defects at the NC interface are randomly distributed in space (powder pattern) so that no difference in the spectra between parallel or perpendicular magnetic field orientations is observed. Thereby, bulk Si/matrix interface defects (anisotropic spectra) and NC interface defects (isotropic spectra) can be distinguished [102].

ESR spectra are commonly represented as the first derivative dP_{μ}/dB as a function of **B** and for analysis of the defect species a simulation has to be carried out. The quantification of the defect density (i.e., the amount of spins in the sample volume) is usually accomplished by means of a comounted calibrated marker sample with a well known spin density. The sample material has to be placed into a microwave cavity whose dimensions are determined by the wavelength of the microwave radiation. Often the cavity volume is just a couple of mm^3 so that the samples have to be cut and stacked over each other. To increase the amount of measureable spins in the cavity the volume of the Si NC films should be maximized by choosing the thinnest possible substrate. In the case of Si substrates, of course p-Si has to be chosen since otherwise the signal of the donor electron in n-Si will dominate the spectrum which masks any defect signals.

4.5.2 *Classification of Point Defects*

Point defects occur at three locations: in the matrix surrounding the Si QDs, in the interior of the QDs, and at the interface between QD and matrix. Obviously, defects within the QD and at the interface are the most critical ones since their presence is likely to interfere with the QD properties. Point defects located in the matrix might be tolerable if either their spatial position is sufficiently far away from the QD or if their electronic states are far beyond the QD band edges. States within the QD band gap provided, point defects are highly efficient recombination centers. In the case of indirect semiconductors like Si, the exciton lifetime is quite long (μs to ms timescale) because the electron and hole have not sufficient overlap in momentum space to recombine directly. As a generic structural feature point defects are highly localized in real space.

According to Heisenberg's uncertainty principle this demands a virtually arbitrary state in momentum space and provides hence any momentum required for the recombination of the exciton. In a defective QD the exciton recombines very fast on a timescale of picoseconds, which is many orders of magnitude faster than the intrinsic recombination and explains the infinitesimal radiative recombination probability. Though many point defects act as nonradiative recombination centers also radiative defects (localized centers) exist.

4.5.2.1 Defects in the Si/SiO_2 system

The Si NC/SiO_2 system benefits from the profound knowledge of the interface properties developed in the course of metal-oxide-semiconductor field-effect transistor (MOS-FET) technology. It represents therefore the best understood of all three interfaces and is used here as a case study.

In the Si/SiO_2 system six major defect centers were identified, of which three are interface defects. The SiO_2 specific defects are EX and E'. EX is modeled as an electron delocalized over four oxygen atoms (backbonded to Si atoms) at the site of a Si vacancy (see Fig. 4.6a) and occurs in densities of up to 10^{18} cm^{-3}, depending on the oxidation conditions of the Si wafer [103, 104]. The EX center does not (as far as currently known) interact with light and can be efficiently deactivated by annealing in H_2. E' centers are a group of point defects that usually occur upon irradiation of SiO_2 with UV light, X-rays, or ionizing radiation. The E' center is a dangling bond located at a Si atom which is back-bonded to three oxygen atoms or, in other words, an oxygen vacancy (see Fig. 4.6b) [105]. Depending on the precise structure, E' centers in amorphous SiO_2 are labeled with Greek subscripts. E'_γ is the most prominent defect in a-SiO_2 and occurs in thermal oxide in densities in the lower 10^{17} cm^{-3} range [106]. Besides, E' centers are rather high-energy defects and interact with light in the range of 6 eV so that even in the case of their presence in the sample no interference with the Si NC PL is expected [105].

Another potential volume defect in the Si NC/SiO_2 system is the D center, which is a characteristic defect in amorphous or

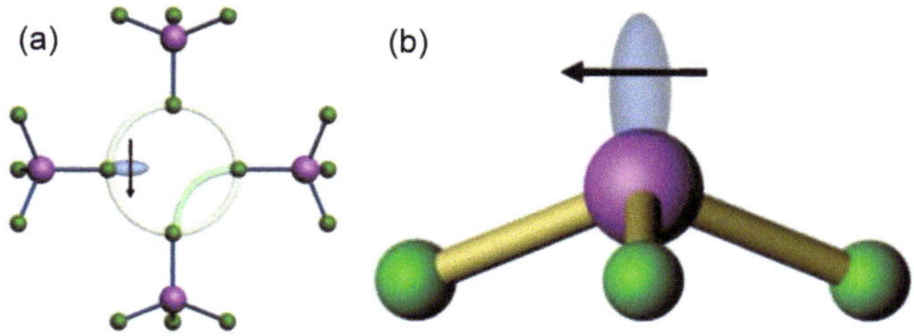

Figure 4.6 Schematics of the SiO_2-specific point defects (a) EX and (b) E' (purple spheres: Si; green spheres: O; arrow: unpaired electron). From Ref. [110].

polycrystalline Si. The D center is a randomly oriented dangling bond located at a Si atom which is backbonded to three Si atoms [107, 108]. Its presence in Si NC/SiO_2 samples annealed at 1100°C was disproven, which supports the assumption that high-temperature annealing forms a highly crystalline Si QD material [109].

The dangling bonds at the Si/SiO_2 interface are labeled as P_b centers and are also configured as unpaired electron at a Si atom which is backbonded to three Si atoms but the DB extends into the SiO_2. Depending on the orientation of the Si crystal three different P_b-type centers occur: P_b, P_{b0}, P_{b1} [111]. The P_b and P_{b0} center are chemically identical ($Si_3 \equiv Si\bullet$, where • denotes the dangling bond) but occur in structurally slightly different configurations [112]. In the P_{b1} center the Si atom having the dangling bond is backbonded by a strained Si–Si bond ($\equiv Si\sim Si\bullet = Si_3$, where $\sim$denotes the strained bond) [113]. Schematics of the P_b-type centers are also shown in Fig. 4.7. For the three fundamental orientations of the Si wafer the occurrence of P_b-type centers is:

- {100}: P_{b0} and P_{b1}
- {110}: P_b
- {111}: P_b

In the case of oxide embedded Si NCs it has not been possible to separate the signals of P_b and P_{b0} so that the notation $P_{b(0)}$ was introduced [102]. According to the ESR literature the natural DB densities at the thermal SiO_2/Si interface for optimum oxidation temperatures are $[P_b] \approx 5 \times 10^{12}$ cm^{-2} and $[P_{b0}]$, $[P_{b1}] \approx 1 \times$

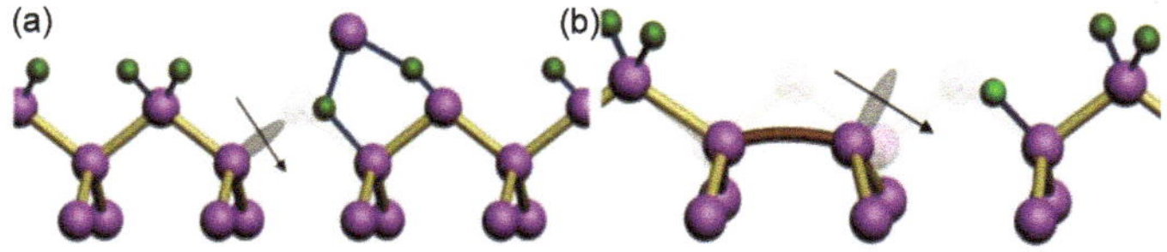

Figure 4.7 Schematics of the (100)-Si-/SiO_2-specific P_b-type point defects (a) P_{b0} and (b) P_{b1} (purple spheres: Si; green spheres: O; arrow: unpaired electron). From Ref. [110].

10^{12} cm^{-2}. The lower density of the (100)-specific defects explains why this wafer orientation is preferred in microelectronic device fabrication. Postoxidation annealing in H_2 ambient passivates the DBs very efficiently and decreases their densities to negligible levels [114]. Interestingly, experimental evidence suggests that the P_{b1} defect is not an electrically active center since it does not have electronic levels in the Si band gap [115, 116]. On the other hand, data was presented that allocates the P_{b1} levels slightly below midgap [117]. In turn, the measurement technique used in that study was shown to be potentially invasive [118], so that the question of the role of the P_{b1} center remains under debate. In general, it has to be noted that though ESR provides comprehensive insight in the nature of paramagnetic defects the potential presence and impact of nonparamagnetic (ESR-invisible) defects has to be kept in mind.

4.5.2.2 Defects in the Si/Si_3N_4 system

In Si_3N_4 two major DBs were found: The K center and the N center. The K center is configured as $N_3 \equiv Si\bullet$ (where • denotes the dangling bond), that is, the DB is localized on a Si atom which is backbonded to three N atoms [119, 120]. Typical K center densities measured by ESR were reported as (2–5) $\times 10^{17}$ cm^{-3} [121, 122], however, for high-temperature-annealed and therefore virtually H-free samples even 2×10^{18} cm^{-3} were observed [123]. In the N center the DB is localized on the N atom which is backbonded to two Si atoms: $\bullet N{=}Si_2$ [124, 125]. In analogy to the E' center in oxide, the N center is often observed after UV irradiation with densities of up to 1×10^{18} cm^{-3} [126]. The energy band diagram developed by Warren et al.

[127] reveals that the N center is a shallow defect in Si_3N_4 with a level only ~0.5 eV above the nitride VB. In contrast, the K center is a midgap state in Si_3N_4 located ~2.6 eV above the nitride VB edge. It might therefore act as a recombination center, if it is located in the vicinity of a nitride-embedded Si NC.

The DBs at the Si/Si_3N_4 interface are labeled as P_{bN} centers and are configured in the same way as the P_b center in oxide ($Si_3{\equiv}Si\bullet$) [122, 128]. Irrespective of the Si wafer orientation only the P_{bN} center has been observed but the densities differ dramatically: $[P_{bN(100)}] \approx (5–7) \times 10^{11}$ cm^{-2} [122] compared to $[P_{bN(111)}] \approx (7–32) \times 10^{12}$ cm^{-2} [128]. Apparently, the (100)-Si/Si_3N_4 interface has intrinsically a quite good interface quality, whereas the (111)-Si/Si_3N_4 interface exhibits DB densities up to six times higher than the (111)-Si/SiO_2 interface. Furthermore, thermally nitrided (111)-Si wafers were shown to have highest P_{bN} densities after prolonged inert gas annealings that cause the total effusion of hydrogen from the films [129]—exactly this is the case of the fabrication method for nitride embedded Si NCs.

4.5.2.3 Defects in the Si/SiC system

For carbide embedded Si NCs no detailed ESR studies have been conducted so far (to the knowledge of the author). The interface defects of bulk Si and a-SiC or c-SiC or respectively the volume defects in SiC are only scarcely investigated and not easily adoptable to the Si NC system. DBs located on a Si atom (P_{SiC}) as well as on a C atom (P_{CC}) in the range of 10^{18} cm^{-3} were reported for LPCVD grown 3C–SiC [130]. In Refs. [131, 132] several different DBs in SiC nanoparticles were measured and labeled DI–DIV. In crystallized SRC films grown by photo-CVD a DB density of even 2×10^{19} cm^{-3} was measured [133]. In accordance to the P_b-type DBs the P_{bC} center was introduced as $Si_3{\equiv}C\bullet$ [134, 135]. The defect situation in the SiC matrix is further complicated by the occurrence of DBs on either the Si or the C atoms depending on the precise composition [136]. Furthermore, the DB density of 3C–SiC even depends on the surface reconstruction of the crystallites [137]. Recapitulating from Section 4.3 the inevitable crystallization of the SiC matrix during annealing into 2–3 nm β-SiC NCs and the typical Si NC size of 5 nm, numerous

SiC NCs surround a single Si NC. Therefore, not only point defects will play a role but also grain boundaries and dislocations.

4.5.3 *Influence of Interface Defects on PL*

4.5.3.1 Interaction of defects with PL in SiO_2-embedded Si NCs

As explained at the beginning of Section 4.5.2 DB defects represent ultimate PL quenching centers [18], hence their occurrence will dominate the PL behavior of the Si NCs. By means of ultrafast PL spectroscopy, with excitation pulses and time resolved detection resolution in the fs range, the dynamics of the early stages of exciton relaxation and recombination can be studied [138, 139]. Further details on the ultrafast spectroscopy can be found in Chapter 5.

Initially, the quantification of the DBs in a size-controlled Si NC sample revealed on average 70% defective NCs, that is, a minority of the NC ensemble is PL active [102, 109]. Later the influence of the high-temperature annealing ambient (N_2 or Ar) as well as the possibility to passivate the DBs with hydrogen was studied. It turns out that the annealing in N_2 atmosphere is superior to Ar since it is able to passivate a significant amount of DBs [140]. For temperatures exceeding $\sim$900°C N_2 is not inert anymore toward the Si/SiO_2 interface and builds up a monolayer of interfacial N atoms which are partially bonded to available DBs—an effect well known from the bulk Si/thermal oxide interface [141, 142]. The same feature was also found for the Si NC/SiO_2 interface and even a very small density of K centers (a typical Si_3N_4 defect) was measured [17]. The impact of passivation annealings in H_2 is tremendous and decreases the DB densities sometimes even below the sensitivity limits of ESR [109] while the PL intensities rise usually by several hundred percent.

The PL peak energy is also influenced by annealings in N_2 and H_2. For N_2 compared to Ar annealing a blue shift (N blue shift) is observed which is pronounced for the smaller Si NCs (2–3 nm) as shown in Fig. 4.8. This effect can be explained by the influence of the polarity of the surface terminating groups (cf. Section 4.1). In contrast, the H_2 passivation causes a red shift of the PL peak which is based on the preferential emission enhancement of the larger

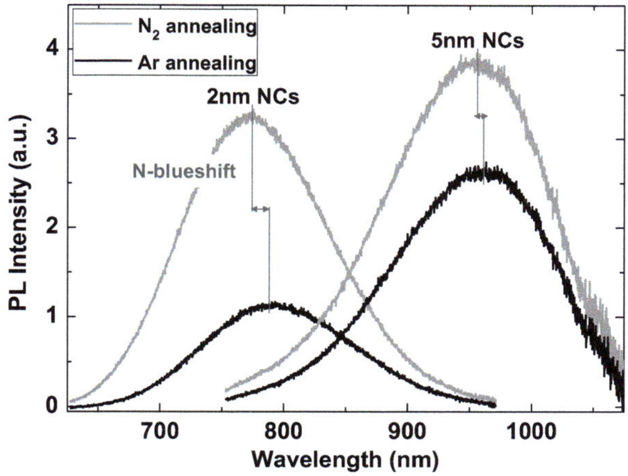

Figure 4.8 PL spectra of size-controlled 2 and 5 nm Si NCs annealed in Ar and N_2; the NC-size-dependent N blue shift is indicated by the vertical lines.

NCs within the narrow size distribution [143]. Within a given Si NC ensemble the larger NCs are more prone to be defective since their surface area is larger. As a consequence the PL spectrum of the unpassivated sample is dominated by the emission from the smaller fraction of the NC ensemble. When the defects are almost entirely passivated by H_2, the PL spectrum is composed of all NCs of the ensemble, but now with a higher contribution of the larger NCs which emit at slightly lower energies. As shown in Fig. 4.9 the H red shift increases with increasing NC size distribution so that it can also be regarded as a measure of size control [140].

In Fig. 4.10 the NC size dependence of the DB density is shown. Whereas for the unpassivated sample set the amount of DBs per NC clearly increases with NC size (lower panel), the concentration of DBs per effective NC interface area is, within error bars, a constant function of NC size (upper panel). That means the amount of DBs scales solely with the interface area of the Si NC: The larger the NC, the higher the probability for a DB defect [144]. Having a closer look at the DB densities, it turns out that $(2.3 \pm 0.8) \times 10^{12}$ P_b-type defects per cm^2 of NC interface area are present—a value well in between the typical DB densities of (111)- and (100)-bulk Si interfaces (cf. Section 4.5.2.1). This result clarifies that

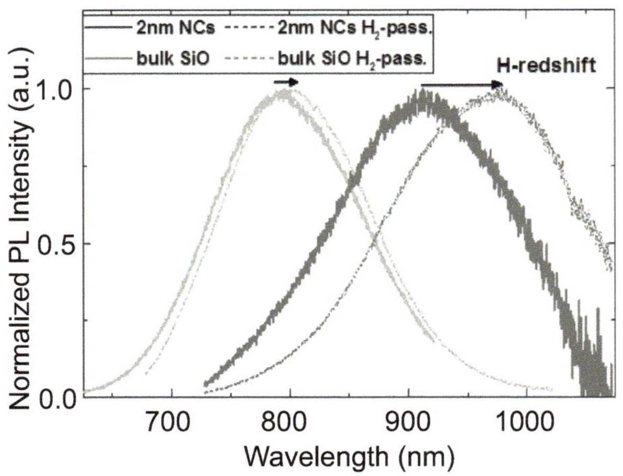

Figure 4.9 PL spectra of a size-controlled 2 nm Si NC sample compared with an annealed SiO bulk film without NC size control; the increased H red shift is indicated by arrows.

the nanoscopic Si/SiO_2 interface is of the same quality as the bulk interface, which is not trivial considering the highly bent surface especially of the smallest Si NCs. A possible explanation can be given following an argumentation of Stesmans et al. [121]: The Si–O bond angle is quite flexible and capable to compensate interfacial stress that arises from the curved interface. H_2 passivation decreases the DB density by a little more than 1 order of magnitude to $(2.0 \pm 0.7) \times 10^{11}$ cm^{-2}, accordingly the ratio of defective NCs decreases to ~6%. It has to be noted that this size dependent DB study [144] was performed on Ar annealed samples to investigate the pristine defect configuration, undisturbed by the influence of interfacial N atoms. Following the evidence given above (N_2 annealing decreases the DB density by ~50% irrespective of H_2 passivation [17, 140]), a fraction of P_b-defective NCs of only ~3% would be expected for N_2 annealed and H_2 passivated samples.

Due to their facet orientation specific occurrence the ratio of $P_{b(0)}$ and P_{b1} can be used to estimate the morphology of the surface terminating planes of the Si NCs. The ratio $R = [P_{b(0)}]/[P_{b1}]$ is for all NC sizes about $R = 1.2 \pm 0.25$. Following the argumentation in Ref. [109] and assuming only a mix of low index facets the idealized

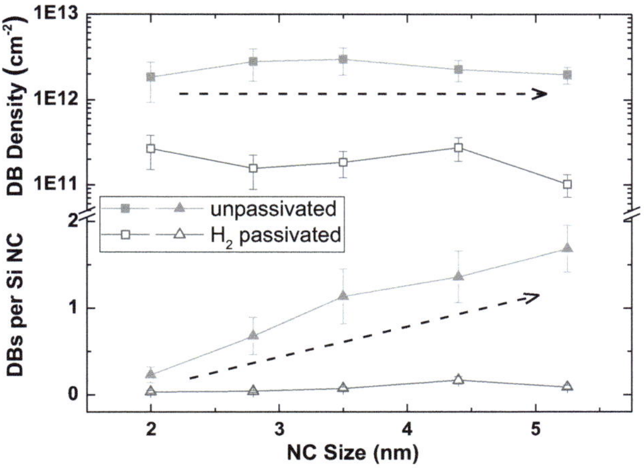

Figure 4.10 P_b-type DB density per Si NC and per effective interface area before and after H_2 passivation as a function of Si NC size.

NC morphology corresponds therefore to a [100] truncated (111) octahedron (cf. inset of Fig. 4.11).

To estimate the impact of the DB defects on the PL of Si NCs, a Poissonian distribution is postulated to calculate the probability $P_{\mathrm{DB}}(k)$ of k defects per Si NC:

$$P_{\mathrm{DB}}(k) = \frac{e^{-n_{\mathrm{DB}}} \cdot n_{\mathrm{DB}}^{k}}{k!} \tag{4.6}$$

where n_{DB} is the average number (expected value) of DB defects per NC. As established before only defect-free NCs (i.e., $k = 0$) have a noninfinitesimal PL emission probability. Therefore, the probability of defect-free, luminescent Si NCs can be derived from:

$$P_{\mathrm{DB}}(k = 0) \quad = \quad e^{-n_{\mathrm{DB}}} \tag{4.7}$$

If the average defect densities of two samples A and B are known, the ratio of the PL intensities I_{PL} can be calculated by:

$$n_{\mathrm{DB,B}} - n_{\mathrm{DB,A}} = \quad \ln\left(\frac{I_{\mathrm{PL,A}}}{I_{\mathrm{PL,B}}}\right) \tag{4.8}$$

An interesting question that remains to be answered is the question of the PL quenching activity of P_{b1}. From a simple PL and ESR analysis it was suggest that the $P_{b(0)}$ DBs are the dominating PL quenching centers [140], however, the sample set with only two size-

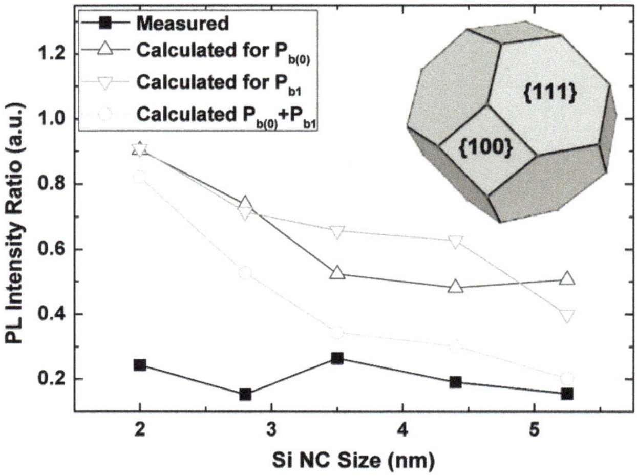

Figure 4.11 Comparison of measured and calculated PL intensity ratios of unpassivated/H_2-passivated samples using Poisson statistics for $P_{b(0)}$ and P_{b1} separately and combined. The inset shows the [100] truncated (111) octahedron.

controlled NC samples was rather small and does not provide good statistics. For the bulk interface the situation has been investigated in detail in terms of ESR and electrical measurements (cf. Section 4.5.2.1), however, an optical study is not feasible with bulk Si in contrast to Si NCs. Using the linear relation between the I_{PL} ratio and the ratio of defect-free probabilities before and after H_2 passivation:

$$\frac{P_{DB}(k=0)}{P_{DB,H_2}(k=0)} = \frac{I_{PL}}{I_{PL,H_2}} \tag{4.9}$$

the measured and the calculated I_{PL} ratios can be compared. The results shown in Fig. 4.11 provide two findings: (i) the Poisson statistics fits much better with the PL measurements for Si NCs $\geq$3.5 nm and (ii) using the single DB species ($P_{b(0)}$ and P_{b1}) no correlation to the measured PL can be obtained. Only with the densities of both DB species together the experimentally observed PL values can be approached. Hence both P_b-type centers seem to act as luminescence quenching centers. The severe disagreement for NCs $<$ 3 nm is puzzling since the measured PL intensity of the passivated samples would have to be much smaller to match the Poisson model. In other words the small Si NCs emit more than they are supposed to from the DB distribution point of view. More realistic is the assumption that in the unpassivated state more not paramagnetic, but H_2-passivatable defects could be present preferentially on smaller Si NCs. Also, a significant source of error arises from the estimation of the Si NC density.

In some papers the values of electrically active interface traps (D_{it} quantified by C–V or DLTS) are compared to the P_b densities (quantified by ESR) and a factor of 2 is found [116, 145]. This must not be misinterpreted as an evidence for a significant number of diamagnetic defect centers. In this case the terminology has to be taken very serious: *defect center density* refers to the amount of physically present defects (as measured by ESR), whereas *defect level density* refers to the electrically observed number of traps. Since the P_b defect is amphoteric it has two charge transition levels (one $e^- \rightarrow$ no e^-, one $e^- \rightarrow$ two e^-) and hence a sweep over the Si band gap reveals two D_{it} peaks [145]. In other words each defect center has two defect levels so that the P_b centers themselves represent already the vast majority of electrically active interface traps.

However, the role of not ESR active (diamagnetic) defects cannot be neglected especially in terms of PL quenching centers which are not necessarily electrical traps. Several options exist in this case: (i) PL quenching and H_2 passivatable, (ii) PL quenching and not H_2 passivatable, and (iii) not PL quenching. Whereas the latter case is of minor importance, especially those defects that are ESR invisible and not eliminable by H_2 elude themselves from any experimental access. In Ref. [10] two specific distorted bonds were identified by means of simulation and shown to create states in the NC band gap: Si–Si and bridging Si–O–Si bonds at the Si NC surface. In that paper these centers were shown to limit the PL QY and to cause subgap absorption. In addition, the detailed investigation of the dynamics and kinetics of the nonradiative recombination in H_2 passivated (i.e., almost DB-free) Si NCs revealed a temperature activated process [146]. Taking into account the migration of excitons [147] between adjacent Si NCs each defective NC (especially if slightly larger in size and hence with a slightly lower band gap) can also annihilate excitons that were originally excited in defect-free NCs.

In conclusion, the available data allows for a quite comprehensive overview of the interaction of DB defects with the PL. However, since ESR, our main technique to observe the point defects and their nanoscopic surrounding, requires paramagnetic states, a hardly accessible parallel world of diamagnetic centers might be present.

4.5.3.2 Interaction of defects with PL in Si_3N_4-embedded Si NCs

The PL of Si NC/Si_3N_4 samples was attributed by many authors to QC [48, 96, 148, 149] but also other radiative recombination paths via defects [47, 150, 151] or band tail states were suggested [123, 152]. In these papers rarely unambiguous evidence of QC was provided and often the well known luminescence of the Si_3N_4 matrix itself (around 2 eV) was not sufficiently discussed. In contrast, time resolved PL measurements revealed lifetimes on the ns or subnanosecond timescale [150, 153, 154] which is a clear indication against radiative exciton recombination in indirect Si NCs. In our recent work, the PL of size-controlled Si NC/Si_3N_4 samples was identified to originate solely from the Si_3N_4 matrix [69]. Moreover, the PL peak blue shift with decreasing NC size which was often observed and misinterpreted as QC effect, is demonstrated by transfer matrix (TMM) simulations to be an interference artifact [14, 69]. In this context and together with data given in Section 4.5.2.2, the reason for the absence of PL from nitride embedded Si NCs has to be discussed. First of all, the two fundamental Si interface orientation exhibit two tremendously different P_{bN}-DB densities: $[P_{bN(100)}] \approx (5–7) \times 10^{11}$ cm^{-2} [122] versus $[P_{bN(111)}] \approx (7–32) \times 10^{12}$ cm^{-2} [128]. Considering the virtually invariant Si–N bond angle (accompanied by a very rigid structure of the Si_3N_4 matrix) and the curvature of the nanoscopic Si NC interface, no significant stress relief by bond angle distortion can be expected [121]. Therefore, the upper limits of the experimentally measured P_{bN} densities have to be considered: $[P_{bN(100)}] = 7 \times 10^{11}$ cm^{-2} and $[P_{bN(111)}] = 3 \times 10^{13}$ cm^{-2}. Assuming furthermore the idealized morphology of a [100] truncated (111) octahedron introduced in the previous section, the average defect densities and defect-free probabilities $P_{bN}(0)$ given in Table 4.1 can be derived for some typical Si NC sizes. Under the questionable assumption that the DB density remains constant with

Table 4.1 Calculated defects per NC and probability for a defect-free NC for Si_3N_4-embedded Si NCs, assuming an idealized truncated octahedron shape and literature values for P_{bN}-DB densities

NC size	Surface area (nm^2)	P_{bN} per NC	$P_{PbN}(0)$
2.5	18.6	4.4	1.3×10^{-2}
3.5	36.5	8.5	1.9×10^{-4}
4.5	60.3	14.1	7.3×10^{-7}
5.5	90.0	21.1	6.9×10^{-10}

NC size even for small nitride embedded NCs the maximum $P_{bN}(0)$ is ~1%. The probability for PL active NCs of larger dimensions is infinitesimal. For comparison the $P_{DB}(0)$ values for oxide embedded Si NCs after 1 h annealing and H_2 passivation are beyond 90% and even the very weakly luminescent samples fabricated by an only few second rapid thermal annealing (RTA) have $P_{DB}(0) \approx$ 2%–4% [155]. If at all only very small Si NCs in Si_3N_4 have a chance to emit a measureable amount of PL, in the experiments however, no trace of PL with Si NC origin was found [69]. Though the P_{bN} defects represent a key element of the optical properties in the Si NC/Si_3N_4 system, also the band tail states, which are well known to protrude deep into the Si_3N_4 gap, have to be considered. In principle, the ability to confine an exciton in a Si NC can be attenuated or even lost, if the band tails of the matrix approach the band edges of the Si NC.

4.5.4 *Influence of Interface Defects on Electrical Transport*

Now that the interaction of DB defects with the PL has been discussed in detail, it should be pointed out that the DBs also contribute substantially to the electrical transport and charging behavior of Si NC/SiO_2 samples. In brief, the electric field is able to cause a charge separation in the amphoteric P_b defect via a two-step band-to-band transition [156]. As a consequence charge carriers are created in the Si NC system even if the SL layers are separated from the wafer and the gate contact by thick SiO_2 barriers that prevent any carrier injection from the contacts [156]. Further details of the electrical transport in Si NCs arrays can be found in Chapter 7.

4.6 Conclusions

The fabrication methods for Si NCs are versatile and well established. The phase separation of Si-rich dielectric matrices upon thermal treatment as well as the formation of Si nanoclusters was studied in detail over the past two decades. The increase of the crystallization temperature of nanoscale silicon requires an annealing at temperatures above 1000°C which has to be carried out in a nonoxidizing ambient to prevent the destruction of Si NCs. An intrinsic feature of the self-organized formation of Si NCs from phase separation in a dielectric matrix is the rather broad log-normal size distribution that is accompanied by further structural disorder from agglomeration and the formation of highly nonspherical nanoclusters. As a consequence, such Si NC samples exhibit a broad ensemble of band gaps, exciton lifetimes, absorption cross sections, inter-NC coupling, etc., which makes the analysis of quantum effects tedious. In contrast, the deposition of nm thin Si-rich and stoichiometric multilayers allows for a control of the NC size and the restriction to quasispherical structures via a simple deposition parameter. Furthermore, the excess Si concentration as well as the stoichiometric barrier thickness (also deposition parameters) can be used to adjust the inter-NC spacing and thereby coupling effects in all 3 dimensions. Various measurement methods can be used to study the formation and the structural properties of Si NC (e.g., FTIR, Raman, TEM, XRD, etc.).

PL spectroscopy represents a very powerful technique to study the Si NCs, provided the samples exhibit a sufficient luminescence QY. However, it has to be pointed out that PL spectra have to be analyzed carefully and critically—not every effect is related to QC just because QDs are in the sample. Potential sources of error and misinterpretation involve defect luminescence, artifacts by interference, artifacts by over excitation (Auger quenching), or insufficient knowledge about the structural and chemical composition (impurities). In general, spectral PL can be complemented with time-resolved and/or temperature-dependent PL measurements to obtain further evidence of the PL origin. The fabrication and measurement of reference samples (e.g., stoichiometric matrix material without QDs that underwent the same annealing process) is an imperative for every study.

The major antagonists of PL are nonradiative defects at the Si QD/interface or within the QD. Whereas the latter case plays a dominant role only for a-Si QDs, the interface defects are inevitable features also for crystalline Si NCs. Among the interface defects Si dangling bonds are predominant and were well studied by ESR. Besides structural information about the defects (which is mainly of academic interest) the analysis of the DB densities and the effect of H_2 passivation are crucial for the actual investigation of Si NC properties. The by far lowest interface defect densities are obtained for oxide embedded Si NCs. Using a postannealing in H_2 samples with a fraction of less than 5% DB-defective Si NC can be obtained.

The emission of PL from Si NCs is based on the radiative recombination of quantum-confined excitons. To obtain sufficient QC the band offsets have to be sufficiently high. Caused by structural disorder in the dielectric material adjacent to the QD (variation in bond angles and bond lengths) an exponentially decaying DOS(E) within the nominal band gap can occur. If these band tail states protrude rather deep into the band gap, nominally sufficient band offsets might be superimposed so that no efficient QC can occur. This might be the case for Si_3N_4 or SiC whereas for SiO_2 no such problems are expected and encountered.

Balancing all properties of the three major matrix materials SiC, Si_3N_4, and SiO_2 no definitive decision can be made. Whereas SiC is supposed to allow for the best charge carrier transport through a Si NC array due to its low band offsets, the inevitable crystallization of the matrix before the Si crystallization involves major structural drawbacks such as an increased defect probability (grain boundaries, dislocations, etc.). The rather rigid structure of Si_3N_4 causes very high DB densities at the Si interface, so that optical and electrical properties are defect dominated. SiO_2 has clearly the best structural parameters and its interface to Si is known to be one of the best in nature (besides epitaxially grown systems). This results in superior optical quality and, for instance, ensemble-QY values on the order of magnitude of 10%. Band offsets of at least 3 to 4 eV between Si and SiO_2 CBs/VBs are the major obstacle if current transport is taken into account.

Finally, the influence of the dielectric matrix on the electronic structure of the Si NCs and the resulting optical and electrical properties has to be underlined. In some sense, QDs are more interface than volume since a significant fraction of the QD atoms is coordinated to the surrounding matrix. It was shown that, for instance, the polarity of the matrix atoms interacts with the band structure of the Si NCs and that for small NCs the influence of the matrix exceeds the impact of the QC.

References

1. Yoffe AD, Low-dimensional systems: quantum size effects and electronic properties of semiconductor microcrystallites (zero-dimensional systems) and some quasi-two-dimensional systems, *Adv. Phys.* **42**, 173 (1993).
2. Trwoga PF, Kenyon AJ, Pitt CW, Modeling the contribution of quantum confinement to luminescence from silicon nanoclusters, *J. Appl. Phys.* **83**, 3789 (1998).
3. Beard MC, Knutsen KP, Yu P, Luther JM, Song Q, Metzger WK, Ellingson RJ, Nozik AJ, Multiple exciton generation in colloidal silicon nanocrystals, *Nano Lett.* **7**, 2506 (2007).
4. Ekimov AI, Hache F, Schanne-Klein MC, Ricard D, Flytzanis C, Kudryavtsev IA, Yazeva TV, Rodina AV, Efros AL, Absorption and intensity-dependent photoluminescence measurements on CdSe quantum dots: assignment of the first electronic transitions, *J. Opt. Soc. Am. B* **10**, 100 (1993).
5. Shaklee KL, Nahory RE, Valley-orbit splitting of free excitons? The absorption edge of Si, *Phys. Rev. Lett.* **24**, 942 (1970).
6. Luo JW, Stradins P, Zunger A, Matrix-embedded silicon quantum dots for photovoltaic applications: a theoretical study of critical factors, *Energy Environ. Sci.* **4**, 2546 (2011).
7. Pavesi L, Will silicon be the photonic material of the third millenium?, *J. Phys.: Condens. Matter* **15**, R1169 (2003).
8. Mangolini L, Jurbergs D, Rogojina E, Kortshagen U, Plasma synthesis and liquid-phase surface passivation of brightly luminescent Si nanocrystals, *J. Lumin.* **121**, 327 (2006).
9. Valenta J, Fucikova A, Vacha F, Adamec F, Humpolickova J, Hof M, Pelant I, Kusova K, Dohnalova K, Linnros J, Light-emission performance of

silicon nanocrystals deduced from single quantum dot spectroscopy, *Adv. Funct. Mater.* **18**, 2666 (2008).

10. Lee BG, Hiller D, Luo JW, Semonin OE, Beard MC, Zacharias M, Stradins P, Strained interface defects in silicon nanocrystals, *Adv. Funct. Mater.* **22**, 3223 (2012).
11. Kovalev D, Heckler H, Polisski G, Koch F, Optical properties of Si nanocrystals, *Phys. Status Solidi B* **215**, 871 (1999).
12. Heitmann J, Muller F, Yi L, Zacharias M, Kovalev D, Eichhorn F, Excitons in Si nanocrystals: confinement and migration effects, *Phys. Rev. B* **69**, 195309 (2004).
13. Valenta J, Juhasz R, Linnros J, Photoluminescence spectroscopy of single silicon quantum dots, *Appl. Phys. Lett.* **80**, 1070 (2002).
14. Dyakov SA, Zhigunov DM, Hartel A, Zacharias M, Perova TS, Timoshenko VY, Enhancement of photoluminescence signal from ultrathin layers with silicon nanocrystals, *Appl. Phys. Lett.* **100**, 061908 (2012).
15. Wolkin MV, Jorne J, Fauchet PM, Allan G, Delerue C, Electronic states and luminescence in porous silicon quantum dots: the role of oxygen, *Phys. Rev. Lett.* **82**, 197 (1999).
16. König D, Rudd J, Green MA, Conibeer G, Role of the interface for the electronic structure of Si quantum dots, *Phys. Rev. B* **78**, 035339 (2008).
17. Hiller D, Goetze S, Munnik F, Jivanescu M, Gerlach JW, Vogt J, Pippel E, Zakharov N, Stesmans A, Zacharias M, Nitrogen at the Si-nanocrystal/SiO2 interface and its influence on luminescence and interface defects, *Phys. Rev. B* **82**, 195401 (2010).
18. Delerue C, Allan G, Lannoo M, Theoretical aspects of the luminescence of porous silicon, *Phys. Rev. B* **48**, 11024 (1993).
19. Averboukh B, Huber R, Cheah KW, Shen YR, Qin GG, Ma ZC, Zong WH, Luminescence studies of a Si/SiO2 superlattice, *J. Appl. Phys.* **92**, 3564 (2002).
20. Hayne M, Maes J, Bersier S, Henini M, Müller-Kirsch L, Heitz R, Bimberg D, Moshchalkov VV, Pulsed magnetic fields as a probe of self-assembled semiconductor nanostructures, *Phys. B* **346**, 421 (2004).
21. Godefroo S, Hayne M, Jivanescu M, Stesmans A, Zacharias M, Lebedev OI, van Tendeloo G, Moshchalkov VV, Classification and control of the origin of photoluminescence from Si nanocrystals, *Nat. Nanotechnol.* **3**, 174 (2008).
22. Canham LT, Silicon quantum wire array fabrication by electrochemical and chemical dissolution of wafers, *Appl. Phys. Lett.* **57**, 1046 (1990).

23. Lehman V, Gösele U, Porous silicon formation: a quantum wire effect, *Appl. Phys. Lett.* **58**, 856 (1991).
24. Shimizu-lwayama T, Fujita K, Nakao S, Saitoh K, Fujita T, ltoh N, Visible photoluminescence in Si+ implanted silica glass, *J. Appl. Phys.* **75**, 7779 (1994).
25. Mutti P, Ghislotti G, Bertoni S, Bonoldi L, Cerofolini GF, Meda L, Grilli E, Guzzi M, Room-temperature visible luminescence from silicon nanocrystals in silicon implanted Si02 layers, *Appl. Phys. Lett.* **66**, 851 (1995).
26. Baru VG, Chernushich AP, Luzanov VA, Stepanov GV, Zakharov LY, O'Donnell KP, Bradley IV, Melnik NN, Optical properties of Si nanocrystals prepared by magnetron sputtering, *Appl. Phys. Lett.* **69**, 4148 (1996).
27. Seifarth H, Grötzschel R, Markwitz A, Matz W, Nitzsche P, Rebohle L, Preparation of Si02 films with embedded Si nanocrystals by reactive r.f. magnetron sputtering, *Thin Solids Films* **330**, 202 (1998).
28. Cheong WS, Hwang NM, Yoon DY, Observation of nanometer silicon clusters in the hot-filament CVD process, *J. Cryst. Growth* **204**, 52 (1999).
29. Kenyon AJ, Trwoga PF, Pitt CW, Rehm G, The origin of photoluminescence from thin films of silicon-rich silica, *J. Appl. Phys.* **79**, 9291 (1996).
30. Rinnert H, Vergnat M, Marchal G, Burneau A, Strong visible photoluminescence in amorphous SiOx and SiOx:H thin films prepared by thermal evaporation of SiO powder, *J. Lumin.* **80**, 445 (1999).
31. Kahler U, Hofmeister H, Visible light emission from Si nanocrystalline composites via reactive evaporation of SiO, *Opt. Mater.* **17**, 83 (2001).
32. Kusova K, Cibulka O, Dohnalova K, Pelant I, Valenta J, Fucikova A, Zidek K, Lang J, Englich J, Matejka P, Stepanek P, Bakardjieva S, Brightly luminescent organically capped silicon nanocrystals fabricated at room temperature and atmospheric pressure, *ACS Nano* **4**, 4495 (2010).
33. Mastronardi ML, Hennrich F, Henderson EJ, Maier-Flaig F, Blum C, Reichenbach J, Lemmer U, Kübel C, Wang D, Kappes MM, Ozin GA, Preparation of monodisperse silicon nanocrystals using density gradient ultracentrifugation, *J. Am. Chem. Soc.* **133**, 11928 (2011).
34. Knipping J, Wiggers H, Rellinghaus B, Roth P, Konjhodzic D, Meier C, Synthesis of high purity silicon nanoparticles in a low pressure microwave reactor, *J. Nanosci. Nanotechnol.* **4**, 1039 (2004).

35. Mangolini L, Thimsen E, Kortshagen U, High-yield plasma synthesis of luminescent silicon nanocrystals, *Nano Lett.* **5**, 655 (2005).
36. Anderson IE, Shircliff RA, Macauley C, Smith DK, Lee BG, Agarwal S, Stradins P, Collins RT, Silanization of low-temperature-plasma synthesized silicon quantum dots for production of a tunable, stable, colloidal solution, *J. Phys. Chem. C* **116**, 3979 (2012).
37. Kusova K, Ondic L, Klimesova E, Herynkova K, Pelant I, Danis S, Valenta J, Gallart M, Ziegler M, Hönerlage B, Gilliot P, Luminescence of free-standing versus matrix-embedded oxide-passivated silicon nanocrystals: the role of matrix-induced strain, *Appl. Phys. Lett.* **101**, 143101 (2012).
38. Mangolini L, Jurbergs D, Rogojina E, Kortshagen U, High efficiency photoluminescence from silicon nanocrystals prepared by plasma synthesis and organic surface passivation, *Phys. Status Solidi C* **3**, 3975 (2006).
39. Kovalev D, Timoshenko VY, Künzner N, Gross E, Koch F, Strong explosive interaction of hydrogenated porous silicon with oxygen at cryogenic temperatures, *Phys. Rev. Lett.* **87**, 068301 (2001).
40. Clément D, Diener J, Gross E, Künzner N, Timoshenko VY, Kovalev D, Highly explosive nanosilicon-based composite materials, *Phys. Status Solidi A* **202**, 1357 (2005).
41. Conibeer G, Green M, Corkish R, Cho Y, Cho EC, Jiang CW, Fangsuwannarak T, Pink E, Huang Y, Puzzer T, Trupke T, Richards B, Shalav A, Lin KL, Silicon nanostructures for third generation photovoltaic solar cells, *Thin Solid Films* **511**, 654 (2006).
42. Persson C, Lindefelt U, Detailed band structure for 3C-, 2H-, 4H-, 6H-SiC, and Si around the fundamental band gap, *Phys. Rev. B* **54**, 10257 (1996).
43. Hill NA, Whaley KB, Size dependence of excitons in silicon nanocrystals, *Phys. Rev. Lett.* **75**, 1130 (1995).
44. van Buuren T, Dinh LN, Chase LL, Siekhaus WJ, Terminello LJ, Changes in the electronic properties of Si nanocrystals as a function of particle size, *Phys. Rev. Lett.* **80**, 3803 (1998).
45. Seguini G, Schamm-Chardon S, Pellegrino P, Perego M, The energy band alignment of Si nanocrystals in SiO2, *Appl. Phys. Lett.* **99**, 082107 (2011).
46. Yi LX, Heitmann J, Scholz R, Zacharias M, Phase separation of thin SiO layers in amorphous SiO/SiO2 superlattices during annealing, *J. Phys. Condens. Matter* **15**, S2887 (2003).

47. Jiang L, Zeng X, Zhang X, The effects of annealing temperature on photoluminescence of silicon nanoparticles embedded in SiNx matrix, *J. Non-Cryst. Solids* **357**, 2187 (2011).
48. Molinari M, Rinnert H, Vergnat M, Evolution with the annealing treatments of the photoluminescence mechanisms in a-SiNx:H alloys prepared by reactive evaporation, *J. Appl. Phys.* **101**, 123532 (2007).
49. Künle M, Kaltenbach T, Löper P, Hartel A, Janz S, Eibl O, Nickel KG, Si-rich a-SiC:H thin films: structural and optical transformations during thermal annealing, *Thin Solid Films* **519**, 151 (2010).
50. Song D, Cho EC, Conibeer G, Huang Y, Flynn C, Green MA, Structural characterization of annealed Si1−xCx/SiC multilayers targeting formation of Si nanocrystals in a SiC matrix, *J. Appl. Phys.* **103**, 083544 (2008).
51. López-Vidrier J, Hernández S, Samà J, Canino M, Allegrezza M, Bellettato M, Shukla R, Schnabel M, Löper P, López-Conesa L, Estradé S, Peiró F, Janz S, Garrido B, Structural, optical and electrical properties of silicon nanocrystals embedded in SixC1−x/SiC multilayer systems for photovoltaic applications, *Mater. Sci. Eng. B* **178**, 639 (2013).
52. Löper P, Canino M, Qazzazie D, Schnabel M, Allegrezza M, Summonte C, Glunz SW, Janz S, Zacharias M, Silicon nanocrystals embedded in silicon carbide: Investigation of charge carrier transport and recombination, *Appl. Phys. Lett.* **102**, 033507 (2013).
53. Hartel AM, Hiller D, Gutsch S, Löper P, Estradé S, Peiró F, Garrido B, Zacharias M, Formation of size-controlled silicon nanocrystals in plasma enhanced chemical vapor deposition grown SiOxNy/SiO2 superlattices, *Thin Solid Films* **520**, 121 (2011).
54. Daldosso N, Das G, Larcheri S, Mariotto G, Dalba G, Pavesi L, Irrera A, Priolo F, Iacona F, Rocca F, Silicon nanocrystal formation in annealed silicon-rich silicon oxide films prepared by plasma enhanced chemical vapor deposition, *J. Appl. Phys.* **101**, 113510 (2007).
55. Chen XY, Lu YF, Wu YH, Cho BJ, Tang LJ, Lu D, Dong JR, Correlation between optical properties and Si nanocrystal formation of Si-rich Si oxide films prepared by plasma-enhanced chemical vapor deposition, *Appl. Surf. Sci.* **253**, 2718 (2006).
56. Criado D, Alayo MI, Pereyra I, Fantini MCA, Structural analysis of silicon oxynitride films deposited by PECVD, *Mater. Sci. Eng. B* **112**, 123 (2004).
57. Fazio B, Vulpio M, Gerardi C, Liao Y, Crupi I, Lombardo S, Trusso S, Neri F, Residual crystalline silicon phase in silicon-rich-oxide films

subjected to high temperature annealing, *J. Electrochem. Soc.* **149**, G376 (2002).

58. Danko VA, Indutnyi IZ, Lysenko VS, Maidanchuk IY, Minko VI, Nazarov AN, Tkachenko AS, Shepelyavyi PE, Kinetics of structural and phase transformations in thin SiOx films in the course of a rapid thermal annealing, *Semiconductors* **39**, 1197 (2005).
59. Nesbit LA, Annealing characteristics of Si rich SiO2 films, *Appl. Phys. Lett.* **46**, 38 (1985).
60. Sarikov A, Litovchenko V, Lisovskyy I, Maidanchuk I, Zlobin S, Role of oxygen migration in the kinetics of the phase separation of non-stoichiometric silicon oxide films during high-temperature annealing, *Appl. Phys. Lett.* **91**, 133109 (2007).
61. Irrera A, Iacona F, Crupi I, Presti CD, Franzò G, Bongiorno C, Sanfilippo D, Di Stefano G, Piana A, Fallica PG, Canino A, Priolo F, Electroluminescence and transport properties in amorphous silicon nanostructures, *Nanotechnology* **17**, 1428 (2006).
62. Anthony R, Kortshagen U, Photoluminescence quantum yields of amorphous and crystalline silicon nanoparticles, *Phys. Rev. B* **80**, 115407 (2009).
63. Torchynska TV, Quantum emission efficiency of nanocrystalline and amorphous Si quantum dots, *Phys. E* **44**, 56 (2011).
64. Buffat P, Borel JP, Size effect on the melting point of gold particles, *Phys. Rev. A* **13**, 2287 (1976).
65. Persans PD, Ruppert A, Abeles B, Crystallization kinetics of amorphous Si/SiO2 superlattice structures, *J. Non-Cryst. Solids* **102**, 130 (1988).
66. Spinella C, Lombardo S, Priolo F, Crystal grain nucleation in amorphous silicon, *J. Appl. Phys.* **84**, 5383 (1998).
67. Zacharias M, Streitenberger P, Crystallization of amorphous superlattices in the limit of ultrathin films with oxide interfaces, *Phys. Rev. B* **62**, 8391 (2000).
68. Scardera G, Puzzer T, Perez-Wurfl I, Conibeer G, The effects of annealing temperature on the photoluminescence from silicon nitride multilayer structures, *J. Cryst. Growth* **310**, 3680 (2008).
69. Hiller D, Zelenina A, Gutsch S, Dyakov SA, López-Conesa L, López-Vidrier J, Estradé S, Peiró F, Garrido B, Valenta J, Kořínek M, Trojánek F, Malý P, Schnabel M, Weiss C, Janz S, Zacharias M, Absence of quantum confinement effects in the photoluminescence of Si3N4–embedded Si nanocrystals, *J. Appl. Phys.* **115**, 204301 (2014).

70. Künle M, Janz S, Nickel KG, Heidt A, Luysberg M, Eibl O, Annealing of nm-thin Si1−xCx/SiC multilayers, *Sol. Energy Mater. Sol. C.* **115**, 11 (2013).
71. Campbell IH, Fauchet PM, The effects of microcrystal size and shape on the one phonon Raman spectra of crystalline semiconductors, *Solid State Commun.* **58**, 739 (1986).
72. Crowe IF, Halsall MP, Hulko O, Knights AP, Gwilliam RM, Wojdak M, Kenyon AJ, Probing the phonon confinement in ultrasmall silicon nanocrystals reveals a size-dependent surface energy, *J. Appl. Phys.* **109**, 083534 (2011).
73. Hessel CM, Reid D, Panthani MG, Rasch MR, Goodfellow BW, Wei J, Fujii H, Akhavan V, Korgel BA, Synthesis of ligand-stabilized silicon nanocrystals with size-dependent photoluminescence spanning visible to near-infrared wavelengths, *Chem. Mater.* **24**, 393 (2012).
74. Hernández S, Martínez A, Pellegrino P, Lebour Y, Garrido B, Jordana E, Fedeli JM, Silicon nanocluster crystallization in SiOx films studied by Raman scattering, *J. Appl. Phys.* **104**, 044304 (2008).
75. Iacona F, Bongiorno C, Spinella C, Boninelli S, Priolo F, Formation and evolution of luminescent Si nanoclusters produced by thermal annealing of SiOx films, *J. Appl. Phys.* **95**, 3723 (2004).
76. López-Vidrier J, Hernández S, Hartel AM, Hiller D, Gutsch S, Löper P, López-Conesa L, Estradé S, Peiró F, Zacharias M, Garrido B, Structural and optical characterization of size controlled silicon nanocrystals in SiO2/SiOxNy multilayers, *Energy Proc.* **10**, 43 (2011).
77. Queeney KT, Weldon MK, Chang JP, Chabal YJ, Gurevich AB, Sapjeta J, Opila RL, Infrared spectroscopic analysis of the Si/SiO2 interface structure of thermally oxidized silicon, *J. Appl. Phys.* **87**, 1322 (2000).
78. Zimina A, Eisebitt S, Eberhardt W, Heitmann J, Zacharias M, Electronic structure and chemical environment of silicon nanoclusters embedded in a silicon dioxide matrix, *Appl. Phys. Lett.* **88**, 163103 (2006).
79. Lopez-Vidrier J, Hernandez S, Hiller D, Gutsch S, López-Conesa L, Estradé S, Peiró F, Zacharias M, Garrido B, Annealing temperature and barrier thickness effect on the structural and optical properties of silicon nanocrystals/SiO2 superlattices, *J. Appl. Phys.* **116**, 133505 (2014).
80. Hiller D, Goetze S, Zacharias M, Rapid thermal annealing of size-controlled Si nanocrystals: Dependence of interface defect density on thermal budget, *J. Appl. Phys.* **109**, 054308 (2011).

81. Wan Z, Huang S, MA Green, Conibeer G, Rapid thermal annealing and crystallization mechanisms study of silicon nanocrystal in silicon carbide matrix, *Nanoscale Res. Lett.* **6**, 129 (2011).
82. Rölver R, Winkler O, Först M, Spangenberg B, Kurz H, Light emission from Si/SiO2 superlattices fabricated by RPECVD, *Microelectron. Reliab.* **45**, 915 (2005).
83. Kachurin GA, Cherkova SG, Marin DV, Yankov RA, Deutschmann M, Formation of light-emitting Si nanostructures in SiO2 by pulsed anneals, *Nanotechnology* **19**, 355305 (2008).
84. Kachurin GA, Tyschenko IE, Zhuravlev KS, Pazdnikov NA, Volodin VA, Gutakovsky AK, Leier AF, Skorupa W, Yankov RA, Visible and near-infrared luminescence from silicon nanostructures formed by ion implantation and pulse annealing, *Nucl. Instrum. Methods B* **122**, 571 (1997).
85. Kahler U, Hofmeister H, Size evolution and photoluminescence of silicon nanocrystallites in evaporated SiOx thin films upon thermal processing, *Appl. Phys. A* **74**, 13 (2002).
86. Shimizu-Iwayama T, Kurumado N, Hole DE, Townsend PD, Optical properties of silicon nanoclusters fabricated by ion implantation, *J. Appl. Phys.* **83**, 6018 (1998).
87. Müller T, Heinig KH, Möller W, Size and location control of Si nanocrystals at ion beam synthesis in thin SiO2 films, *Appl. Phys. Lett.* **81**, 3049 (2002).
88. Zacharias M, Heitmann J, Yi LX, Wildanger E, Scholz R (2003) *Towards the First Silicon Laser*, eds. Pavesi L, Gaponenko S, and Dal Negro, L. (Springer, Dordrecht), p. 134.
89. Yurtsever A, Weyland M, Muller DA, Three-dimensional imaging of nonspherical silicon nanoparticles embedded in silicon oxide by plasmon tomography, *Appl. Phys. Lett.* **89**, 151920 (2006).
90. Frank J (1992) *Electron Tomography: Three-Dimensional Imaging with the Transmission Electron Microscope* (Plenum, New York).
91. Jiang CW, Green MA, Silicon quantum dot superlattices: modeling of energy bands, densities of states, and mobilities for silicon tandem solar cell applications, *J. Appl. Phys.* **99**, 114902 (2006).
92. Bulutay C, Interband, intraband, and excited-state direct photon absorption of silicon and germanium nanocrystals embedded in a wide band-gap lattice, *Phys. Rev. B* **76**, 205321 (2007).

93. Löper P, Müller R, Hiller D, Barthel T, Malguth E, Janz S, Goldschmidt JC, Hermle M, Zacharias M, Quasi-Fermi-level splitting in ideal silicon nanocrystal superlattices, *Phys. Rev. B* ***84***, 195317 (2011).
94. Puthen Veettil B, *Modelling and Characterization of Carrier Transport through Nanostructures*, PhD thesis, UNSW, Australia (2012), Ch. 4.3.3 and Ch. 4.3.2, http://handle.unsw.edu.au/1959.4/51640.
95. Zacharias M, Heitmann J, Scholz R, Kahler U, Schmidt M, Bläsing J, Size-controlled highly luminescent silicon nanocrystals: a SiO/SiO2 superlattice approach, *Appl. Phys. Lett.* **80**, 661 (2002).
96. So YH, Gentle A, Huang S, Conibeer G, Green MA, Size dependent optical properties of Si quantum dots in Si-rich nitride/Si3N4 superlattice synthesized by magnetron sputtering, *J. Appl. Phys.* **109**, 064302 (2011).
97. Di D, Perez-Wurfl I, Conibeer G, Green MA, Formation and photoluminescence of Si quantum dots in SiO2/Si3N4 hybrid matrix for all-Si tandem solar cells, *Sol. Energy Mater. Sol. C.* **94**, 2238 (2010).
98. Di D, Xu H, Perez-Wurfl I, Green MA, Conibeer G, Optical characterisation of silicon nanocrystals embedded in SiO2/Si3N4 hybrid matrix for third generation photovoltaics, *Nanoscale Res. Lett.* **6**, 612 (2011).
99. Ding K, Aeberhard U, Astakhov O, Köhler F, Beyer W, Finger F, Carius R, Rau U, Silicon quantum dot formation in SiC/SiOx hetero-superlattice, *Energy Proc.* **10**, 249 (2011).
100. Ding K, Aeberhard U, Beyer W, Astakhov O, Köhler F, Breuer U, Finger F, Carius R, Rau U, Annealing induced defects in SiC, SiOx single layers, and SiC/SiOx hetero-superlattices, *Phys. Status Solidi A* **209**, 1960 (2012).
101. Wan Z, Patterson R, Huang S, Green MA, Conibeer G, Ultra-thin silicon nitride barrier implementation for Si nano-crystals embedded in amorphous silicon carbide matrix with hybrid superlattice structure, *Eur. Phys. Lett.* **95**, 67006 (2011).
102. Stesmans A, Jivanescu M, Godefroo S, Zacharias M, Paramagnetic point defects at SiO2/nanocrystalline Si interfaces, *Appl. Phys. Lett.* **93**, 023123 (2008).
103. Stesmans A, New intrinsic defect in as-grown thermal SiO2 on (111)Si, *Phys. Rev. B* **45**, 9501 (1992).
104. Stesmans A, Scheerlinck F, Generation aspects of the delocalized intrinsic EX defect in thermal SiO2, *J. Appl. Phys.* **75**, 1047 (1994).
105. Skuja L, Optically active oxygen-deficiency-related centers in amorphous silicon dioxide, *J. Non-Cryst. Solids* **239**, 16 (1998).

106. Stesmans A, Afanas'ev VV, Interlayer-related paramagnetic defects in stacks of ultrathin layers of SiOx, Al2O3, ZrO2, and HfO2 on (100)Si, *J. Appl. Phys.* **97**, 033510 (2005).
107. Biegelsen DK, Stutzmann M, Hyperfine studies of dangling bonds in amorphous silicon, *Phys. Rev. B* **33**, 3006 (1986).
108. Umeda T, Yamasaki S, Isoya J, Tanaka K, Electron-spin-resonance center of dangling bonds in undoped a-Si:H, *Phys. Rev. B* **59**, 4849 (1999).
109. Jivanescu M, Stesmans A, Zacharias M, Inherent paramagnetic defects in layered Si/SiO2 superstructures with Si nanocrystals, *J. Appl. Phys.* **104**, 103518 (2008).
110. Jivanescu M, *Intrinsic Point Defects in a-SiO2 with Embedded Si Nanoparticles Probed by ESR*, PhD thesis, KU Leuven, Belgium (2010), Ch. 1.2.1.
111. Poindexter EH, MOS interface states: overview and physicochemical perspective, *Semicond. Sci. Technol.* **4**, 961 (1989).
112. Helms CR, The silicon-silicon-dioxide system: its microstructure and imperfections, Poindexter EH, *Rep. Prog. Phys* **57**, 791 (1994).
113. Stesmans A, Nouwen B, Afanas'ev VV, Pb1 interface defect in thermal (100)Si/SiO2: 29Si hyperfine interaction, *Phys. Rev. B* **58**, 15801 (1998).
114. Stesmans A, Interaction of Pb defects at the (111)Si/SiO2 interface with molecular hydrogen: simultaneous action of passivation and dissociation, *J. Appl. Phys.* **88**, 489 (2000).
115. Stesmans A, Afanas'ev VV, Electrical activity of interfacial paramagnetic defects in thermal (100)Si/SiO2, *Phys. Rev. B* **57**, 10030 (1998).
116. Stesmans A, Afanas'ev VV, Undetectability of the Pb1 point defect as an interface state in thermal (100)Si/SiO2, *J. Phys. Condes. Matter* **10**, L19 (1998).
117. Campbell JP, Lenahan PM, Density of states of Pb1 Si/SiO2 interface trap centers, *Appl. Phys. Lett.* **80**, 1945 (2002).
118. Stesmans A, Afanas'ev VV, Invasive nature of corona charging on thermal Si/SiO2 structures with nanometer-thick oxides revealed by electron spin resonance, *Appl. Phys. Lett.* **82**, 2835 (2003).
119. Lenahan PM, Curry SE, First observation of the 29Si hyperfine spectra of silicon dangling bond centers in silicon nitride, *Appl. Phys. Lett.* **56**, 157 (1990).

120. Warren WL, Lenahan PM, Electron-nuclear double-resonance and electron-spin-resonance study of silicon dangling-bond centers in silicon nitride, *Phys. Rev. B* **42**, 1773 (1990).

121. Stesmans A, van Gorp G, ·Si≡Si3 defect at thermally grown (111)Si/Si3N4 interfaces, *Phys. Rev. B* **52**, 8904 (1995).

122. Aubert P, von Bardeleben HJ, Delmotte F, Cantin JL, Hugon MC, Electron-paramagnetic-resonance study of the (100)Si/Si3N4 interface, *Phys. Rev. B* **59**, 10677 (1999).

123. Savall C, Bruyere JC, Krautwurm J, Correlations between ESR and photoluminescence in slightly hydrogenated silicon nitride, *J. Phys. D* **28**, 565 (1995).

124. Warren WL, Lenahan PM, Curry SE, First observation of paramagnetic nitrogen dangling-bond centers in silicon nitride, *Phys. Rev. Lett.* **65**, 207 (1990).

125. Warren WL, Lenahan PM, Kanicki J, Electrically neutral nitrogen dangling-bond defects in amorphous hydrogenated silicon nitride thin films, *J. Appl. Phys.* **70**, 2220 (1991).

126. Warren WL, Kanicki J, Rong FC, Poindexter EH, Paramagnetic point defects in amorphous silicon dioxide and amorphous silicon nitride thin films, *J. Electrochem. Soc.* **139**, 880 (1992).

127. Warren WL, Kanicki J, Robertson J, Lenahan PM, Energy level of the nitrogen dangling bond in amorphous silicon nitride, *Appl. Phys. Lett.* **59**, 1699 (1991).

128. Stesmans A, van Gorp G, Observation of the localized Si dangling-bond Pb defect at the Si/Si3N4 interface, *Phys. Rev. B* **39**, 2864 (1989).

129. Stesmans A, The ·Si≡Si3 defect at various (111)Si/SiO2 and (111)Si/Si3N4 interfaces, *Surf. Sci. Tech.* **4**, 1000 (1989).

130. Nagasawa H, Yamaguchi Y, Izumi T, Tonosaki K, Heteroepitaxial growth and ESR evaluation of 3C-SiC, *Appl. Surf. Sci.* **70**, 543 (1993).

131. Charpentier S, Kassiba A, Emery J, Cauchetier M, Investigation of the paramagnetic centres and electronic properties of silicon carbide nanomaterials, *J. Phys. Condens. Mater.* **11**, 4887 (1999).

132. Kassiba A, Makowska-Janusik M, Boucle J, Bardeau JF, Bulou A, Herlin N, Mayne M, Armand X, Stoichiometry and interface effects on the electronic and optical properties of SiC nanoparticles, *Diam. Rel. Mater.* **11**, 1243 (2002).

133. Shevaleevskii OI, Tsvetkov AA, Larina LL, Myong SY, Lim KS, Paramagnetic structural defects and conductivity in hydrogenated

nanocrystalline carbon-doped silicon films, *Semiconductors* **38**, 528 (2004).

134. Cantin JL, von Bardeleben HJ, Shishkin Y, Ke Y, Devaty RP, Choyke WJ, Identification of the carbon dangling bond center at the 4H-SiC/SiO2 interface by an EPR study in oxidized porous SiC, *Phys. Rev. Lett.* **92**, 015502 (2004).
135. Macfarlane PJ, Zvanut ME, Characterization of paramagnetic defect centers in three polytypes of dry heat treated, oxidized SiC, *J. Appl. Phys.* **88**, 4122 (2000).
136. Robertson J, The electronic and atomic structure of hydrogenated amorphous Si–C alloys, *Philos. Mag. B* **66**, 615 (1992).
137. Kitabatake M, Greene JE, Structure of 3 $\times$ 2, 5 $\times$ 2, and 7 $\times$ 2 reconstructed 3C-SiC(001) surfaces obtained during epitaxial growth: molecular dynamics simulations, *Appl. Phys. Lett.* **69**, 2048 (1996).
138. Malý P, Trojánek F, Kudrna J, Hospodková A, Banás S, Kohlová V, Valenta J, Pelant I, Picosecond and millisecond dynamics of photoexcited carriers in porous silicon, *Phys. Rev. B* **54**, 7929 (1996).
139. Trojánek F, Neudert K, Malý P, Dohnalová K, Pelant I, Ultrafast photoluminescence in silicon nanocrystals studied by femtosecond up-conversion technique, *J. Appl. Phys.* **99**, 116108 (2006).
140. Hiller D, Jivanescu M, Stesmans A, Zacharias M, Pb(0) centers at the Si-nanocrystal/SiO2 interface as the dominant photoluminescence quenching defect, *J. Appl. Phys.* **107**, 084309 (2010).
141. Raider SI, Gdula RA, Petrak JR, Nitrogen reaction at a silicon-silicon dioxide interface, *Appl. Phys. Lett.* **27**, 150 (1975).
142. Green ML, Sorsch T, Feldman LC, Gusev EP, Garfunkel E, Lu HC, Gustafsson T, Ultrathin SiOxNy by rapid thermal heating of silicon in N2 at T=760–1050°C, *Appl. Phys. Lett.* **71**, 2978 (1997).
143. Cheylan S, Elliman RG, Effect of hydrogen on the photoluminescence of Si nanocrystals embedded in a SiO2 matrix, *Appl. Phys. Lett.* **78**, 1225 (2001).
144. Jivanescu M, Hiller D, Zacharias M, Stesmans A, Size dependence of Pb-type photoluminescence quenching defects at the Si nanocrystal interface, *Eur. Phys. Lett.* **96,** 27003 (2011).
145. Poindexter EH, Gerardi GJ, Rueckel ME, Caplan PJ, Johnson NM, Biegelsen DK, Electronic traps and Pb centers at the Si/SiO2 interface: bandgap energy distribution, *J. Appl. Phys.* **56**, 2844 (1984).

146. Hartel AM, Gutsch S, Hiller D, Zacharias M, Intrinsic non-radiative recombination in ensembles of silicon nanocrystals, *Phys. Rev. B* **87**, 035428 (2013).

147. Pavesi L, Influence of dispersive exciton motion on the recombination dynamics in porous silicon, *J. Appl. Phys.* **80**, 216 (1996).

148. Kim TW, Cho CH, Kim BH, Park SJ, Quantum confinement effect in crystalline silicon quantum dots in silicon nitride grown using SiH4 and NH3, *Appl. Phys. Lett.* **88**, 123102 (2006).

149. Nguyen PD, Kepaptsoglou DM, Ramasse QM, Olsen A, Direct observation of quantum confinement of Si nanocrystals in Si-rich nitrides, *Phys. Rev. B* **85**, 085315 (2012).

150. Deshpande SV, Gulari E, Brown SW, Rand SC, Optical properties of silicon nitride films deposited by hot filament chemical vapor deposition, *J. Appl. Phys.* **77**, 6534 (1995).

151. Ko C, Han M, Shin HJ, Generation of optically active states in a-SiNx by thermal treatment, *J. Lumin.* **131**, 1434 (2011).

152. Kistner J, Chen X, Weng Y, Strunk HP, Schubert MB, Werner JH, Photoluminescence from silicon nitride: no quantum effect, *J. Appl. Phys.* **110**, 023520 (2011).

153. Ma LB, Song R, Miao YM, Li CR, Wang YQ, Cao ZX, Blue-violet photoluminescence from amorphous Si-in-SiNx thin films with external quantum efficiency in percentages, *Appl. Phys. Lett.* **88**, 093102 (2006).

154. Dal Negro L, Yi JH, Kimerling LC, Hamel S, Williamson A, Galli G, Light emission from silicon-rich nitride nanostructures, *Appl. Phys. Lett.* **88**, 183103 (2006).

155. Hiller D, Gutsch S, Hartel AM, Löper P, Gebel T, Zacharias M, A low thermal impact annealing process for SiO2-embedded Si nanocrystals with optimized interface quality, *J. Appl. Phys.* **115**, 134311 (2014).

156. Gutsch S, Laube J, Hartel AM, Hiller D, Zakharov N, Werner P, Zacharias M, Charge transport in Si nanocrystal/SiO2 superlattices, *J. Appl. Phys.* **113**, 133703 (2013).

Chapter 5

Excited-State Relaxation in Group IV Nanocrystals Investigated Using Optical Methods

František Trojánek,[a] Petr Malý,[a] and Ivan Pelant[b]

[a] *Faculty of Mathematics and Physics, Charles University in Prague, Ke Karlovu 3, 121 16 Prague 2, Czech Republic*

[b] *Institute of Physics AS CR, v.v.i., Cukrovanická 10, 162 53 Prague 6, Czech Republic*

frantisek.trojanek@mff.cuni.cz

5.1 Introduction

The electronic and optical properties of silicon nanostructures are of fundamental importance for many prospective applications, including photovoltaic devices, fluorescence labeling of live cells and targeted drug delivery, light sources for silicon photonics, and silicon nanocrystal (SiNC)-based memories. More than two-decade worldwide research in this field has established a global scheme of electronic excitation decay in luminescent SiNCs: Upon creating an electron–hole (e/h) pair, no matter whether optically or via electric injection, energy relaxation of both free electrons and holes sets in, followed usually by localizing of the photocarriers in surface-related

Nanotechnology and Photovoltaic Devices: Light Energy Harvesting with Group IV Nanostructures
Edited by Jan Valenta and Salvo Mirabella

ISBN 978-981-4463-63-8 (Hardcover), 978-981-4463-64-5 (eBook)
www.panstanford.com

states. Radiative recombination of these trapped electrons and holes results in long-lived orange-red luminescence radiation, frequently denoted as the S(low)-band or sometimes also "the excitonic band." The observed long luminescence decay time (of the order of 10–100 μs) reflects the indirect-band-gap-nature of SiNCs, inherited from bulk silicon. In some SiNCs, in addition to the S-band, another luminescence band appears (on the blue wing of the visible region) featuring much faster decay, the so-called F(ast)-band. The F-band is frequently observed in SiNCs derived from porous silicon [1], sometimes in certain chemically synthesized SiNCs [2], and it is reported rather exceptionally also in SiNCs embedded in a SiO_2 matrix [3].

The above scenario, however, may have multiple subtle variations in dependence on NC size, surface passivation, presence or absence of closely spaced other nanoparticles, etc. Besides, under intense excitation the nonradiative recombination of Auger type usually occurs. In this chapter we shall review various processes that may happen during fast photoexcited carrier relaxation in SiNCs before the steady-state luminescence is observed. These processes can be regarded from two competitive points of view:

- One is interested in light emission functionality of SiNCs with future prospects in silicon nanophotonics light sources. In this case, fast relaxation (thermalization) of excitons (electron–hole pairs) is beneficial as a rule, followed by high-efficiency radiative decay. Nonradiative recombination and energy transfer (exciton migration) to nearby SiNCs in a dense system of these nanoparticles can be regarded as rather undesirable steps.
- One is interested in photovoltaic use of optically excited SiNCs. Then neither fast relaxation of hot excitons nor effective radiative exciton annihilation is of primary interest. On the contrary, instead of relaxation, excitation energy conversion into the useful form of electron–hole pairs injected from a given SiNC into a nearby one or extracted from NCs to the host matrix (and, eventually, into conductive electrodes) should be the requested virtue of the SiNCs ensemble. This is especially important when

high-energy photons are applied to excite the NCs so that multiple-exciton generation is achieved [4–6].

We shall attempt to classify the relaxation phenomena in SiNCs according to their rate, that is, whether they occur on the femtosecond, picosecond or nanosecond scales. First of all, we shall describe relevant experimental methods.

5.2 Experimental Methods

The remarkable progress in laser and optics technology during past 30 years resulted in laser systems providing ultrashort pulses of optical radiation with a high intensity and broad spectral tunability. The pulses as short as a few femtoseconds are available in the visible spectral range which is important for investigation of optical transitions in silicon nanostructures. This, in combination with very good stability, reliability, and commercial availability of to-date lasers, has opened the way to improvement of ultrafast laser experiments and their accessibility to a wider scientific community. Three main methods have been established as standard tool for investigation of charge carrier dynamics in semiconductor nanostructures, namely, pump and probe techniques (time-resolved transmission and reflection), time-resolved photoluminescence (PL) measurements, and the laser-induced transient grating technique.

5.2.1 *Pump and Probe Technique*

In the pump and probe technique two time-synchronized optical pulses are used to study optical response of the sample. The two pulses are obtained usually by the amplitude splitting of a primary laser beam (pulse). Two beams intersect under a small angle in the sample studied (see Fig. 5.1).

The defined time interval τ between the pump and probe pulses can be adjusted by changing the optical path of the pulses using an optical delay line (which is typically a computer driven linear translation stage with a retroreflector) (cf. Fig. 5.2). The wavelengths of the pump and probe pulses can be usually tuned independently.

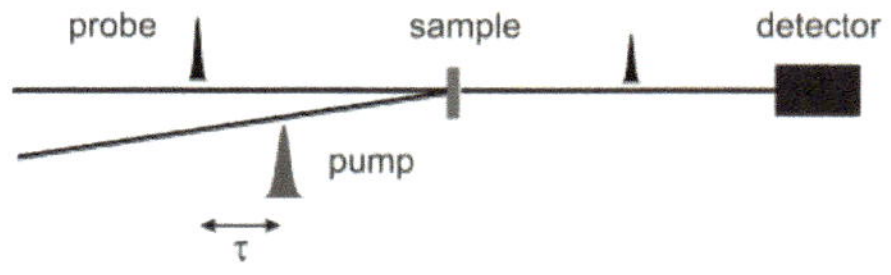

Figure 5.1 Principle of the pump and probe transmission technique.

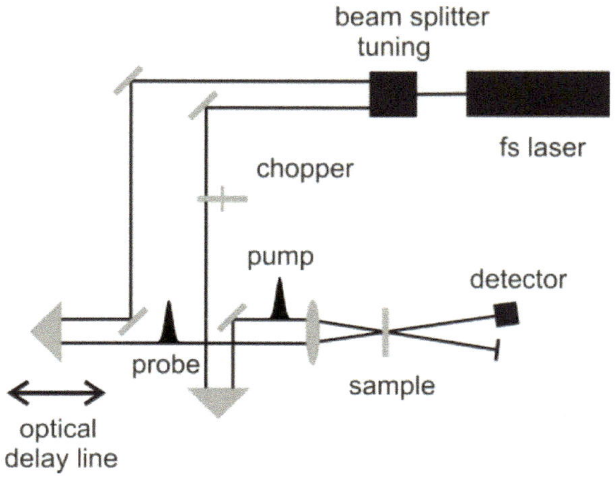

Figure 5.2 Experimental setup of the pump and probe technique.

In a routine setup the pump pulse has a high intensity and is tuned to the absorption region of the sample so that it creates real excitations in the material as molecular excited states or free electron–hole pairs in semiconductors. The pulse-induced changes in the population of energy states lead to the changes in the absorption of the probe pulse. A photodetector is used to measure the energy of the probe pulse after passing the excited region of the sample as a function of the time delay τ (positive or negative) between the pump and probe pulses. There is no condition on the time resolution of the detector (usually a standard photodiode can be used) as the time resolution of the technique is given in principle by the time width of the used pulses and by the step of the translation stage. The result of such a measurement is presented most often as the normalized transient transmission

$$\frac{\Delta T(\tau)}{T} = \frac{T_{\mathrm{E}}(\tau) - T_0}{T_0}$$

where $T_{\mathrm{E}}(\tau)$ and T_0 is the transmission of the probe pulse measured with and without the pump pulse, respectively.

The transient transmission is monitored in the spectral window corresponding to the probe pulse spectrum. Depending on the particular setup, the wavelength of the probe pulse can be tuned using nonlinear optical transformations. In this case, the time evolution of the transient transmission spectra in fairly broad interval can be measured. Moreover, also the polarization state of the light pulses can be controlled by polarizers and/or phase plates.

The sensitivity of the technique, understood as the minimum ratio $\frac{\Delta T(\tau)}{T}$ that can be measured, is an important parameter of given setup. At the first sight, one could expect to achieve a sufficient signal by increasing the pump pulse energy. However, the investigated phenomena might depend strongly on excitation level. For example, in semiconductor NCs the photoexcited carrier recombination strongly depends on the number of photoexcited electron–hole pairs per NC and often single-pair regime is required. The other issue is the repetition rate of the pulses used. Higher rates (typically 80 MHz) for which the signal-to-noise-ratio is often very good correspond to the time delay between successive pump pulses of about 12 ns. However, in many cases this time does not exceed the time constants of decay processes studied. In these cases a reduced repetition rate is required.

The measured transient transmission reflects both the changes in sample surface reflection and volume extinction (absorption plus scattering). The technique can be modified to measure transient reflection, or simultaneously both the transient transmission and reflection. In many cases the modulation of transmission due to reflection changes can be neglected. Introducing the extinction coefficient ε at the probe-pulse wavelength by the relation $T_{\mathrm{E},0} = \exp(-\varepsilon_{\mathrm{E},0} d)$, where d is the sample thickness, the normalized transient transmission is

$$\frac{\Delta T(\tau)}{T} = \exp[-(\varepsilon_{\mathrm{E}} - \varepsilon_0)d] - 1$$

and, after keeping only the first term of approximation of the exponential function for a small extinction change,

$$\frac{\Delta T(\tau)}{T} \approx -(\varepsilon_{\mathrm{E}} - \varepsilon_0)d.$$

In many cases the extinction corresponds to the absorption and absorption changes are thus monitored. For investigation of charge carrier dynamics in silicon nanostructures, it is convenient to tune the probe pulse wavelength into the transparent region of unexcited sample so that the excited state absorption is measured ($\varepsilon_0 = 0$). It is often directly proportional to the number of photoexcited carriers ($\varepsilon_{\mathrm{E}} \propto n$) [7], in which case the time evolution of the transient transmission monitors directly the dynamics of population of photoexcited charge carriers (the transmission decreases after excitation in accord with a negative sign),

$$\frac{\Delta T(\tau)}{T} \propto -n(\tau).$$

5.2.2 *Up-Conversion Technique*

The time-resolved PL can be measured by directly monitoring the emitted light signal with, for example, fast photodiodes, photomultipliers, a gated charge-coupled device (CCD), or a streak camera. However, the best time resolution (tens or hundreds of femtoseconds, limited basically by the laser pulse duration) can be achieved by techniques of optical gating. One of them, which is frequently used, is the up-conversion technique based on the nonlinear optical sum frequency generation.

The principle of the up-conversion can be understood from the scheme in Fig. 5.3. The measured luminescence light (frequency

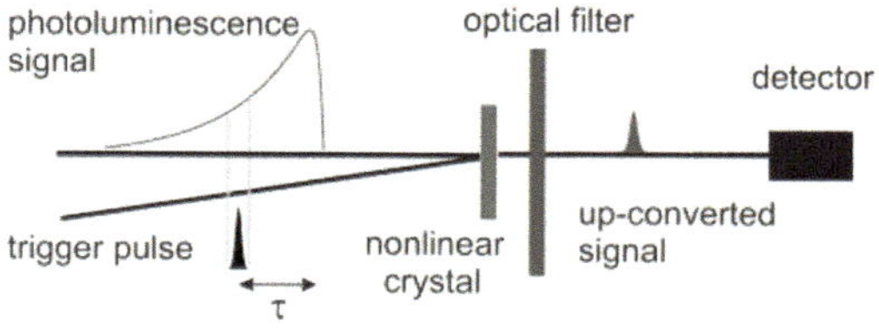

Figure 5.3 Principle of the up-conversion technique.

ω_{LUM}) is spatially overlapped with a laser triggering pulse (frequency ω_{TP}) in a nonlinear crystal. If both the light signals overlap also temporarily and under fulfillment of specific conditions (phase matching, polarization), the up-converted light pulse at the sum frequency $\omega_{UP} = \omega_{LUM} + \omega_{TP}$ is generated. The up-converted signal behind the crystal can be spatially and spectrally filtered by pinholes, filters and/or a monochromator, and detected, for example, by a photon-counting photomultiplier or a CCD camera. The magnitude of the up-converted signal is directly proportional to the magnitude of the PL intensity profile overlapping with the switching laser pulse. The time evolution of the PL signal can be therefore measured by changing the time delay between the ultrashort switching pulse and the longer PL signal (more exactly, the time of light excitation of the sample) (see Fig. 5.4). The sum-frequency generation in a nonlinear crystal operates as an ultrafast optical gate driven by the triggering pulse. For an efficient up-conversion the phase-matching condition is to be fulfilled which means that the wavevector of the up-converted light is equal to the sum of those of the triggering and PL lights. This condition is conveniently adjusted, for example, by rotating the nonlinear crystal. In case of a spectrally broadband PL its spectrum can be measured at given time delay by simultaneous adjustments of the crystal angle and monochromator grating position.

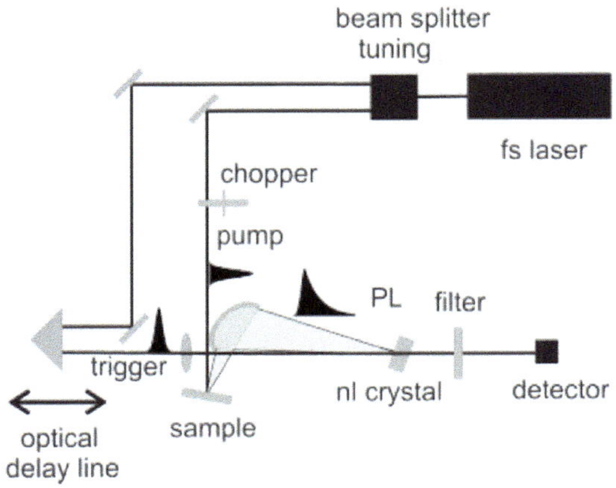

Figure 5.4 Experimental setup of the up-conversion technique.

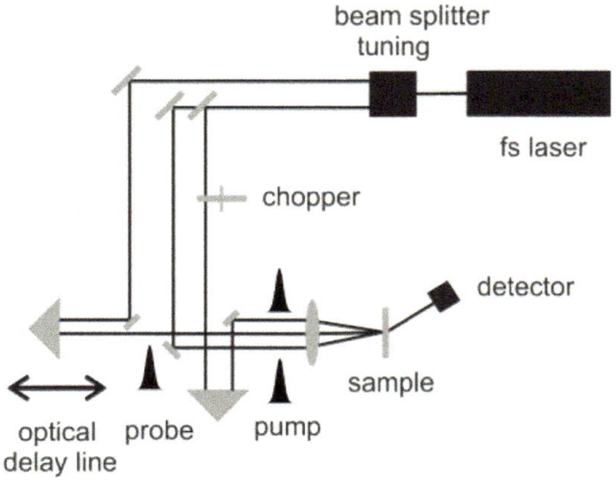

Figure 5.5 Experimental setup of the transient grating technique.

5.2.3 *Transient Grating Technique*

The transient grating (also laser-induced grating, LIG) technique is one of methods of four-wave mixing where four waves interact in matter. Four-wave mixing can be used for investigation of ultrafast coherent processes. In semiconductor research, the transient grating technique is used mostly for measurement of carrier diffusion. In this modification—see Fig. 5.5—of the technique two ultrashort pump pulses overlap temporarily and spatially in the sample. The two pump beams with wavelength λ_P intersect under angle ϑ, producing an interference pattern with the period

$$\Lambda = \frac{\lambda_P}{2\sin(\vartheta/2)}.$$

For symmetrical geometry of the sample, that is, when the normal to the sample surface corresponds to the axis of incident beams, the light intensity is spatially modulated as $I = 2I_0(1 + \cos(2\pi x/\Lambda))$, where I_0 is intensity of each beam and x stands for a spatial coordinate in the plane of the two beams, perpendicular to their axis. If the wavelength λ_P is tuned into the absorption region of the sample, the density of photoexcited carriers with the same profile is created (under assumption of single-photon absorption and no saturation). This leads to a periodical spatial modulation of complex index of refraction (both of refractive index and absorption) which, in the first approximation, is proportional to the number of photoexcited carriers [8, 9]. In this way an optical grating is created. The third, a probe pulse, can be diffracted at this grating. Usually the efficiency of the first-order diffraction as a function of the time delay between the pump and probe pulses is measured.

For a thin grating, the diffraction efficiency is proportional to the refractive index or absorption change squared [8, 9]. The diffraction efficiency decays in time due to "smoothing" of the grating which can be caused by two simultaneous effects, (i) recombination of photoelectrons with photoholes (lifetime τ) and (ii) lateral photocarrier diffusion (diffusion coefficient D). The spatiotemporal evolution of the carrier population $N(x, t)$ can be described by equation

$$\frac{\partial N(x,t)}{\partial t} = D\frac{\partial^2 N(x,t)}{\partial x^2} - \frac{N(x,t)}{\tau}.$$

Solving this equation with initial condition of periodically modulated carrier population one obtains an exponential decay of diffraction efficiency $\propto \exp(-t/\tau_D)$ with the decay time τ_D given by

$$\frac{1}{2\tau_D} = \frac{1}{\tau} + \frac{4\pi^2 D}{\Lambda^2}. \quad (5.1)$$

In a standard experiment the grating dynamics is measured under different angles ϑ, that is, under different grating periods Λ. For smaller grating periods the grating decay is faster because of raising the importance of the photocarrier lateral diffusion. The evaluation of the experiment is based on plotting $(2\tau_D)^{-1}$ against $4\pi^2/\Lambda^2$. A linear plot is obtained, the slope of which yields the carrier diffusion coefficient D.[9] Moreover, the intercept of the plotted straight line with the y axis gives the reciprocal value of the carrier lifetime τ.

5.2.4 *Time-Resolved Terahertz Spectroscopy*

Of special importance for noncontact investigation of photoexcited electronic system in nanoparticles and its ultrafast relaxation appears time-resolved terahertz spectroscopy (TRTS). In principle, this experimental method is a modification of the standard optical pump and probe technique described above. In a similar way, carriers are excited in the sample by an ultrashort optical pump pulse. The sample response is monitored using a time-delayed THz pulse [10]. The THz ($\nu = 10^{12}$ Hz) probe pulses can be generated by technique of optical rectification. This is a second order optical nonlinear process that can be perceived as the degenerate case of difference frequency generation for two identical frequencies. In a suitable nonlinear medium (without center of inversion), a short laser pulse creates a short-lived DC polarization of the same duration. Provided the laser pulse is of picosecond or subpicosecond duration, the DC polarization pulse represents a source of electromagnetic radiation, emitted in the form of a pulse that contains a very small number of frequency cycles in the terahertz range. As a rule, a ZnSe crystal excited with 800 nm, 100 fs laser pulses is used for generating the THz pulses.

What is essential now is that the THz probe pulses "feel" photocarrier motion on the length scale $l = \sqrt{D\tau}$ driven by carrier diffusion coefficient D and time interval $\tau \simeq \nu^{-1}$. By considering D in bulk silicon to be of the order of 10 cm^2/s, we get immediately $l = \sqrt{D\tau} \approx 30$ nm. Therefore, probing length of this method fits perfectly typical NC dimensions and covers possibly also their close surroundings, making TRTS an important tool for investigating trapping of photoexcited carriers at NC surface/interface states with subpicosecond time resolution.

5.3 Femtosecond Phenomena

It has been widely accepted nowadays that absorption of the excitation radiation and the creation of electron–hole pairs occur in the NC's core. On the other hand—given the small volume of NCs which are attractive for photonic, photovoltaic, and biological applications—the photocarriers can very quickly diffuse from the core toward the NC's surface.[a] The "hot" carriers may relax by

---

[a]This statement, very frequently used, is oversimplified and can be misleading. One should use a more exact expression: The photocarrier wavefunctions, initially delocalized uniformly over the nanocrystal volume, get partially localized closer to the surface. The photocarriers then feel, more or less, the influence of surface passivation species.

transforming part of their excitation energy to lattice vibrations. An insight into this relaxation processes can be obtained with the aid of femtosecond PL spectroscopy.

Reported experiments of this type are not numerous, which is obviously determined by a rather specific and not easy-to-manage experimental technique (Section 5.2). First we shall discuss the particular case of SiNCs excited with 400 nm (3.1 eV) pulses of about 100 fs duration [11–13]. Let us think about how and where, within the first Brillouin zone, the excitation valence-to-conduction-band transitions are to be realized. Bulk silicon is an indirect band-gap semiconductor and, as recent density functional theory (DFT) calculations suggest [14], SiNCs inherit this property down to about 2 nm in size unless they are modified in a specific way. Quantum confinement may open the bulk $\sim$1.17 eV indirect band gap up to $\sim$2 eV. Optical transitions across such a gap are still easily accessible by the 3.1 eV photons. Excitation path thus seemingly takes place via indirect phonon-assisted transitions.

However, due to complexity of the silicon band structure, there is an interesting interplay between the direct and indirect gaps. While the indirect band gap increases as a result of quantum confinement, the direct one decreases [15, 16]. This strongly affects pertinent oscillator strengths and, in particular, lowers the value of the direct band gap down from the bulk value of 3.32 eV. It is then anticipated that 3.1 eV photon energy fits well into the direct (no-phonon) absorption transitions in the center of the first Brillouin zone, the Γ-point (Fig. 5.6). Then the photocreated electrons find themselves in a highly nonequilibrium state (Γ_{15}) and tend to "fall down" toward the indirect (Δ_1) conduction band minimum close to the X-point rapidly (Fig. 5.6).

This temporal evolution of excitation energy is reflected in time-resolved PL spectroscopy. At the moment, let us disregard possible recombination processes taking place around the Γ-point. Ultrafast PL studies carried out by the up-conversion technique (Section 5.2) feature an emission band slightly blue-shifted with respect to steady-state (time-integrated) PL—see Fig. 5.7, inset. Besides, these femtosecond experiments reveal an efficient photocarrier thermalization via phonon emission, manifesting itself through a very fast rise of PL signal (corresponding to the time resolution of

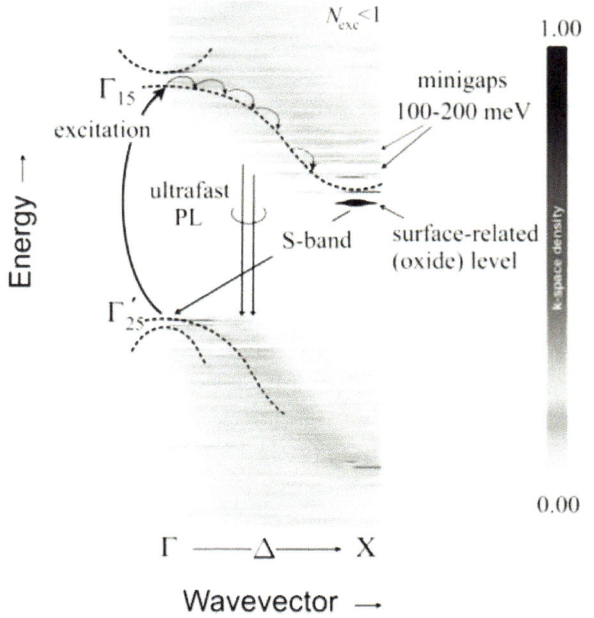

Figure 5.6 Schematics of photoexcitation and ultrafast photoluminescence transitions in SiNCs sized approximately 2–4 nm. Minigaps (sketched here mainly in conduction band energies [14]) entail slowing down the electron thermalization (dashed arrows); consequently no-phonon radiative transitions may happen along the $\Gamma \rightarrow X$ direction, owing to the blurring of electron and hole states in **k**-space. The dashed curves are drawn to evoke memories of the bulk silicon energy band structure, the principal features of which are inherited by these SiNCs.

the up-conversion method). The states which the PL originates from are populated within 300 fs after photoexcitation [11]. This ultrafast rise indicates a fairly high value of $\geq$3.8 eV/ps for the initial energy loss rate per electron–hole pair.

The smooth curve in Fig. 5.7 represents fitting of the experimental data with a two-exponential decay function, convoluted with the time response of the experiment. The time constants extracted in this way are 400 fs and 16 ps (these values are slightly sample and wavelength dependent, ranging around 250–1000 fs and 3–20 ps, respectively). Now, the principal questions to be answered read:

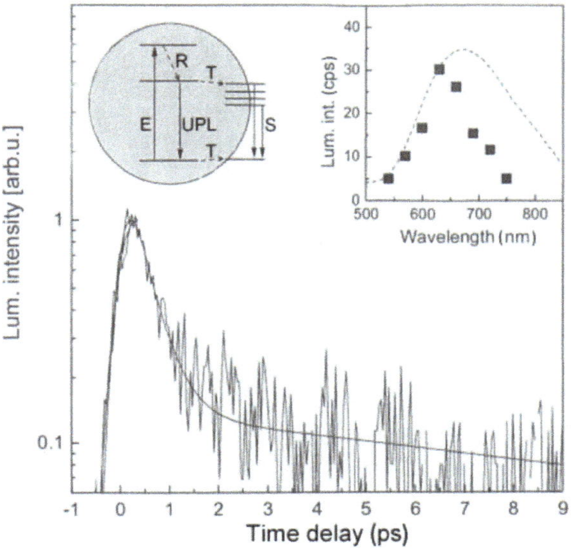

Figure 5.7 Room-temperature ultrafast photoluminescence dynamics of SiNCs excited by 400 nm femtosecond pulses. Smooth curve: a two-exponential fit. Insets: (right) comparison of the ultrafast (zero time, full symbols) and time-integrated emission spectra (dashed curve); (left) simplified scheme of relaxation and recombination channels: E = excitation, R = relaxation, UPL = ultrafast radiative recombination due to no-phonon direct recombination of the not-fully-relaxed core "exciton" (see Fig. 5.6), T = carrier trapping to the surface-related states at X, and S = slow phonon-assisted radiative recombination of relaxed excitons (S-band). After Ref. [11].

1) Why is the ultrafast PL blue-shifted?
2) For what reason does this blue-shifted light emission decay so extremely fast?[a]

Seemingly the luminescence could originate in direct recombination of hot electron–hole pairs in the Γ-point. This would answer easily the first question. However, a big energetic difference between the relevant photon energy (~2 eV) and the Γ_{15}–Γ'_{25} energy interval (~3 eV) testifies against such interpretation. This is supported by another observation: Ensembles of smaller SiNCs

[a] As a rule, luminescence decay time in semiconductors falls into the nanosecond–microsecond range.

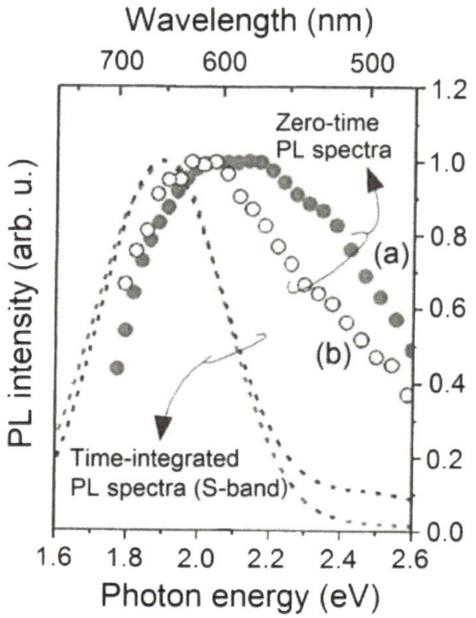

Figure 5.8 Ultrafast (zero-time) photoluminescence spectra of two ensembles of SiNCs: (a) 2.2–2.6 nm (solid symbols) and (b) ~3.5 nm (open symbols). Red-shifted time-integrated spectra are drawn by dashed curves. Room temperature, after Ref. [13].

(curve a in Fig. 5.8) have this emission situated at distinctly shorter wavelengths compared with larger SiNCs (curve b in Fig. 5.8). Given the above-discussed descent of the Γ_{15}-state with shrinking NC size, one should expect quite opposite behavior. Neither light emission during the electron thermalization from Γ_{15} to Δ_1 can explain the questions (1) and (2) in a straightforward manner because (i) the spectrum should have been considerably larger and (ii) the recombination would have been much slower because of predominantly indirect character.

Recent theoretical calculations of energy band structure of SiNCs [14] seem to allow submitting plausible interpretation of the ultrafast PL (Fig. 5.6). Of principal importance here appears widening (blurring) of electronic energy levels in **k**-space together with appearance of relatively wide forbidden minigaps occurring along the conduction and/or valence band (depending on surface passivation) dispersion. Widths of these minigaps (up to ~200 meV) are larger than LO-/TO-phonon energy (~63 meV), consequently they can serve as a phonon bottleneck [7], retarding the thermalization of photoelectrons (photoholes). The population of photocarriers, temporarily piled up in the bottleneck, then recombines radiatively via *direct* (*no-phonon*) transitions; this is facilitated because of the above-mentioned widening of electronic wavefunctions in **k**-space. Relevant transitions are labeled as "ultrafast PL" in Fig. 5.6. The fast decay of this luminescence is mediated by delayed phonon emission across the minigaps. Because the thermalized photocarriers (e.g., electrons "down" at the X-point) usually interact with surface-related states, in simplistic terms it can be said that the decay is caused *by quenching due to carrier surface trapping*. The characteristic time of this trapping (decay of the ultrafast PL is ~400 fs, Fig. 5.7) agrees reasonably with previously reported electronic surface trapping time in CdSe NCs (170 fs) [17]. In larger SiNCs (diameter >4 nm) the relaxation rates are somewhat lower (10^{11}–10^{12} s^{-1}), as calculated recently in a detailed theoretical study of the phonon-assisted intraband relaxation processes of hot electrons and holes by Moskalenko et al. [18] and shown experimentally by Lioudakis et al. [19].

It is obvious that the ultrafast PL is blue-shifted with respect to the time-integrated or c.w. excited luminescence labeled as S-band in Fig. 5.6. The S-band, characterized by a long decay time (10–100 μs), is due to indirect phonon-assisted electron–hole (X–Γ) recombination [20] in cooperation with surface-induced localized levels within the band gap; a classical example of such an electronic state is the oxygen-related level located close below the conduction band edge [21]. However, other surface species, in particular various grafted alkyl groups, can participate in the S-band luminescence, too (e.g., Ref. [20], or see the dashed-line-drawn spectra in Fig. 5.8. One of them belongs to oxide-capped SiNCs, while the other one to alkyl-capped SiNCs [13]).

Two remarks may be of interest. First, the ultrafast PL under femtosecond pumping in SiNCs can result in room-temperature optical amplification due to stimulated emission (optical gain) [22]. A light pulse can be amplified by propagation through the previously photoexcited region of silicon nanocrystalline material.

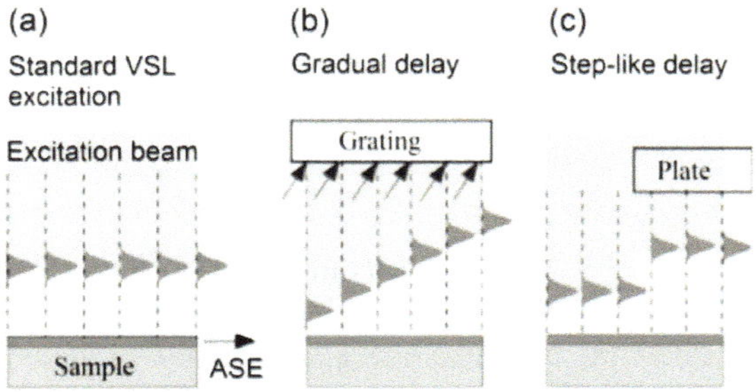

Figure 5.9 Experimental setup for investigating optical gain during ultrafast electron–hole recombination (modified variable-stripe-length [VSL] method). (a) Schematics of the standard VSL pump over the whole stripe length, (b) gradual femtosecond pump pulse delay, and (c) step-like pump pulse delay to compensate for the ultrafast depopulation of the upper level. ASE = amplified spontaneous emission. After Ref. [22].

In case of perpendicular geometry of the pump beam (see Fig. 5.9a) the carriers are excited simultaneously within the whole volume of the sample, and the propagating pulse "feels" amplification at different photon energies in different spatial positions due to a fast relaxation of carriers described above. More effective amplification concentrated to a narrow spectral interval can be achieved in a properly modified pump setup of variable pump delay as shown in Figs. 5.9b and 5.9c.

Second, one could argue that all the results discussed up to now in this section can be explained (at least qualitatively) in a completely different way, namely, in the framework of energy/exciton ultrafast diffusion between closely spaced NCs. The underlying mechanism would consist in interplay of the quantum confinement effect with a nonnegligible size distribution of SiNCs ensemble: The short laser pulse brings all the NCs to excited electronic state, however, fast transfer of excitation energy from smaller NCs (wide band gap) to larger ones (narrower band gap) follows. This is obviously reflected in the overall appearance of the fast band-to-band luminescence, if one accepts that the diffusing excitons recombine radiatively on their "travel" over NCs. In this case the luminescence undergoes a fast red shift and, eventually, it is transformed into the standard S-band.

Indeed, such an interpretation cannot be rejected a priori, even if two experimental facts serve as arguments against it. First of all, no continuous red shift of the ultrafast PL has been observed; this light emission disappears without any noticeable spectral shift while the S-band becomes growing slowly at the same time. The second argument reads: the experimental results in Figs. 5.7 and 5.8 are basically identical, one of them being acquired in relatively densely packed SiNCs in an SiO_2-based matrix (Fig. 5.7), while the second one was measured on a colloidal suspension containing low density of SiNCs ($\leq 10^{18}$ cm^{-3}, Fig. 5.8). The mean internanocrystal separation $\geq$ 8 nm in the colloids hardly allows efficient exciton diffusion between $\sim$3 nm NCs. Nevertheless, because the colloidal suspensions under question were prepared from a mechanically pulverized porous silicon layer, the presence of larger crumbs (10–100 nm) containing interconnected SiNCs was quite possible.

Consequently, it cannot be excluded that the exciton and/or photocarrier transfer between NCs may contribute to the ultrafast excitation energy relaxation, depending strongly on the type of samples, the density of SiNCs and the magnitude of laser pump fluence. This brings us to a brief discussion of experimental results brought up by the THz spectroscopy. Reports on application of this method are not numerous at present, but they confirm the reality of long-range carrier transport, following femtosecond carrier injection [23]. The authors of Ref. [23] investigated a series of nanocomposites (SiNCs/SiO_2 solid matrix) with varying density of SiNCs—from NC volume fraction $\rho_{Si} = 16\%$ to 80%. (Theoretical three-dimensional percolation threshold for the onset on a conductive path in similar systems is known to be $\rho_{Si} = 33\%$; above this threshold the composite should be "metallic" and below "insulating.") The samples were pumped with 400 nm, 100 fs laser pulses in an experimental setup similar to that described in Section 5.2. The time-delayed THz probe monitored ultrafast changes of the AC conductivity of the samples in a reflection mode. Theoretical modeling of the experimental observation was performed in the framework of a phenomenological Drude–Smith model, in which the plasma frequency ω_p and the photocarrier scattering time τ played—among other quantities—the role of fit parameters. The essence of the obtained results can be inferred in Fig. 5.10.

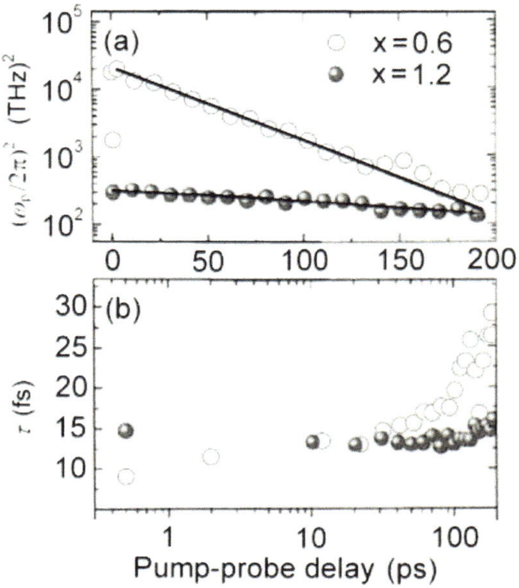

Figure 5.10 Dynamics of the (a) square of the plasma frequency ω_p^2 and (b) carrier-scattering time τ in composites with silicon volume fraction lower (full symbols) and higher (open symbols) than the percolation threshold. After Ref. [23].

It is a common knowledge that the squared plasma frequency is proportional to the carrier density [24]. In Fig. 5.10a, this density follows an exponential decay with recombination time of 39 ps in a composite above the percolation threshold (open circles); the recombination time increases to 220 ps in a sample with silicon volume fraction below the percolation threshold (full circles). This rapid increase in carrier lifetime can be explained by the breakup of percolation path as the Si volume fraction drops below the metal–insulator transition. The carriers then remain isolated in their parent NC. In Fig. 5.10b one can recognize two completely different behaviors. In the sample "above percolation threshold" an obvious increase in carrier scattering time τ occurs (from ~10 fs to ~30 fs) with increasing time delay between the laser pump and the THz probe. This can be properly understood when expressing τ via an electron mean free path $l_{mf} = v_{th}\tau$, using a thermal velocity $v_{th} = \sqrt{3k_B T/m^*} = 2.3 \times 10^5$ m/s, assuming $m^* = 0.26\ m_e$ and $T = 300$ K. The increase in l_{mf} then occurs from ~3 nm to ~10 nm as it is energetically favorable for charges to migrate from smaller to larger particles (possibly aggregates). The value of 10 nm is close to the bulk Si electron at comparable charge carrier densities of 10^{18}–10^{19} cm^{-3}, when the electron mobility is approximately 300 cm^2/Vs. On the other hand, in the sample "below the percolation threshold" the charges are localized to their parent nanoparticles and the scattering time τ, set by the mean particle diameter, remains constant.

These conclusions seem to be partly in line with ultrafast decay of femtosecond-laser-induced transient grating in silicon-quantum-dot-based optical waveguides [25]. Here, the LIG created in a thin layer of SiNCs (fabricated by Si^+ ion implantation into an Infrasil slab followed by annealing) exhibits very short decay time (picoseconds) that was found to decrease with decreasing grating constant Λ. However, unlike the above case of THz measurements, the standard model of photocarrier diffusion between NCs was not able to explain the observation because of necessity to apply unrealistically high carrier diffusion constant to fit the experimental results. Instead, the authors invoked *exciton* diffusion between nanoparticles and/or enhanced exciton radiative decay rate in a cavity represented by the periodically modulated planar structure (Purcell effect).

Final remark in this section concerns germanium (as another important element of the group IV) and femtosecond carrier relaxation/recombination in Ge nanoparticles. Research of GeNCs has been rather rare in comparison with silicon. There are only few reports on ultrafast spectroscopy of GeNCs as the experiments are more complicated due to the spectral positions of fundamental transitions in the near infrared region. The picture of carrier dynamics seems to be similar to that of SiNCs, but further research is needed to obtain its unambiguous picture. As an example we can mention the study of ultrafast carrier dynamics in above-band-gap energy states (visible spectral range) of GeNCs which revealed a bulk-like band structure for NC sizes smaller than the exciton Bohr radius [26] or femtosecond pump and probe measurements of efficient Auger recombination in GeNCs [27]. Picosecond dynamics of electron and holes with femtosecond time resolution in vertically aligned germanium nanowires (mean diameter of 18–30 nm) was observed, using the optical pump and probe technique, by Prasankumar et al. [28]. The lifetime of both electrons and holes decreased with decreasing nanowire diameter, demonstrating the importance of surface effects.

Interestingly enough, recently an increased research activity is devoted to femtosecond events connected with the relaxation processes of nonequilibrium carriers in *bulk* Ge which play currently an important role in development of active optical devices for CMOS-compatible photonics. Germanium, even if being indirect-gap material like silicon, is sometimes called "quasidirect" semiconductor because the energetic separation between the absolute conduction band minimum in the L-point ($2\pi/a(1/2, 1/2, 1/2)$) and the local Γ-point valley in the center of the Brillouin zone is very small: 136 meV at 300 K (Fig. 5.11). Consequently, an endeavor can be traced back to the sixties of the last century to modify the germanium band structure so that no-phonon direct radiative recombination at the Γ-point can be achieved and employed to realization of a germanium laser. This goal has been attained, indeed, by Michel's group in MIT in 2010 [29]. They applied two concurrent physical effects on a thin

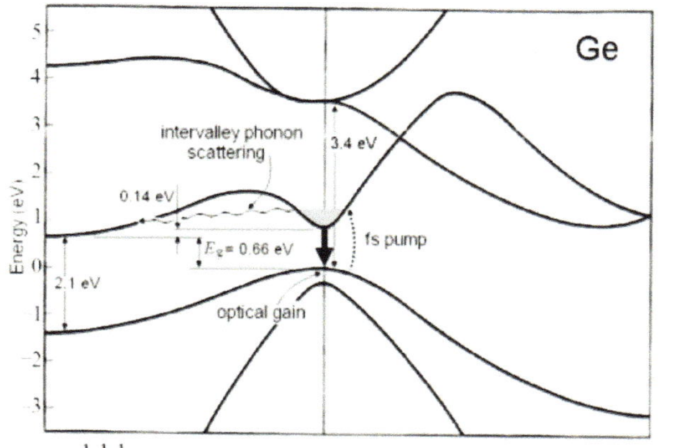

Figure 5.11 Schematics of the germanium band structure showing transitions that govern optical gain from the direct gap.

Ge film in order to increase electronic population of the Γ-valley and to observe lasing at ~1600 nm:

- Biaxial tensile strain
- Heavy n^+-doping

Tensile strain raises the L-valley up, while pushing the Γ-valley down, making the energy separation between the valleys smaller and, consequently, reducing the unfavorable $\Gamma \rightarrow L$ intervalley electron scattering. The heavy *n*-doping simultaneously shifts the Fermi level up into the Γ-valley facilitating for external pumping to produce an inverse population in the Γ-point.

Optical gain observed experimentally under steady-state optical pumping was ~50 cm^{-1} [30]. It is obvious, however—considering the optical gain in Ge being dominated by the direct gap recombination—that one would expect to observe a higher optical gain under ultrafast pumping compared to steady state one because, in the former case, all the electrons in the Γ-valley will participate in the stimulated transitions, before they are scattered into the L-valleys and thus lost for light amplification. The lifetime of the $\Gamma \rightarrow L$ intervalley electron scattering is ~230 fs in bulk Ge. Upon ultrafast carrier injection, probing light pulses shorter than this scattering time and applied synchronously with the pump pulses, feel *inherent* optical gain from the direct gap, without being reduced by the $\Gamma \rightarrow L$ electron scattering.

Broadband femtosecond (pulse width <80 fs) transmittance spectroscopy, recently realized using a modified pump and probe experiment in the wavelength range of 1500–1700 nm [31] revealed the inherent direct gap gain to be $\geq$1300 cm^{-1}, that is, 25× greater than the steady-state optical gain (see Fig. 5.12). This value of gain is comparable to III–V semiconductors. It implies that the performance of Ge lasers has hidden reserves and can be considerably improved by engineering the $\Gamma \rightarrow L$ intervalley scattering [31].

5.4 Picosecond and Nanosecond Phenomena

The discussion on ultrafast relaxation phenomena in SiNCs presented in the previous section was limited, even if not stated

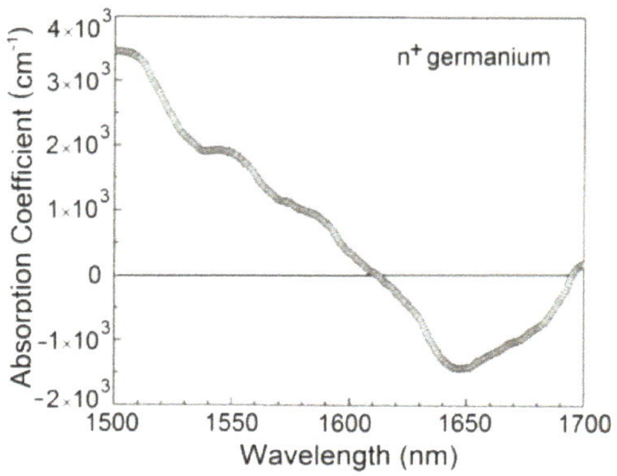

Figure 5.12 Femtosecond absorption spectrum of a n^+-Ge thin film under 1.2×10^{19} cm^{-3} carrier injection. The negative absorption coefficient at $\lambda = 1600$–1700 nm reveals optical amplification in the direct band gap (positive optical gain $\approx$1300 cm^{-1}). Room temperature, after Ref. [31].

explicitly, to the case of weak photoexcitation. This can be characterized by average population less than a single created exciton (electron–hole pair) per NC, $N_{\text{exc}} < 1$. Let us have a look at Fig. 5.6 again and let us consider additional effects which can take place when photocarriers are generated in a SiNC close to the Γ-point under much stronger optical pumping. Then multiple excitons in an NC are easily created ($N_{\text{exc}} > 1$) and, consequently, various interactions between electrons and holes may happen. These interactions exert influence on hot-carrier relaxation and recombination paths and manifest themselves usually on time scales 10–100 ps, either in time-resolved PL or via transient photoinduced absorption.

One of the processes to be considered is carrier–carrier scattering, for example, the conduction electrons close to the bottleneck region (e_1, e_2) can scatter—see Fig. 5.13—so that one of them falls down to the conduction band edge and the released energy is transferred to the second electron promoting it back near to the Γ_{15}-point. Another process is the Auger recombination [16] in which one conduction electron (e_3) recombines with a hole in the valence band transferring the energy to another electron (e_4). In this way, the nonequilibrium carriers generated by the laser pump not only

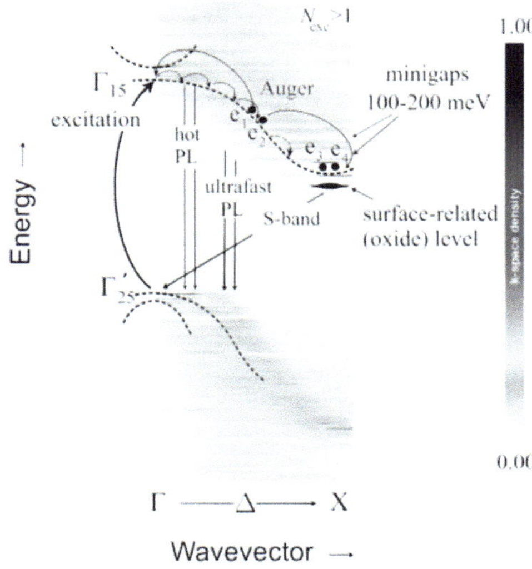

Figure 5.13 Relaxation processes taking place in SiNCs in the case of high-density excitation ($N_{exc} > 1$), as drawn in the silicon band structure along the Γ–Δ–X direction. Nonradiative Auger recombination of pairs of electrons e_1, e_2 or e_3, e_4 contributes to filling back the Γ-states of the conduction band by nonequilibrium carriers. Adapted from Ref. [14].

lose their energy by electron–phonon interaction but also can gain back the excess energy in the bands.

Enhancement of the no-phonon radiative recombination channel happens as a natural consequence of this re-excitation. This channel, labeled "hot PL" in Fig. 5.13, originates in radiative recombination of nonequilibrium electrons (and holes) distributed over many blurred energy levels around the Γ-point and downward along the Γ–Δ–X direction. It is then expected that this photoluminescence band should (i) feature rather large spectral width and (ii) be blue-shifted with respect to the S-band. Figure 5.14 confirms these expectations. Further evidence for the origin of this emission band comes from its spectral shift with reducing the NC size: the band undergoes a considerable red shift when the NC size is decreased from 5.5 nm to 2.5 nm [16]. This behavior is quite opposite to what is exhibited by the S-band and reflects perfectly the peculiarity of the band structure of SiNCs, namely, the interplay between the direct and indirect gaps discussed in Section 5.3.

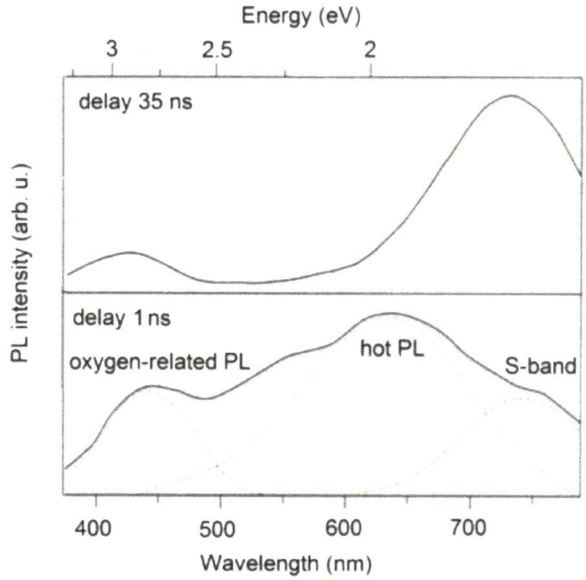

Figure 5.14 Time-resolved photoluminescence spectra from an ensemble of SiNCs with 2.5 nm average diameter embedded in a SiO_2 thin film under intense pumping (5 ns pulses, $N_{exc} > 1$). The spectrum taken at 1 ns "delay" (i.e., during the laser pulse) exhibits a relatively intense band at ~630 nm due to no-phonon hot-carrier recombination—see the "hot PL" arrows in Fig. 5.13. The curve peaked at ~750 nm represents the standard phonon-assisted S-band with microsecond decay. At the delay of 35 ns the hot photoluminescence is no more present. For the band at ~420 nm see text. Adapted from Ref. [16].

The emission band peaked at ~420 nm in Fig. 5.14 deserves also a short discussion now. Its decay in time is somewhat longer (nanosecond up to tens of nanoseconds) than that of the hot PL (10–100 ps), as evidenced by the upper panel of Fig. 5.14. This indicates that the origin of the relevant radiative recombination is not immediately governed by the Auger processes conditioned by multiple-exciton generation. Indeed, this band occurs under weak photopumping ($N_{exc} < 1$), too, and has been tentatively ascribed to transitions within an oxygen-related luminescence center [16], thus being strongly affected by the NC's surface. It is tempting to identify this luminescence with the F-band, currently observed between 400–460 nm in many types of SiNCs, including porous silicon and even organically capped silicon nanoparticles submitted to oxidation [32]. However, alternative interpretations of the F-band have been proposed. In particular, one of them appoints its origin to the core of a subensemble of small silicon nanoparticles [33]. Another extreme is the attribution of the F-band solely to defects in the oxide shell of nanoparticles, when one refers frequently to a similar blue luminescence emitted by defect states in SiO_2 [34]. Even more, recently a paper by Dasog et al. [35] appeared, raising a hypothesis that the blue emission characterized by nanosecond dynamics is due to the presence of trace nitrogen (and oxygen) contamination of Si nanoparticles. In summary, there are probably a variety of luminescence channels which may contribute to the appearance of the blue F-band.

Closer experimental insight into dynamics and other features of Auger-type processes in SiNCs with $N_{\mathrm{exc}} > 1$ is provided by a paper by Trojánek et al. [36]. We shall mention one particular result here—PL study under 35 ps, 532 nm pumping. The corresponding excitation photon energy of $\sim$2.33 eV is not sufficient to excite electrons up to the Γ_{15}-extreme of the conduction band, therefore, basically only the Δ-valley should be populated with electrons and a slow luminescence S-band originating in $\Delta \rightarrow \Gamma'_{25}$ recombination is expected alone to occur in luminescence spectrum. In fact, however, two components are observed in PL, fast and slow (Fig. 5.15); the slow one has emission spectrum identical with that of steady-state PL S-band, indeed, which is not surprising. The fast component deserves more attention: it is very broad and considerable part of the luminescence signal is situated energetically *above* the excitation (anti-Stokes emission). Analysis of the luminescence decay revealed that the decay time found from a single exponential fit is 105 ps and that amplitude of the fast component scales quadratically with the excitation energy density (Fig. 5.16). The decay time of 105 ps agrees very well with the above cited characteristic times of the processes governed by Auger-type recombination. The anti-Stokes behavior of the fast component can be appropriately explained as a consequence of radiative recombination of Auger electrons (re-excited from the

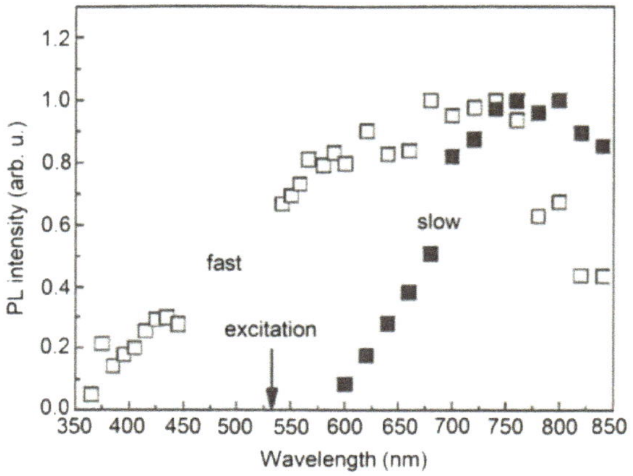

Figure 5.15 Spectra of the fast and slow photoluminescence components in ~3 nm SiNCs under 532 nm, 35 ps excitation. Room temperature, after Ref. [36].

Δ-valley to the Γ-states) with holes residing at the valence band top Γ'_{25}. The quadratic scaling of the fast component reflects the Auger recombination time τ_A dependence on the photoexcited electron density n, namely, $\tau_A \propto 1/n^2$ (for NCs containing two electron–hole pairs $n = 2/V$ holds where V is the NC volume).[a]

All in all, the features of the fast PL component displayed in Figs. 5.15 and 5.16 confirm the effectiveness of Auger excitation of nonequilibrium carriers in SiNCs with $N_{exc} > 1$. It is worth noting yet that these results were obtained in a series of samples constituted by SiNCs (with average diameter of ~3 nm) embedded in a SiO_2 matrix. Variable density of SiNCs had no impact on picosecond dynamics of PL and transient absorption, indicating that internanocrystal exciton and/or photocarrier transfer did not participate in the relevant type of experiment.

Final remark of this section will be concerned (again) with germanium. In the preceding section we touched on the issue of the zone-center direct-gap recombination in germanium thin films and we discussed related engineering of the intervalley

[a]Possible two-photon excitation processes, characteristic also by quadratic pump intensity dependence, were excluded by means of auxiliary experiments.

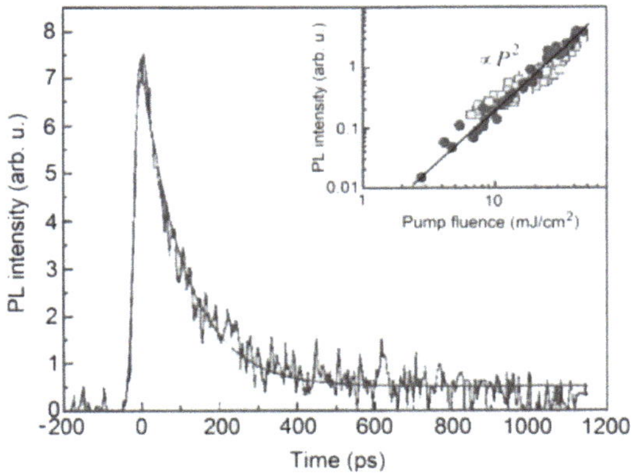

Figure 5.16 Decay of the photoluminescence fast component from Fig. 5.15 (at 600 nm) excited by 532 nm, 35 ps pulses, as measured by a streak camera. Inset: Pump intensity dependence of the amplitude of the fast component. The solid line is a quadratic function $I_{\mathrm{PL}} \propto aP^2$. Room temperature, after Ref. [36].

$\Gamma \leftrightarrow$ L scattering. Here, it is of interest to refer to a recent paper by Terada et al. [37] demonstrating room-temperature direct-gap electroluminescence at ~1590 nm (0.78 eV) with *nanosecond* dynamics from *n*-type bulk Ge, even in the absence of built-in strain. The authors achieved pulse modulation electroluminescence at 10 MHz (limited by the bandwidth of pump pulse generator) and explained their observation by thermal intervalley scattering which pumps electrons into the direct Γ-valley from the lower lying L-valleys that take up electrons thermally released from donors. This thermal electron pumping from the indirect L-valleys may compete (or add in favor) with strain-engineered direct-indirect crossover and, consequently, increase the chance for realization of Ge-based electroluminescent photonic device, working in the telecommunication window.

The recombination of photoexcited carriers extends up to microsecond–millisecond time range. It occurs predominantly between electrons and holes trapped in spatially separated sites and intersite hopping/tunneling has to precede the recombination.

These processes fall out of the scope of this chapter devoted to ultrafast dynamics.

Acknowledgments

This work was financially supported by the EU project NASCEnT (FP7-245977) and by the Grant Agency of the Czech Republic (Grants No. 13-12386S and P108/12/G108).

References

1. Dohnalová K, Žídek K, Ondič L, Kůsová K, Cibulka O, Pelant I (2009) Optical gain at the F-band of oxidized silicon nanocrystals, *Journal of Physics D: Applied Physics*, **42**(13), 135102 (5pp).
2. Pi XD, Liptak RW, Deneen Nowak J, Wells NP, Carter CB, Campbell SA, Kortshagen U (2008) Air-stable full-visible-spectrum emission from silicon nanocrystals synthesized by an all-gas-phase plasma approach, *Nanotechnology*, **19**(24), 245603 (5pp).
3. de Boer WDAM, Timmerman D, Gutiérrez Roldán MA, Molina SI, Gregorkiewicz T (2012) Photoluminescence excitation spectroscopy of Si nanocrystals in SiO2, *ECS Transactions*, **45**(5), 3–8.
4. Timmerman D, Izeddin I, Stallinga P, Yassievich IN, Gregorkiewicz T (2008) Space-separated quantum cutting with silicon nanocrystals for photovoltaic applications, *Nature Photonics*, **2**(2), 105–109.
5. Timmerman D, Valenta J, Dohnalová K, de Boer WDAM, Gregorkiewicz T (2011) Step-like enhancement of luminescence quantum yield of silicon nanocrystals, *Nature Nanotechnology*, **6**(11), 710–713.
6. Govoni M, Marri I, Ossicini S (2012) Carrier multiplication between interacting nanocrystals for fostering silicon-based photovoltaics, *Nature Photonics*, **6**(10), 672–679.
7. Pelant I, Valenta J (2012) *Luminescence Spectroscopy of Semiconductors*, Oxford University Press, Oxford.
8. Hönerlage B (2000) Méthodes expérimentales de l'optique non linéaire, in *L'optique non Linéaire et ses Matériaux* (Lévy R, Jonathan JM, eds), Collection de la Société Française d'Optique Vol. 6, EDP Sciences, Les Ulis, pp. 89–155.

9. Eichler HJ, Gunter P, Pohl DW (1986) *Laser-Induced Dynamic Grating*, Springer, Berlin.
10. Titova LV, Cocker TL, Wang X, Meldrum A, Hegmann FA (2012) Ultrafast carrier dynamics in silicon nanocrystal films, *ECS Transactions*, **45**(5), 21–29.
11. Trojánek F, Neudert K, Malý P, Dohnalová K, Pelant I (2006) Ultrafast photoluminescence in silicon nanocrystals studied by femtosecond up-conversion technique, *Journal of Applied Physics*, **99**(11), 116108 (3pp).
12. Sykora M, Mangolini L, Schaller RD, Kortshagen U, Jurgbergs D, Klimov VI (2008) Size-dependent intrinsic radiative decay rates of silicon nanocrystals at large confinement energies, *Physical Review Letters*, **100**(6), 067401 (4pp).
13. Žídek K, Trojánek F, Malý P, Ondič L, Pelant I, Dohnalová K, Šiller L, Little R, Horrocks BR (2010) Femtosecond luminescence spectroscopy of core states in silicon nanocrystals, *Optics Express*, **18**(24), 25241–25249.
14. Hapala P, Kůsová K, Pelant I, Jelínek P (2013) Theoretical analysis of electronic band structure of 2-to-3-nm Si nanocrystals, *Physical Review B*, **87**(19), 195420 (13pp).
15. Prokofiev AA, Moskalenko AS, Yassievich IN, de Boer WDAM, Timmerman D, Zhang H, Buma WJ, Gregorkiewicz T (2009) Direct bandgap optical transitions in Si nanocystals, *JETP Letters*, **90**(12), 758–762.
16. de Boer WDAM, Timmerman D, Dohnalová K, Yassievich IN, Zhang H, Buma WJ, Gregorkiewicz T (2010) Red spectral shift and enhanced quantum efficiency in phonon-free photoluminescence from silicon nanocrystals, *Nature Nanotechnology*, **5**(12), 878–884.
17. Schoenlein RW, Mittleman DM, Shiang JJ, Alivisatos AP, Shank CV (1993) Investigation of femtosecond electronic dephasing in CdSe nanocrystals using quantum-beat-suppressed photon echoes, *Physical Review Letters*, **70**(7), 1014–1017.
18. Moskalenko AS, Berakdar J, Poddubny AN, Prokofiev AA, Yassievich IN, Goupalov SV (2012) Multiphonon relaxation of moderately excited carriers in Si/SiO2 nanocrystals, *Physical Review B*, **85**(8), 085432 (17pp).
19. Lioudakis E, Othonos A, Nassiopoulou AG (2007) Ultrafast transient photoinduced absorption in silicon nanocrystals: coupling of oxygen-related states to quantized sublevels, *Applied Physics Letters*, **90**(17), 171103 (3pp).

20. Hannah DC, Yang J, Podsiadlo P, Chan MK, Demortière A, Gosztola DJ, Prakapenka VB, Schatz GC, Kortshagen U, Schaller RD (2012) On the origin of photoluminescence in silicon nanocrystals: pressure-dependent structural and optical studies, *Nano Letters*, **12**(8), 4200–4205.
21. Wolkin MV, Jorne J, Fauchet PM, Allan G, Delerue C (1999) Electronic states and luminescence in porous silicon quantum dots: the role of oxygen, *Physical Review Letters*, **82**(1), 197–200.
22. Žídek K, Pelant I, Trojánek F, Malý P, Gilliot P, Hönerlage B, Oberlé J, Šiller L, Little R, Horrocks BR (2011) Ultrafast stimulated emission due to quasidirect transitions in silicon nanocrystals, *Physical Review B*, **84**(8), 085321 (9pp).
23. Cooke DG, Meldrum A, Jepsen U (2012) Ultrabroad terahertz conductivity of Si nanocrystal film, *Applied Physics Letters*, **101**(21), 211107 (4 pp).
24. Peyghambarian N, Koch SW, Mysyrowicz A (1993) *Introduction to Semiconductor Optics*, Prentice Hall, Englewood Clifs.
25. Pelant I, Tomašiunas R, Sirutkaitis V, Valenta J, Ostatnický T, Kůsová K, Elliman RG (2008) Ultrafast decay of femtosecond laser-induced grating in silicon-quantum-dot-based optical waveguides, *Journal of Physics D: Applied Physics*, **41**(1), 015103 (5pp).
26. Tognini P, Stella A, De Silvestri S, Nisoli M, Stagira S, Cheyssac P, Kofman R (1999) Ultrafast carrier dynamics in germanium nanoparticles, *Applied Physics Letters*, **75**(2), 208–210.
27. Robel I, Gresback R, Kortshagen U, Schaler RD, Klimov V (2009) Universal size-dependent trend in Auger recombination in direct-gap and indirect-gap semiconductor nanocrystals, *Physical Review Letters*, **102**(17), 177404 (4pp).
28. Prasankumar RP, Choi S, Trugman SA, Picraux ST, Taylor AJ (2008) Ultrafast electron and hole dynamics in germanium nanowires, *Nano Letters*, **8**(6), 1619–1624.
29. Liu JF, Sun X, Camacho-Aguilera R, Kimerling LC, Michel J (2010) Ge-on-Si laser operating at room temperature, *Optics Letters*, **35**(5), 679–681.
30. Liu JF, Sun X, Kimerling LC, Michel J (2009) Direct-gap optical gain of Ge on Si at room temperature, *Optics Letters*, **34**(11), 1738–1740.
31. Wang X, Kimerling LC, Michel J, Liu J (2013) Large inherent optical gain from the direct gap transitions of Ge thin films, *Applied Physics Letters*, **102**(13), 131116 (4pp).

32. Hua F, Erogbogbo F, Swihart MT, Ruckenstein E (2006) Organically capped silicon nanocrystals with blue photoluminescence prepared by hydrosilylation followed by oxidation, *Langmuir*, **22**(9), 4363–4370.

33. Valenta J, Fucikova A, Pelant I, Kůsová K, Dohnalová K, Aleknavičius A, Cibulka O, Fojtík A, Kada G (2008) On the origin of the fast photoluminescence band in small silicon nanoparticles, *New Journal of Physics*, **10**(7), 073022 (6pp).

34. Skuja L (1992) Isoelectronic series of twofold coordinated Si, Ge and Sn atoms in glassy SiO_2: a luminescence study, *Journal of Non-Crystalline Solids*, **149**(1–2), 77–95.

35. Dasog M, Yang Z, Regli S, Atkins TM, Faramus A, Singh MP, Muthuswamy E, Kauzlarich SM, Tilley RD, Veinot JGC (2013) Chemical insight into origin of red and blue photoluminescence arising from freestanding silicon nanocrystals, *ACS Nano*, **7**(3), 2676–2685.

36. Trojánek F, Neudert K, Bittner M, Malý P (2005) Picosecond photoluminescence and transient absorption in silicon nanocrystals, *Physical Review B*, **72**(7), 075365 (6pp).

37. Terada Y, Yasutake Y, Fukatsu S (2013) Time-resolved electroluminescence of bulk Ge at room temperature, *Applied Physics Letters*, **102**(4), 041102 (3pp).

Chapter 6

Carrier Multiplication in Isolated and Interacting Silicon Nanocrystals

Ivan Marri,[a] Marco Govoni,[b] and S. Ossicini[a]
[a]*Department of Sciences and Methods for Engineering, University of Modena and Reggio Emilia, Italy*
[b]*Department of Chemistry, University of California Davis, USA*
marri@unimo.it, mgovoni@ucdavis.edu

6.1 Introduction

An important challenge of the modern scientific research is oriented in promoting the establishment of cheap and renewable energy sources. The most appealing and promising technology is solar based, that is, photovoltaics (PVs). Recently, thanks to the impressive results achieved in the field of nanotechnologies and the advent of new nanomaterials, new nanocrystal (NC)-based solar cells have been proposed. The employment of NCs in solar cell devices can, in principle, lead to photoconversion efficiency higher than the one obtainable in single junction systems, also when low-grade (inexpensive) materials, with low production costs and low-energy consumption, are adopted. In these systems the possibility to control optoelectronic properties by size, shape, and compositions

Nanotechnology and Photovoltaic Devices: Light Energy Harvesting with Group IV Nanostructures
Edited by Jan Valenta and Salvo Mirabella

ISBN 978-981-4463-63-8 (Hardcover), 978-981-4463-64-5 (eBook)
www.panstanford.com

manipulations, and to exploit new nondissipative recombination mechanisms, represents one of the most used routes to develop and create new materials that can be integrated into existing devices in order to extend the portion of sunlight frequency available for photon-to-current conversion. A detailed analysis of both radiative and nonradiative recombination mechanisms as well as of the electron–phonon scattering processes allows us to identify microscopic parameters that can be tuned to improve solar cell performances and to design innovative devices with properties modelled to satisfy specific requirements for solar energy applications. In this context, numerical calculations can be used to give a detailed description of electronic and optical excitations in both k-dispersive and low-dimensional nanosystems, with an accuracy that complements experimental observations. The possibility offered by theoretical simulations to isolate single decay paths and to quantify their relevance is fundamental to both understand microscopic properties of quantum dot (QD)-based solar cell devices and to support the design of new PV devices. In low-dimensional systems, quantum confinement is responsible for a significant enhancement of carrier–carrier Coulomb interaction that is the main mechanism at the base of both the carrier multiplication (CM) (also called multiple-exciton generation [MEG]) and the Auger recombination (AR) effects. CM is a Coulomb driven nonradiative recombination mechanism that results in the generation of multiple electron–hole (e–h) pairs after absorption of a single photon. In this process an excited electron (hole) decays to a lower energy state in the conduction band (valence band) by transferring its excess energy to (at least) one electron that is excited across the band gap (from an occupied state in the valence band to an empty state in the conduction band; see Fig. 6.1). Obviously CM is permitted only when the excess energy of the carrier igniting the process (initial carrier) exceeds the energy gap (E_g) of the system (CM threshold $\geq 2E_g$).

Understanding which conditions yield to an efficient CM dynamics is of fundamental importance in order to harvest photons excess energy and convert it into additional e–h pairs, increasing thus solar cell photocurrent and boosting the maximum theoretical PV efficiency over the so called Shockley–Queisser (SQ) limit (for more details see Chapter 1).

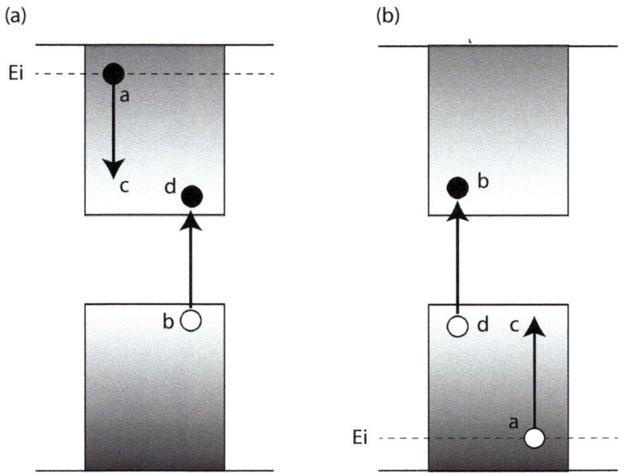

Figure 6.1 CM ignited by electron (left) and hole (right) relaxation mechanisms are depicted in the figure.

Because of the restrictions imposed by energy and momentum conservation and by fast phonon relaxation processes (see Fig. 6.2), CM is often inefficient in bulk semiconductors. At nanoscale, instead, CM is favored by:

(a) the quantum confinement that enhances the carrier–carrier Coulomb interaction [27],
(b) the lack of restrictions imposed by momentum conservation law [6],
(c) the presence of discrete electronic structures that reduce the probability of phonon emission thanks to the so-called phonon-bottleneck effect [10, 30, 34, 37].

It is important to note that the occurrence of thermal relaxation of excited hot carriers through inelastic carrier–phonon scattering (and the subsequent phonon emission) strongly reduces minority carrier lifetimes and adversely affects solar cell performances. At the same time, solar cell quantum efficiency is strongly influenced by the competition between CM relaxation processes and thermalization mechanisms (for a detailed analysis of thermal relaxation processes in silicon NCs see Chapter 5). When the phonon bottleneck dominates the carrier–phonon scattering, new

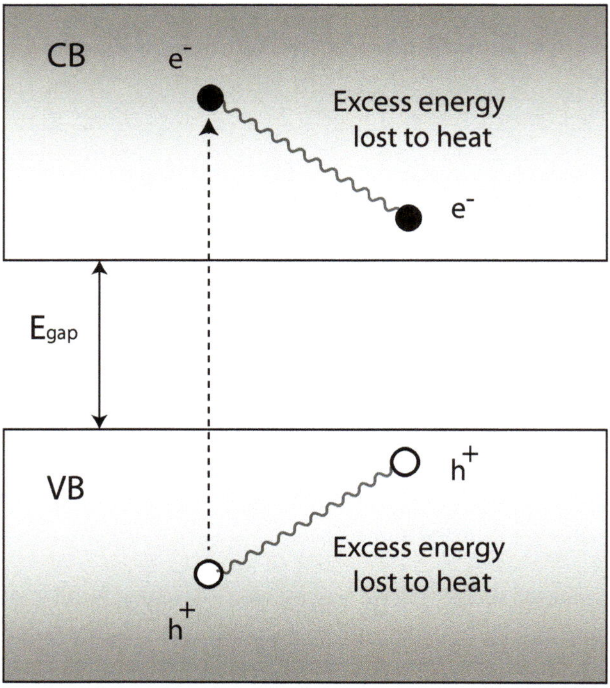

Figure 6.2 A photon of energy greater than the energy gap of the semiconductor is absorbed by the system and an electron is excited from the valence band (VB) to the conduction band (CB). The e–h pair excess energy is then quickly lost into heat through carrier–phonon scattering processes. The useful part of the photon energy is therefore equal to the energy gap of the system.

nondissipative recombination processes (as, for instance, the CM) emerge. In these conditions, CM represents an effective way to minimize the occurrence of energy loss events and thus constitutes a possible route to increasing solar cell performances.

Band-to-band AR is the counterpart of CM. It is one of the most important nonradiative recombination mechanisms in semiconductors, as proven both experimentally [11, 12, 57] and theoretically [9, 20, 24, 36]. AR strongly influences the excess carrier lifetime and therefore the performance of semiconductor-based optoelectronic devices; for instance, it decreases solar cell PV efficiency [26] and, impending the population inversion, strongly reduces optical gain.

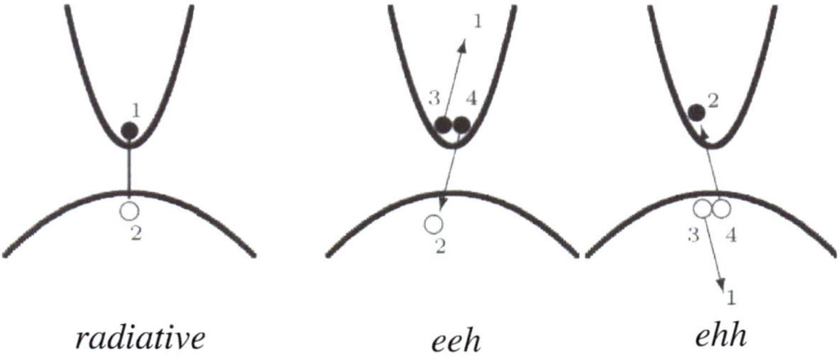

Figure 6.3 Schematic representation of radiative (left) and nonradiative direct Auger (right) band-to-band recombinations: *eeh* and *ehh* processes are depicted.

In *n*-doped (*p*-doped) materials, AR is dominated by electron–electron–hole *eeh* (electron–hole–hole *ehh*) scattering. In this case an electron (hole) in the conduction band (valence band) decay via nonradiative recombination with a valence hole (valence electron), conserving energy and momentum by exciting an electron in the conduction band (see Fig. 6.3). It is evident that AR, without emitting or absorbing photons, does not preserve the number of conduction or valence carriers which are reduced through *eeh* or *ehh* recombinations.

In this chapter we analyze, by first principles calculations, both CM and AR processes. In particular we focus our attention on the physics that is at the heart of CM, analyzing both CM dynamics in isolated and interacting silicon (Si) NCs. AR lifetimes will be then estimated for isolated Si NCs of about 2nm of diameter.

6.2 Carrier Multiplication and Auger Recombination in Low-Dimensional Nanosystems

CM is detected by monitoring the signature of multiexciton decay dynamics using ultrafast transient absorption (TA) spectroscopies. Effects induced by CM on excited-carrier dynamics after single-photon absorption have been observed in a wide range of systems, like, for instance, PbSe and PbS [15, 29, 33, 44, 46, 55], Si [7],

CdSe and CdTe [17, 45, 47], InAs [38, 48], InP [52] and have been interpreted using different theoretical not fully ab initio approaches [2, 3, 16, 39, 49, 50]. Moreover, thanks to the work of Semonin et al. [43], a relevant photocurrent enhancement arising from CM was observed in a PbSe-based QD solar cell, which proved the feasibility of increasing solar cell performances by exploiting high-energy photogenerated carriers. It is evident that, in this context, the possibility to use the nontoxic and largely diffused silicon instead of lead-based materials can give a drastic boost to the future development of QD-based solar cell devices.

Recently, a new CM scheme was hypothesized by Timmerman et al. [53, 54] and Trinh et al. [56] in order to explain measurements conducted on dense arrays of Si NCs (NC–NC separation $\leq$ 1 nm) and obtained from photoluminescence (PL) and induced absorption (IA) experiments, respectively. In the first set of experiments authors proved that, although the excitation cross section is wavelength dependent and increases with reducing the excitation wavelength, the maximum time-integrated PL signal for a given sample saturates at the same level independently of the excitation wavelength or amount of generated e–h pairs per NC after a laser pulse. In this case saturation occurs when every NC absorbs at least one photon. This process was explained by considering a new energy-transfer-based CM scheme, termed "space-separated quantum cutting" (SSQC). CM by SSQC is driven by the Coulomb interaction between carriers of different NCs and differs from traditional CM dynamics because the generation of two e–h pairs after absorption of a single photon occurs in two different and separated Si NCs; a highly excited carrier decays to lower energy states transferring its excess energy to a close NC where an extra e–h pair is generated. Distributing the excitation among several nanostructures, CM via SSQC might therefore be one of the most suitable routes for solar cell loss factor minimization. Experiments conducted by Trinh et al., instead, pointed out the lack of fast decay components in the IA dynamics for high-energy ($h\nu > 2E_{\text{gap}}$) excitations. Again, IA signal measured for high excitation photon energies ($h\nu \approx 2.7E_{\text{gap}}$) was proven to be two times higher than the one recorded at energy below the CM threshold ($h\nu \approx 1.6E_{\text{gap}}$), yielding to a double number of generated excitons when CM is active. These effects were interpreted to be driven by a quantum cutting mechanism that, immediately after the absorption of a single high-energy photon ($h\nu > 2E_{\text{gap}}$) produces the direct formation of two e–h pairs localized onto two separated (and interacting) Si NCs. The measured quantum yield was proven to be very similar to the one detected in the PL experiments conducted by Timmerman et al., pointing out a similar microscopic origin of the recorded PL and IA signals. Similarly, occurrence of efficient CM in film of strongly coupled PbSe QDs with 1,2-ethanediamine (EDA) ligands, with an efficiency that exceeds that for PbSe QDs in solution, was recently proven by M. Aerts et al. [1].

AR is an intrinsic process that dominates multiexcitons non-radiative recombination in NCs. It is the inverse of CM and it follows CM because in this case multiple e–h pairs are generated. AR can be detected by ultrafast spectroscopy techniques, by observing transient transmission and reflectivity of laser light at frequencies under or near the band gap [58], by analyzing radiative recombination dynamics [22, 23] or by photoconductivity measurements [5]. As proved for the first time by Klimov et al. [28], Auger coefficient C_A [20] shows an universal dependence on NCs volume, being proportional to R^3, where R is NC radius [27, 35, 41]. Auger dynamics are very important in QD solar cells; by accelerating energy loss processes they contribute to reduce solar cell performances [14]. The possibility to identify conditions that maximize CM effects and minimize occurrence of AR mechanisms is therefore fundamental in order to improve solar cell performances. In the following chapters, results obtained in the calculation of CM and AR lifetimes for Si NCs will be reported.

6.3 Theory

On the theoretical side, CM dynamics have been investigated using three different approaches:

(a) Coherent superpositions of single- and multiexciton states (strong coupling limit [50]). In this picture absorbed photons instantly generate a coherent superposition of resonant (and almost degenerate) excited single excitons and biexcitons (or multiexcitons). Phonon-induced intraband relaxation merely stabilizes the populations leading to efficient biexciton (multiexciton) production, affected by the fast biexciton (multiexciton) intraband relaxation rate. This model predicts that oscillations between states of varying numbers of (energy allowed) excitons may be observed by ultrafast spectroscopy.

(b) Second-order perturbation theory (weak coupling limit [39, 49]). In this picture the process is simulated using second-order perturbation theory with the perturbation being given by the sum of the electron–photon coupling and the screened Coulomb interaction. In this model electron–electron Coulomb interaction couples a virtual single exciton state to a multiexciton state.

(c) First-order perturbation theory (impact ionization [II] scheme) [2, 16, 40]. CM is here described as a impact ionization process that follows the primary photoexcitation event. In this model the de-excitation of a highly excited-carrier to lower-energy states is followed by the generation of an extra e–h pair.

In our approach CM lifetimes are calculated using the scheme of point (c) in a fully ab initio way, that is, within the density functional theory (DFT), applying first-order perturbation theory (Fermi's Golden Rule) to Kohn–Sham (KS) states. We therefore consider excited electrons (holes) and their relaxation by II. In our model, the decay of an exciton into a biexciton is split into the separated decay dynamics of an electron and a hole (one of the two particles is active while the other is a spectator [40]). The simultaneous involvement of both particles in the process is neglected in the present treatment, but could be included taking into account e–h correlation. The process we consider therefore is the sum of two events, one ignited by an electron (decay of an electron in a negative trion, hole spectator), the other ignited by a hole (decay of a hole in a positive trion, electron spectator). CM rates for mechanisms ignited by electrons (Eq. (6.1)) and holes (Eq. (6.2)) are reported below as a function of the energy of the initial carrier E_i, where $E_i = E_a$.

$$R^{e}_{n_a,\mathbf{k}_a}(E_i) = \sum_{n_c,n_d}^{cond.} \sum_{n_b}^{val.} \sum_{\mathbf{k}_b,\mathbf{k}_c,\mathbf{k}_d}^{1BZ} 4\pi \Big[\mid M_D \mid^2 + \mid M_E \mid^2 + \mid M_D - M_E \mid^2 \Big] \delta(E_a + E_b - E_c - E_d) \quad (6.1)$$

$$R^{h}_{n_a,\mathbf{k}_a}(E_i) = \sum_{n_c,n_d}^{val.} \sum_{n_b}^{cond.} \sum_{\mathbf{k}_b,\mathbf{k}_c,\mathbf{k}_d}^{1BZ} 4\pi \Big[\mid M_D \mid^2 + \mid M_E \mid^2 + \mid M_D - M_E \mid^2 \Big] \delta(E_a + E_b - E_c - E_d) \quad (6.2)$$

where indexes n and $\mathbf{k}$ identify KS states, $1BZ$ is the first Brillouin zone and $\mid M_D \mid$ and $\mid M_E \mid$ are the direct and exchange screened Coulomb matrix elements, respectively. In our simulations, the delta function for energy conservation was implemented in the form of a Gaussian distribution with a full width at half maximum of 0.02 eV, for each of the considered systems. In reciprocal space, M_D and M_E assume the form:

$$\mathrm{M}_D = \frac{1}{V} \sum_{\mathbf{G},\mathbf{G}'} \rho_{n_d,n_b}(\mathbf{k}_d, \mathbf{q}, \mathbf{G}) \mathrm{W}_{\mathbf{GG}'} \rho^{*}_{n_a,n_c}(\mathbf{k}_a, \mathbf{q}, \mathbf{G}') \quad (6.3)$$

and

$$\mathrm{M}_E = \frac{1}{V} \sum_{\mathbf{G},\mathbf{G}'} \rho_{n_c,n_b}(\mathbf{k}_c, \mathbf{q}, \mathbf{G}) \mathrm{W}_{\mathbf{GG}'} \rho^{*}_{n_a,n_d}(\mathbf{k}_a, \mathbf{q}, \mathbf{G}'), \quad (6.4)$$

where both $\mathbf{k}_c + \mathbf{k}_d - \mathbf{k}_a - \mathbf{k}_b$ and $\mathbf{G}$, $\mathbf{G}'$ are vectors of the reciprocal space, $\mathbf{q} = (\mathbf{k_c} - \mathbf{k_a})_{\mathbf{1BZ}}$ and $\rho_{n,m}(\mathbf{k}, \mathbf{q}, \mathbf{G}) = \langle \mathbf{n}, \mathbf{k} | \mathbf{e}^{\mathbf{i}(\mathbf{q}+\mathbf{G})\cdot\mathbf{r}} | \mathbf{m}, \mathbf{k} - \mathbf{q} \rangle$ is the oscillator strength. The Fourier transform of the screened interaction, identified by a matrix in $\mathbf{G}$ and $\mathbf{G}'$ is given by;

$$W_{\mathbf{G},\mathbf{G}'} = \frac{4\pi}{\mid \mathbf{q} + \mathbf{G} \mid^2} \delta_{\mathbf{G},\mathbf{G}'} + \frac{4\pi}{\mid \mathbf{q} + \mathbf{G} \mid^2} \chi_{\mathbf{G},\mathbf{G}'}(\mathbf{q}, \omega = 0) \frac{4\pi}{\mid \mathbf{q} + \mathbf{G}' \mid^2} \quad (6.5)$$

where the polarizability $\chi_{\mathbf{G},\mathbf{G}'}(\mathbf{q}, \omega)$ has been obtained by solving Dyson's equation in the random phase approximation [32] (the presence of off-diagonal terms in the solution of the Dyson's equation is related to the inclusion of local fields). The first term on the right-hand side of Eq. (6.5) denotes the bare interaction, while the second one includes the screening caused by the medium [60]. After convergence tests we adopted, for all considered NCs, an energy cutoff of about 0.5 Hartree in the calculation of the second term of Eq. (6.5). CM lifetimes are then calculated as reciprocal of rates of Eq. (6.1) and (6.2), summing over all possible final states, and are given as a function of the energy of the initial carrier E_i. In our approach, multiexciton configurations are calculated without including many-body corrections. As a consequence Coulomb matrix elements of Eqs. 6.3 and 6.4, already calculated to evaluate CM lifetimes, can be used to estimate AR rates and Auger coefficients as a function of the biexciton energy or as a function of the minority carriers concentration [20]. In this work we will calculate AR lifetimes as a product of an effective Coulomb matrix element and the density of the final states, thus using a standard procedure already adopted for the calculation of CM lifetimes [3]. On the contrary of CM, however, initial states are in this case biexciton states, while final states are single excitons.

6.4 One-Site CM: Absolute and Relative Energy Scale

To study CM in a sparse array of Si NCs, where NC–NC interactions can be neglected, we have considered four different spherical and hydrogenated Si NCs with different diameters: $Si_{35}H_{36}$ (1.3 nm), $Si_{87}H_{76}$ (1.6 nm), $Si_{147}H_{100}$ (1.9 nm) and the $Si_{293}H_{172}$ (2.4 nm) (see Fig. 6.4). In these systems, hydrogen passivation ensures that dangling-bond-related states are not present in the energy gap region. A direct comparison between calculated CM lifetimes for all the considered systems will shed light on the role played by quantum confinement effect on CM dynamics.

Electronic structures have been obtained from first principles using DFT with a norm-conserving pseudopotential plane-wave supercell approach [18]. Local density approximation (LDA) has been used to calculate the exchange-correlation functional. Energy levels have been obtained considering a wavefunction cutoff of 20 Hartree. Calculated KS states have been subsequently used to obtain two particle screened Coulomb matrix elements and, then, CM lifetimes. An exact box-shaped Coulomb cutoff has been adopted to avoid spurious Coulomb interactions between replicas [42]. Calculated CM lifetimes are reported in Fig. 6.5 as a function of

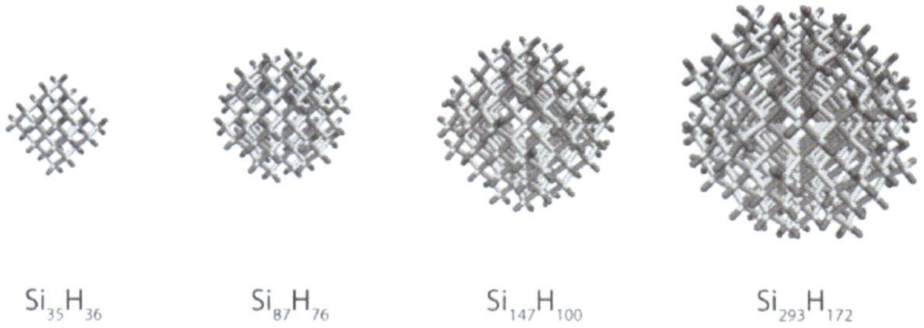

Figure 6.4 Free-standing hydrogenated and spherical Si NCs with a diameter of 1.3, 1.6, 1.9, and 2.4 nm and LDA energy gaps of 3.42, 2.50, 2.21 and 1.70 eV are reported in the figure.

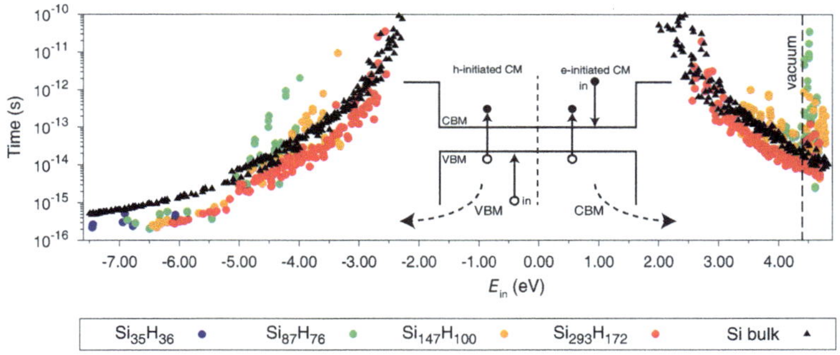

Figure 6.5 Calculated CM rates are reported in the figure (colored dots) and compared with the one obtained for Si bulk (black triangles). Zero is placed at half gap. Reprinted from Ref. [21], Copyright 2012 Nature Publishing.

the energy of the initial carrier (absolute energy scale) and are compared with the one obtained for silicon bulk.

CM lifetimes calculated for mechanisms ignited by electron relaxation are reported on the right part of Fig. 6.5 (positive energies) while CM lifetimes calculated for mechanisms ignited by hole relaxation are reported on the left (negative energies). We found that:

(a) CM is active when the initial carrier excess energy exceeds the energy gap of the system. CM threshold depends on the energy gap of the system and moves toward lower energies with increasing NCs size. CM lifetimes decrease when the energy of initial carrier increases because of the increased density of final states, and move from fraction of nanoseconds (near the CM threshold) to fraction of femtoseconds (fs). For the largest NC ($E_{\text{gap}}^{\text{Si}_{293}\text{H}_{172}} \approx 1.7$ eV), CM lifetimes settle under tenths of picoseconds (ps) when the excess energy of the initial carrier exceeds about 2.2 eV, under hundredths of ps when it exceeds about 3.2 eV.

(b) Far from the activation threshold, CM is proven to be more efficient in Si NCs than in Si bulk. A similar effect is not predicted in direct gap lead-chalcogenide-based materials where CM seems to be more efficient in bulk than in NCs [13]. On the contrary, at low excess energies (i.e., near the CM threshold) CM rates are smaller in Si NCs than in Si bulk because of the smaller density of final states.

(c) For the considered NCs, far from the activation threshold, CM seems to be independent of NCs size.

(d) When ignited by vacuum states, CM processes show lifetimes that strongly oscillate on a large range of values (see transitions calculated for energies above the vertical dashed line of Fig. 6.5) and that depend on the chosen periodic boundary conditions. Inclusion of vacuum states in the calculation of CM lifetimes can lead, therefore, to nonphysical results.

Our simulations not consider neither effects induced by the presence of an embedding matrix nor effects due to the presence of a liquid solvent. For such systems, ab initio analysis of CM relaxation dynamics needs huge computational efforts that go beyond the potentialities offered by modern supercomputer facilities. Inclusion of an external solvent, for instance, requires the implementation of a polarized continuum model (PCM) or the inclusion of a large number of solvent molecules in the simulation box. In both cases calculations become extremely heavy under a computational point of view and not feasible for large NCs.

Point (c) underlines an interesting property of hydrogenated Si NCs. As pointed out by G. Allan and C. Delerue [4], CM rates can be written as the product between the square modulus of an effective Coulomb matrix element and the density of the final states, that is,

$$R(E_i) \propto | W_{\text{eff}}(E_i) |^2 \cdot \rho(E_i). \tag{6.6}$$

It is evident that the maximum CM efficiency can be realized by maximizing both Coulomb interaction (and therefore quantum confinement effects) and density of final states. The latter can be easily calculated imposing energy conservation, while effective Coulomb matrix elements can be extracted from relation (6.6) and from relations (6.1) and (6.2).

Our calculations point out that, far from the activation threshold, an almost exact compensation between $| W_{\text{eff}}(E) |^2$, that increase when NCs size decrease, and $\rho(E)$, that increase when NCs size increase, exists. As a consequence, at high energies, CM lifetimes seem to be independent of the NCs size.

Remarkably, CM in both Si bulk and Si NCs is slower than in Pb-based like-bulk materials. This conclusion is supported by results of Fig. 6.6 where calculated CM lifetimes for both PbS and PbSe bulks are compared with the ones obtained for Si bulk. This condition underlines the necessity of increasing CM efficiency in Si-based nanomaterials, for instance, by applying an external stress [59], by codoping with boron and phosphorous (codoping decreases

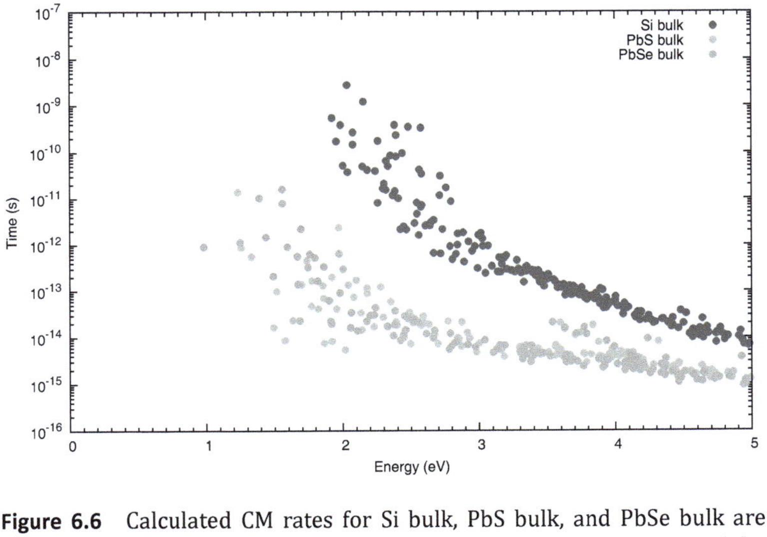

Figure 6.6 Calculated CM rates for Si bulk, PbS bulk, and PbSe bulk are reported in the figure. We consider here only mechanisms ignited by electron relaxation. Zero is place at half gap. It is evident that CM lifetimes of lead-chalcogenide-based materials are smaller than of silicon-based materials.

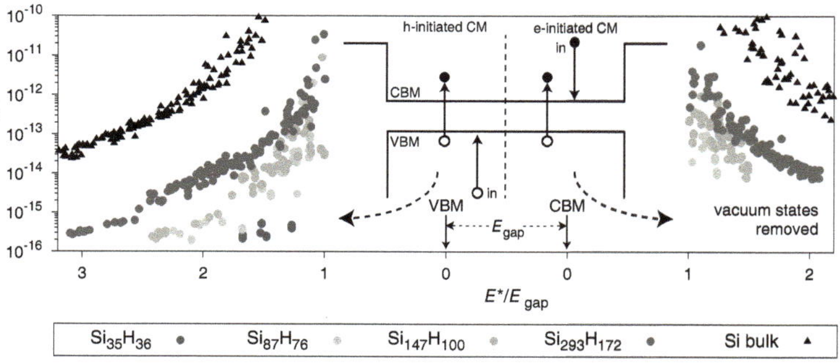

Figure 6.7 CM lifetimes as a function of the ratio between energy of initial carrier E^* and energy gap of the systems are reported in the plot. Zero is placed at half gap. Reprinted from Ref. [21], Copyright 2012 Nature Publishing.

the NC's energy gap and therefore the CM threshold [25]) or by exploiting NC–NC interaction (see Section 6.5).

A strong dependence of CM lifetimes on NCs size is instead observed when a relative energy scale is adopted. In this case CM lifetimes are calculated as a function of the ratio between the excess energy of the initial carrier (energy of the carrier measured from the respective band edge, E^*) and the energy gap of the system E_{gap}.

Results obtained are reported in Fig. 6.7. As argued by M. C. Beard et al. [8] and Delerue et al. [13], the use of an absolute energy scale is probably more appropriated when we investigate microscopic properties of the CM while the use of a relative energy scale results more suitable when implications of CM in real devices are discussed. Results of Fig. 6.7 prove therefore that, for solar cell applications, CM shows benefits induced by the quantum confinement of the carrier density.

By extrapolating effective Coulomb matrix elements from CM calculations, we have estimated Auger lifetimes for all the considered systems. Our calculations point out, for instance, that Auger lifetime asses to about 1 ps when the larger NC is considered. This result is in good agreement with experimental measures by Beard et al. [7], where biexciton lifetimes were estimated for NCs of 3.8, 6.8, and 9.5 nm of diameter. By considering a linear dependence of Auger decay lifetime on the NC volume [27, 35, 41], we can extrapolate a biexciton lifetime of few units of *ps* for NCs of about 2.4 nm of diameter. A similar result was obtained by Klimov et al. [27] for CdSe NCs of 2.4 nm of size, where the biexciton decay lifetime was evaluated to be about 6 ps.

6.5 Two-Site CM: Wavefunction-Sharing Regime

As discussed in Section 6.2, benefits induced on CM dynamics by NCs interplay have been proved by D. Timmerman, M. T. Trinh and A. Aerts [1, 53, 54, 56]. To investigate effects induced by NC–NC interaction on CM dynamics, we have calculated CM lifetimes for systems obtained placing two NCs in the same simulation box, that is, the $Si_{293}H_{172}\times Si_{35}H_{36}$ and the $Si_{293}H_{172}\times Si_{147}H_{100}$ (size of the box 5 nm × 5 nm × 10 nm, NC–NC separation d = 1.0, 0.8, 0.6, 0.4 nm). Having two NCs in the same cell, wavefunctions are now free to delocalize to both NCs. This effect increases when electronic states move to higher energy and when NC–NC separation decreases. A detailed determination of the percentage of localization of wavefunctions can be obtained by evaluating, for all the states of the system, the square modulus of wavefunctions along the *x*, *y*, and *z* directions,

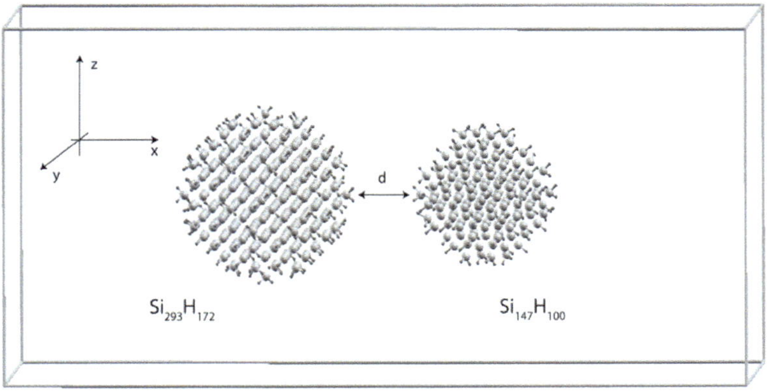

Figure 6.8 A couples of interacting Si NCs are reported in the figure, that is the $Si_{293}H_{172}\times Si_{147}H_{100}$. The size of the simulation box is (5 nm × 5 nm × 10 nm).

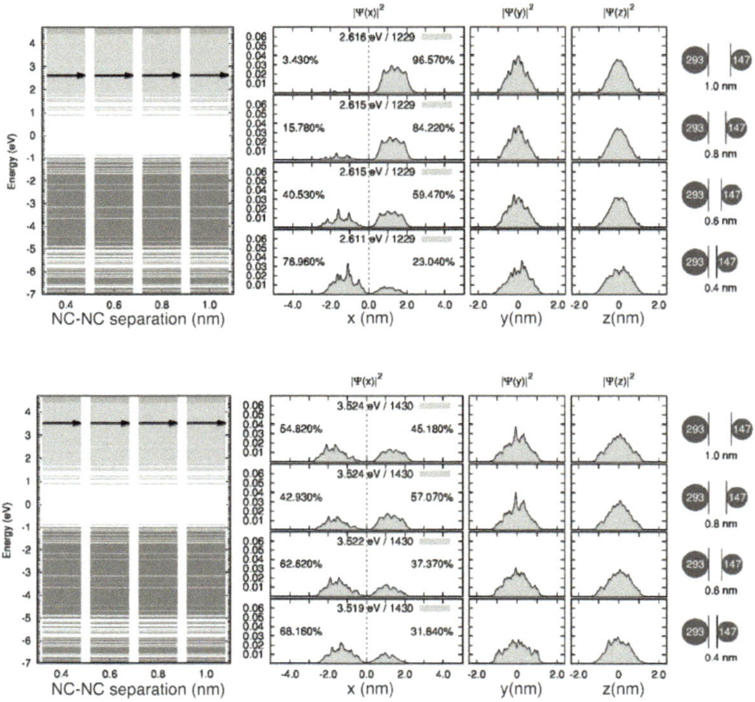

Figure 6.9 Calculated $| \psi(x) |^2$, $| \psi(y) |^2$ and $| \psi(z) |^2$ functions for two selected states of energy 2.61 and 3.52 eV and for a NC–NC separation of 1.0, 0.8, 0.6, and 0.4 nm are reported in the figure. Zero is placed at half gap. Selected states are identified by an horizontal arrow. Axis x, y and z are defined in Fig. 6.8. A different tonality of gray identifies valence band states, conduction band states and vacuum states, respectively.

$$| \psi(x) |^2 = \int_\Omega | \psi(x, y, z) |^2 \, dydz \tag{6.7}$$

$$| \psi(y) |^2 = \int_\Omega | \psi(x, y, z) |^2 \, dxdz \tag{6.8}$$

$$| \psi(z) |^2 = \int_\Omega | \psi(x, y, z) |^2 \, dxdy. \tag{6.9}$$

As an example, we report in Fig. 6.9 $|\psi(x)|^2$, $|\psi(y)|^2$ and $|\psi(z)|^2$ calculated for two arbitrary selected states of the $Si_{293}H_{172} \times Si_{147}H_{100}$ system. It is evident that when NCs are placed in close

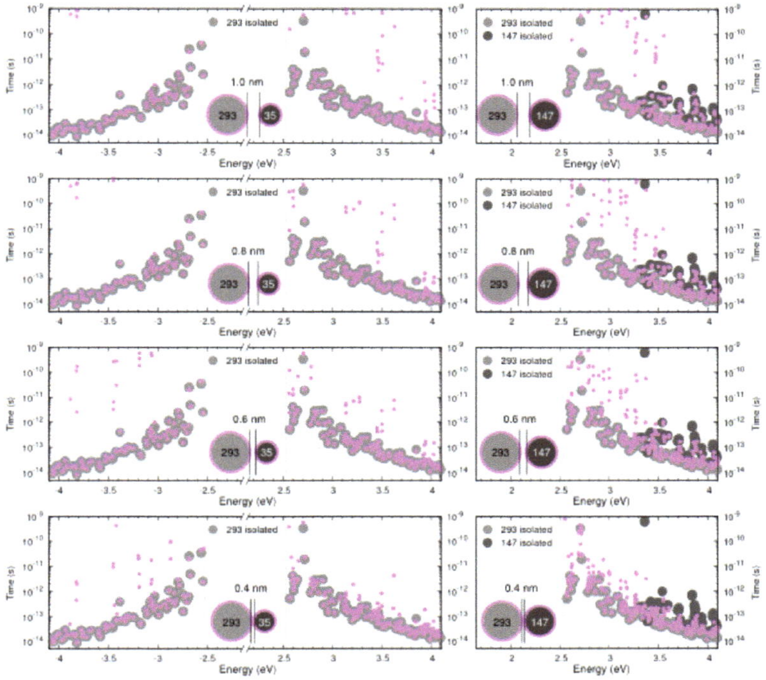

Figure 6.10 Calculated CM lifetimes for a system of two interacting Si NCs are reported in the figure as a function of the energy of the initial state (purple points). For comparison, light and dark gray points represent CM lifetimes for the noninteracting Si NCs.

proximity ($d \leq 1$ nm), electronic wavefunctions can extend to both NCs. To study effects induced on CM dynamics by NCs interplay, we calculate CM lifetimes for both $Si_{293}H_{172} \times Si_{35}H_{36}$ and $Si_{293}H_{172} \times Si_{147}H_{100}$ systems by turning off (step 1) and then by turning on (step 2) NC–NC interaction. Results obtained are reported in Fig. 6.10. We observe that:

(a) When NC–NC interaction is turned off, CM lifetimes are given by the sum of the one calculated for the single, isolated, Si NCs (see Fig. 6.10, gray points). This situation well reproduces CM dynamics in a sparse array of Si NCs where NC–NC interaction can be neglected. In the plot, CM thresholds of single NCs are clearly visible.

(b) When NC–NC interaction is turned on, new CM decay channels (two-site CM effects) appear in the plot (see Fig. 6.10, purple points).

Importantly, when NCs interplay is enhanced, that is the NC–NC separation is reduced from 1.0 to 0.4 nm, the fastest recorded CM transitions do not significantly move toward smaller lifetime values but, instead, a significant increment in the number of fast CM decay processes is observed [61]. As a consequence, by increasing the number of fast CM decay channels, NC–NC interaction improves CM efficiency. Again, when NCs are strongly coupled ($d = 0.4$ nm) tracks of the single CM thresholds disappear and the system shows an unique well defined CM threshold that corresponds to the one of the largest NC (the $Si_{293}H_{172}$). In this situation the system formed by two interacting Si NCs appears as a single and unique quantum system. To define the microscopic origin of CM decay mechanisms, we introduce a color scale and a new parameter (namely spill-out) to define the percentage of localization of initial state. Red points (spill-out $= 0\%$) identify transition ignited by states that are completely localized on the larger NC (the $Si_{293}H_{172}$), blue points (spill-out $= 100\%$) identify transition that are ignited by states localized on the smaller NC. Colors from red to blue (spill-out from 0% to 100%) identify transitions ignited by states delocalized on both NCs. Results obtained are reported in Fig. 6.11. Varying d from 1.0 to 0.4 nm, two-site CM lifetimes significantly decrease up to 3 orders of magnitude. Effects induced by wavefunction sharing are well depicted in the figure. Remarkably, modifications induced in the electronic structure by the change in the NC–NC separation do not significantly influence one-site CM events.

CM in a dense array of NCs can be divided in two components, that is, one-site and two-site events. Two-site events can be divided in two types, that is, SSQC and Coulomb-driven charge transfer (CDCT) decay mechanisms (see Fig. 6.12).

SSQC (see the second panel of Fig. 6.12) is an energy-transfer-based CM mechanism that occurs when a high-energy carriers decays toward the band edge and an extra e–h pair is generated in a nearby NC. CDCT is a charge-transfer-based CM mechanism that occurs, for instance, when a high-energy electron decays in

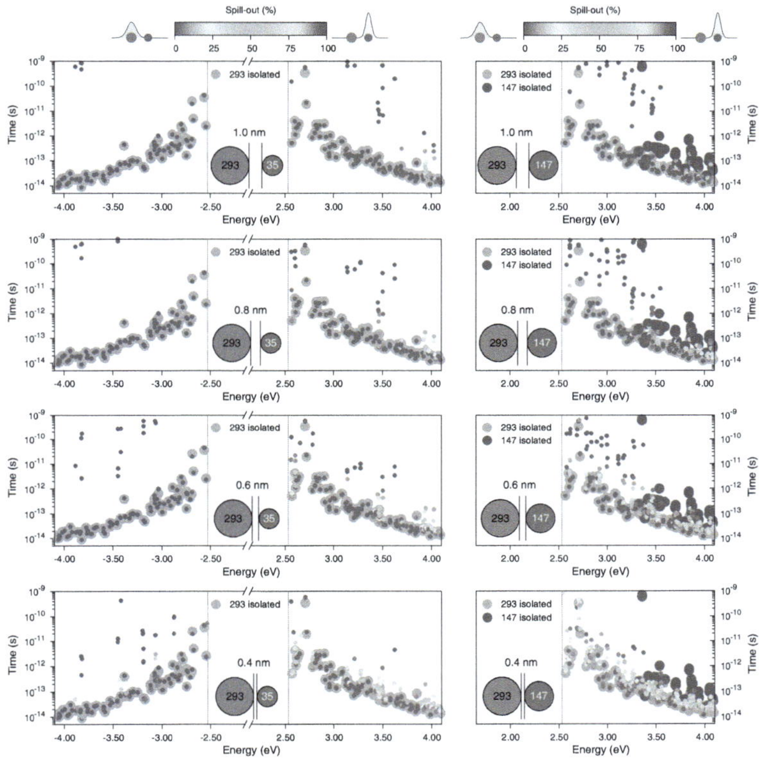

Figure 6.11 Calculated CM lifetimes as a function of the spill-out parameter are reported in the figure. Reprinted from Ref. [21], Copyright 2012 Nature Publishing.

the conduction bands of a nearby NC where an extra e–h pair is generated (CDCT mechanisms ignited by the transfer of a single charge are depicted in the third box of Fig. 6.12). A general definition of one-site CM, CDCT and SSQC valid for all possible configurations (wavefunctions localized on a single NC or shared by two NCs) is reported below:

$$\frac{1}{\tau^{e/h}_{\text{one-site}}} = \sum_{n_b\mathbf{k}_b}\sum_{n_c\mathbf{k}_c}\sum_{n_d\mathbf{k}_d}\Big[s_{n_a\mathbf{k}_a}s_{n_b\mathbf{k}_b}s_{n_c\mathbf{k}_c}s_{n_d\mathbf{k}_d} + (1-s_{n_a\mathbf{k}_a})(1-s_{n_b\mathbf{k}_b})(1-s_{n_c\mathbf{k}_c})(1-s_{n_d\mathbf{k}_d})\Big]\frac{1}{\tau^{e/h}_{(a,b)\to(c,d)}},$$

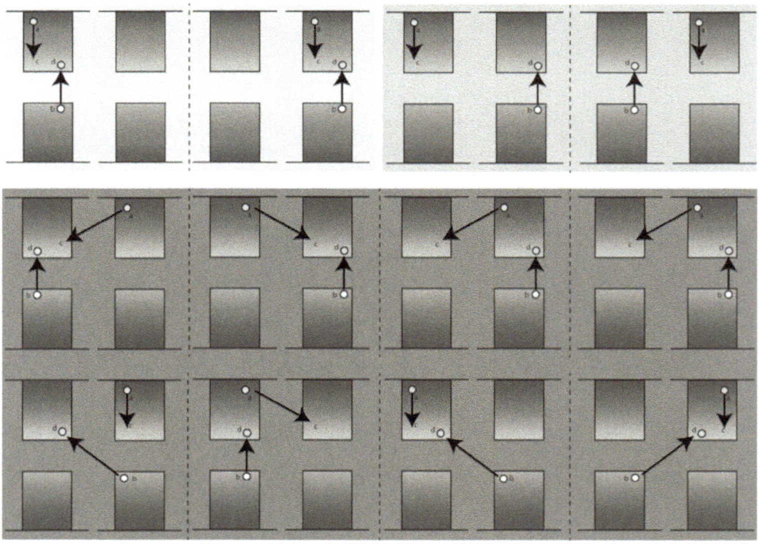

Figure 6.12 One-site CM (white box), SSQC (light gray box) and CDCT (gray box) transitions are depicted in the figure.

$$\frac{1}{\tau^{e/h}_{\mathrm{SSQC}}} = \sum_{n_b\mathbf{k}_b}\sum_{n_c\mathbf{k}_c}\sum_{n_d\mathbf{k}_d}\Big\{\left[(1-s_{n_a\mathbf{k}_a})s_{n_b\mathbf{k}_b} + s_{n_a\mathbf{k}_a}(1-s_{n_b\mathbf{k}_b})\right] \times \left[s_{n_c\mathbf{k}_c}(1-s_{n_d\mathbf{k}_d}) + s_{n_d\mathbf{k}_d}(1-s_{n_c\mathbf{k}_c})\right]\Big\}\frac{1}{\tau^{e/h}_{(a,b)\to(c,d)}}$$

$$\frac{1}{\tau^{e/h}_{\mathrm{CDCT}}} = \frac{1}{\tau^{e/h}_{n_a\mathbf{k}_a}} - \frac{1}{\tau^{e/h}_{\mathrm{SSQC}}} - \frac{1}{\tau^{e/h}_{\text{one-site}}}.$$

where

$$\frac{1}{\tau^{e/h}_{(a,b)\to(c,d)}} = 4\pi\left[\mid \mathrm{M}_D\mid^2 + \mid \mathrm{M}_E\mid^2 + \mid \mathrm{M}_D - \mathrm{M}_E\mid^2\right] \times \delta(E_a + E_b - E_c - E_d)$$

is the total CM rate for the generic single CM decay path $(a, b) \to (c, d)$, $\tau^{e/h}_{n_a\mathbf{k}_a}$ is the total CM lifetime, s_a, s_b, s_c and s_d are the spill-out parameters of a, b, c and d carriers (see Fig. 6.1) and $\tau_{\text{one-site}}$, τ_{SSQC} and τ_{CDCT} denote the one-site CM, SSQC and CDCT lifetimes. Calculated SSQC and CDCT lifetimes are reported in Fig. 6.13 as a function of the energy of the initial carrier for both

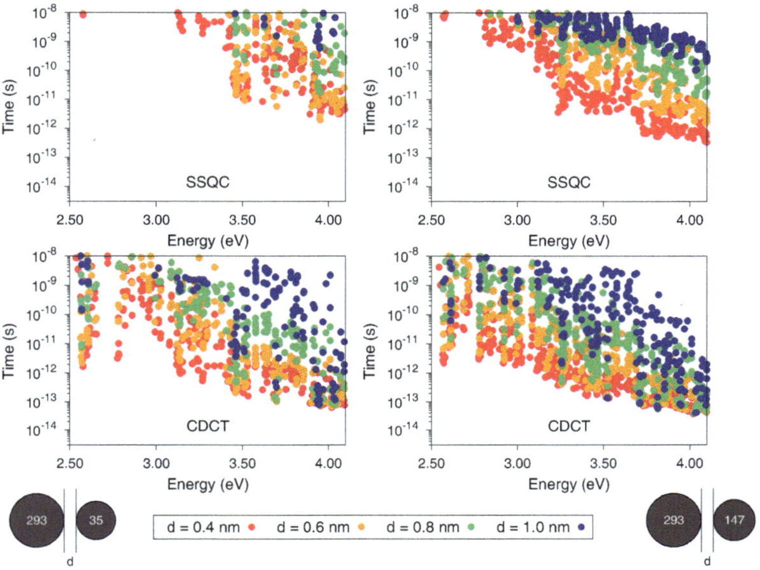

Figure 6.13 Calculated SSQC and CDCT lifetimes are reported in the figure. Here different colors identify different NC–NC separations. Reprinted from Ref. [21], Copyright 2012 Nature Publishing.

the considered systems (we consider only mechanisms ignited by electron relaxation). Here different colors identify different NC–NC separation. Our results point out that two-site CM lifetimes strongly decrease when the energy of the initial carrier (E_i) increase and when the NC–NC separation is reduced. Remarkably, SSQC lifetimes decrease when NCs size increases. A similar behavior is not observed neither for one-site not for CDCT CM transitions. Again, SSQC (and in general the two-site CM mechanisms) can benefit from experimental conditions where the embedding matrix (formation of minibands) or the presence of several interacting NCs (typical condition of three-dimensional realistic systems) is expected to amplify the relevance of SSQC. For these reasons our estimation for the SSQC processes is an upper bound for the lifetime of energy transfer quantum cutting mechanisms. As a consequence in a realistic system SSQC lifetimes can settle at fractions of picoseconds in a large window of energies.

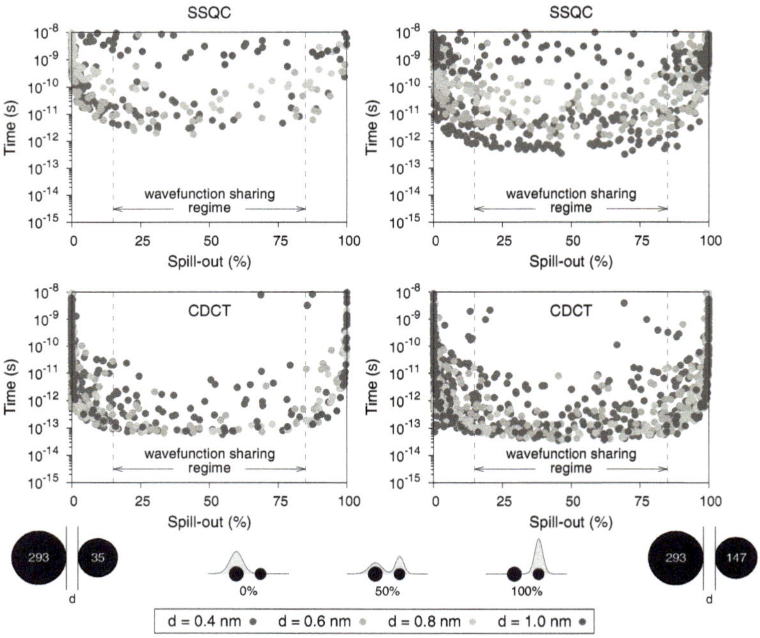

Figure 6.14 Calculated SSQC and CDCT lifetimes are reported in the figure as a function of the spill-out parameter. Reprinted from Ref. [21], Copyright 2012 Nature Publishing.

The results reported in Fig. 6.13 and in Fig. 6.5 suggest the following lifetime hierarchy, that is,

$$\tau_{\text{one-site}} \leq \tau_{\text{CDCT}} \leq \tau_{\text{SSQC}}$$

which points out that one-site CM mechanisms are typically faster than two-site CM mechanisms and that CDCT transitions are faster than SSQC transitions. A direct formation of excitons in neighboring Si NCs after absorption of a single photon is therefore not compatible with our results and cannot be used to interpret experimental evidences by Trinh et al. [56]. Two-site CM mechanisms are always dominated by one-site CM events.

A simple model based on a not-perpetual cyclic procedure of one-site CM, SSQC and Auger exciton recycling has been recently proposed to interpret results by Trinh et al. [21].

To identify the conditions that maximize two-site CM events, we report calculated SSQC and CDCT lifetimes as a function of the spill-out parameter.

It is evident that two-site CM processes are slow when inital state wavefunction is localized onto one single NC (spill-out equal to 0% or 100%). However we observe changes up to 3 orders of magnitude in both SSQC and CDCT lifetimes when the initial state ceases to be completely localized onto one NC and at least the 15% of the wavefunction is shared by two NCs. As a consequence, the maximum efficiency for the two-site CM events is recorded when the initial carrier wavefunction extends to both NCs and the spill-out parameter ranges from 15% to 85%. These conditions define the so-called "wavefunction-sharing regime" [21], where both energy and charge transfer processes are maximized.

6.6 Conclusions

In this chapter we have studied, by first principles, CM processes in both isolated and interacting Si NCs. We have discussed one-site CM events in isolated Si NCs proving benefits induced by quantum confinement of the electronic density. We have also investigated two-site CM mechanisms in strong coupled Si NCs proving that such effects can be divided in two components, that is, SSQC and CDCT. On the basis of our calculations we have shown that one-site CM events are always faster that two-site CM events and that CDCT mechanisms are faster than SSQC processes. Conditions that maximize two-site CM effects have been discussed and a new regime called the wavefunction-sharing regime has been introduced.

Acknowledgments

The authors thank the Super-Computing Interuniversity Consortium CINECA for support and high-performance computing resources under the Italian Super-Computing Resource Allocation (ISCRA) initiative, PRACE for awarding us access to resource IBM BGQ based in

Italy at CINECA, and the European Community's Seventh Framework Programme (FP7/2007-2013; grant agreement 245977).

References

1. M. Aerts et al., Nano Lett. 4485, 11 (2011).
2. G. Allan and C. Delerue, Phys. Rev. B 73, 205423 (2006).
3. G. Allan and C. Delerue, Phys. Rev. B 79, 195324 (2009).
4. G. Allan and C. Delerue, Phys. Rev. B 79, 125340 (2008).
5. A. I. Andrushko, K. M. Salikov and S. V. Slobodchikov, Sov. Phys. Semicond. 20, 255 (1986).
6. M. C. Beard, J. Chem. Phys. Lett. 2, 1282 (2011).
7. M. C. Beard et al., Nano Lett. 7, 2506 (2007).
8. M. C. Beard et al., Nano Lett. 10, 3019 (2010).
9. A. R. Beattie and P. T. Landsberg, Proc. R. Soc. London A 249, 16 (1959).
10. H. Benisty, C. M. Sotomayor-Torre's and C. Weisbuch, Phys. Rev. B 44, 10945 (1991).
11. L. M. Blinov, E. A. Bobrowa, V. V. Vavilon and G. N. Galkin, Sov. Phys. Solid State 9, 2537 (1968)
12. R. Conradt and J. Aengenheister, Solid State Commun. 10, 321 (1972).
13. C. Delerue, G. Allan, J. J. H. Pijpers and M. Bonn, Phys. Rev. B 81, 125306 (2010).
14. A. L. Efros, V. A. Kharchenko and M. Rosen, Solid State Commun. 93, 281 (1995).
15. R. J. Ellingson et al., Nano Lett. 5, 865 (2005).
16. A. Franceschetti, J. M. An and A. Zunger, Nano Lett. 6, 2191 (2006).
17. D. Gachet, A. Avidan, I. Pinkas and D. Oron, Nano Lett. 10, 164 (2010).
18. P. Giannozzi et al., J. Phys. Condens. Matter 21, 395502 (2009).
19. A. Goetzberger, J. Knobloc and B. Voss, *Cristalline Silicon Solar Cells*, John Wiley & Sons, New York, 1994.
20. M. Govoni, I. Marri and S. Ossicini, Phys. Rev. B 84, 075215 (2011).
21. M. Govoni, I. Marri and S. Ossicini, Nat. Photon. 6, 672 (2012).
22. M. I. Guseva, N. I. Zotova, A. V. Koval and D. N. Nasledov, Sov. Phys. Semicond. 9, 591 (1975).
23. S. Hausser, G. Fuchs, A. Hangleiter, K. Streubel and W. T. Tsang, Appl. Phys. Lett. 913 (1990).

24. D. Hill and P. T. Landsberg, Proc. R. Soc. London A 347, 547 (1976); 347, 565 (1976).
25. F. Iori, E. Degoli, E. Luppi, R. Magri, I. Marri, G. Cantele, D. Ninno, F. Trani, and S. Ossicini, J. Luminesc. 121, 335 (2006).
26. M. J. Kerr, A. Cuevas, and P. Campbell, Prog. Photovolt. Res. Appl. 11, 97 (2003).
27. V. I. Klimov et al., Science 290, 314 (2000a).
28. V. I. Klimov, A. A. Mikhailovsky, D. W. McBranch, C. A. Leatherdale and M. G. Bawendi, Science 287, 1011 (2000b).
29. M. Ji, S. Park, S. T. Connor, T. Mokari, Y. Cui and K. J. Gaffney, Nano Lett. 9, 1217 (2009).
30. P. T. Landsberg, *Recombination in Semiconductors*, Cambridge University Press, 1991.
31. O. Lazarenkova and A. Balandin, J. Appl. Phys. 89, 5509 (2001).
32. A. Marini, C. Hogan, M. Grünig and D. Varsano, Compt. Phys. Commun. 180, 1392 (2009).
33. G. Nair, S. M. Geyer, L.Y. Chang and M. G. Bawendi, Phys. Rev. B 78, 125325 (2008).
34. A. J. Nozik, Phys. E: Low-dimensional Syst. Nanostruct. 14, 115 (2002).
35. L. A. Padilha, J. T. Stewart, R. L. Sandberg, W. K. Bae, W. Koh, J. M. Pietryga and V. I. Klimov, Acc. Chem. Res. 46, 1261 (2013).
36. S. Picozzi, R. Asahi, C. B. Geller and A. J. Freeman, Phys. Rev. Lett. 89, 197601 (2002).
37. J. M. Pietryga, K. K. Zhuravlev, M. Whitehead, V. I. Klimov and R. D. Schaller, Phys. Rev. Lett. 101, 217401 (2008).
38. J. J. H. Pijpers et al., J. Phys. Chem. C 111, 4146 (2007).
39. E. Rabani and R. Baer, Chem. Phys. Lett. 496, 227 (2010).
40. E. Rabani and R. Baer, Nano Lett. 8, 4488 (2008).
41. I. Robel, R. Gresback, U. Kortshagen, R. D. Schaller and V. I. Klimov, Phys. Rev. Lett. 102, 177404 (2009).
42. C. A. Rozzi, D. Varsano, A. Marini, E. K. U. Gross and A. Rubio, Phys. Rev. B 73, 205119 (2006).
43. O. E. Semonin et al., Science 334, 1530 (2011).
44. R. D. Schaller and V. I. Klimov, Phys. Rev. Lett. 92, 186601 (2004).
45. R. D. Schaller, M. A. Petruska and V. I. Klimov, Appl. Phys. Lett. 87, 253102 (2005).

46. R. D. Schaller, M. Sykora, J. M. Pietryga and V. I. Klimov, Nano Lett. 6, 424 (2006).
47. R. D. Schaller, M. Sykora, S. Jeong and V. I. Klimov, J. Phys. Chem. B 110, 25332 (2006).
48. R. D. Schaller, J. M. Pietryga and V. I. Klimov, Nano Lett. 7, 3469 (2007).
49. R. D. Schaller, V. M. Agranovich and V. I. Klimov, Nat. Phys. 1, 189, (2005).
50. A. Shabaev, A. L. Efros and A. J. Nozik, Nano Lett. 6, 2856 (2006).
51. W. Shockley and H. J. Queisser, J. Appl. Phys. 32, 510 (1961).
52. S. K. Stubbs et al., Phys. Rev. B 81, 081303 (2010).
53. D. Timmerman, I. Izeddin, P Stallinga, I. N. Yassievich and T. Gregorkiewicz, Nat. Photon. 2, 105 (2008).
54. D. Timmerman, J. Valenta, K. Dohnalova, W. D. A. M. de Boer and T. Gregorkiewicz, Nat. Nanotech. 6, 710 (2011).
55. M. T. Trinh et al., Nano Lett. 8, 1713 (2008).
56. M. T. Trinh et al., Nat. Photon. 6, 316321 (2012).
57. K. G. Svantesson, N. G. Nilsson and L. Huldt, Solid State Commun. 9, 213 (1971).
58. K. L. Vodopyanov, H. Graener, C. C. Phillips and T. J. Tate, Phys. Rev. B 46 194 (1992).
59. S. Wippermann, M. Vörös, D. Rocca, A. Gali, G. Zimanyi and G. Galli, Phys. Rev. Lett. 110, 046804 (2013).
60. I. Marri, M. Govoni and S. Ossicini, Beilstein J. Nanotechnol. 6, 343 (2015).
61. I. Marri, M. Govoni and S. Ossicini, J. Am. Chem. Soc. 136, 13257 (2014).

Chapter 7

The Introduction of Majority Carriers into Group IV Nanocrystals

Dirk König
Australian Centre for Advanced Photovoltaics, University of New South Wales (UNSW), Sydney NSW 2052, Australia
dirk.koenig@unsw.edu.au

7.1 Introduction

Group IV nanocrystals (NCs) are about to revolutionize the electronic devices and gadgets we are using today due to their specific properties. These properties are pivotal in using group IV NCs for new Third Generation solar cells of high efficiency at low material cost, providing renewable energies for future economic growth with minimum climate impact. Group IV NCs advance large-scale integration into regions which provide massive increase in compute performance at much decreased energy consumption.

In order to unfold their full potential, the technology to assign one majority carrier type per Group IV NC must exist so that *n*- and *p*-type (electron- and hole-dominated) NCs can form electronic devices in analogy to existing solid-state electronics like

Nanotechnology and Photovoltaic Devices: Light Energy Harvesting with Group IV Nanostructures
Edited by Jan Valenta and Salvo Mirabella

ISBN 978-981-4463-63-8 (Hardcover), 978-981-4463-64-5 (eBook)
www.panstanford.com

metal-oxide-semiconductor field-effect transistors (MOSFETs) and diodes. This holds true also for such devices working in the ballistic transport regime. Their mass production becomes realistic at device dimensions below 14 nm. This size establishes one group IV NC as the semiconductor volume of an individual electronic device as is the case for MOSFETs with gate length below 14 nm. Until now, there is no clear route how to achieve group IV NC doping with reasonable effort and success rate in large-scale production, or indeed even in research. Hence, new ways must be found for the introduction of majority carriers into Si NCs. This chapter investigates Si as a typical group IV semiconductor. However, derived findings also apply to other group IV NC materials such as germanium (Ge) and Si-Ge alloys (SiGe).

The chapter is organized as follows. Section 7.2 evaluates the theory of conventional NC doping. One major aspect in Section 7.2.1 is the thermodynamics of NC doping which considers the most stable configuration of the NC-dopant-dielectric system. We arrive at very fundamental processes like NC self-purification and the complete bond saturation of dopants as their most stable energetic configuration. Another aspect are the electronic properties of dopants in NCs (assuming they get activated within the NC lattice) discussed in Section 7.2.2. We compare these properties of NCs with the dopant presenting a point defect. High accuracy hybrid density functional theory (h-DFT) calculations are a very powerful simulation tool to understand and predict electronic properties of Si NCs and dopants embedded in dielectrics on the atomic scale. Section 7.2.3 presents h-DFT results of the electronic nature of phosphorous (P) in SiO_2, in $SiO_{0.9}$, at and within completely OH-terminated 1.5 nm Si NCs. Experimental results found in the literature are evaluated in Section 7.3. We divide the survey into doped Si nanovolumes for next-generation ultra-large-scale integration (NG-ULSI) FET devices: Section 7.3.1; free-standing NCs: Section 7.3.2; and embedded NCs formed by segregation anneal from a Si-enriched matrix (Si oxide - SiO_2, nitride - Si_3N_4) as precursor: Section 7.3.3. With these findings, we look into alternative approaches for introducing majority carriers into NCs: Section 7.4. All outcomes are summarized in Section 7.5 together with an outlook on the research field.

7.2 Theory of Conventional Nanocrystal Doping

Before we evaluate NCs, it is instructive to have a brief look at the phenomenology of impurity diffusion as a function of structure size, temperature and associated exposure time.

At high temperatures, impurities in Si are known to diffuse over distances in the cm range. In fact, this behavior is exploited for float-zone refinement of Si ingots which is carried out at temperatures (T) of ca. 1450°C [5, 40]. On the microscopic scale, it has been established in the 1980s that dopants massively diffuse to grain boundaries (GBs) in multicrystalline (mc) Si [39] over a few hundred μm. For poly-Si with grain sizes of about 100 nm this process occurs during solid phase crystallization at $T \approx 600°C$ [72, 84]. This segregation effect is so large that is was investigated for carrier collection from thick μc-Si layers in solar cells. Research in this matter was abandoned after it emerged that the Shockley–Read–Hall (SRH) recombination rate along these grain boundaries is very high due to massive impurity concentrations, incurring enormous carrier losses.

In dielectrics like SiO_2 or Si_3N_4 the segregation of excess Si forming embedded Si NCs is carried out at $T = 1100 \pm 100°C$ for 10 to 180 s by rapid thermal anneal [33] or by conventional furnace anneal for 30 to 120 min [34, 86]. With these temperatures it is fairly clear that conventional doping or indeed any incorporation of foreign atoms other than anions of the embedding dielectric into Si NCs occurs only with a very low probability, except their presence is increased to concentrations of $\geq$ 0.5 mol-%, or the Si excess concentration is increased to levels where an interconnected Si NC network is formed. Former turns the precursor layers for embedded Si NCs into an alloy with different electronic structure, see Sections 7.2.1, 7.2.2 and 7.3.2. Si excess concentrations beyond separate Si NC formation reduce control over structural and electronic properties due to spontaneous crystallization into Si NC networks (porous Si, Si sponge) [56] and reduce or even completely remove quantum properties. Such amorphous materials still may have electronic band gaps exceeding the value of the crystalline (c)-Si bulk phase, but are close to values of a-Si. However, transport properties of Si NC networks are very likely to be inferior to a-Si. The remaining dielectric forms an antidot lattice with the Si NC network whereby carrier transport is hampered and additional defects are introduced. Hence, it is questionable whether practical solar cells based on such materials will be able to compete with low-cost a-Si-based solar cells, which have stable conversion efficiencies of $\eta \approx 10\%$ [26].

Over the last 15 years, ultra-large-scale integration (ULSI) developed from gate lengths of 130 nm to the current 22 nm technology node. Dopant activation in MOSFETs is crucial for device operation. With decreasing device dimensions, the diffusion of dopants as opposed to their sole activation in the Si lattice became increasingly challenging. Dopant activation by annealing was optimized over the last years in order to meet required sharp doping profiles and high dopant activation probabilities. Rapid thermal anneals (RTAs) with $\approx$900 to 1100°C and $t = 10$ to 30 s were used for the 250 nm to 130 nm technology nodes [12]. The 65 nm and 45 nm technology nodes already required a flash anneal for dopant activation without significant dopant relocation by diffusion [25, 51]. In 2010, laser spike annealing was introduced to cope with shrinking device dimensions [8]. Even with such a sophisticated annealing technology at hand, dopant activation with a sharp doping profile (i.e., a shallow junction) is a major challenge for the 22 nm and in particular 15 nm (150 Å) technology nodes [52].

7.2.1 *Thermodynamics: Stable vs. Active Dopant Configurations*

One fundamental property for integrating dopants into a lattice is the required energy: the formation energy E_{form} for an impurity atom to be built into the local lattice. This means for donors with an extra valency to establish bonds to all of its first next-neighbor (1-nn) atoms and for acceptors with one valency less to saturate all of its valency states by forming bonds to 1-nn host atoms. Then, donors have a dangling bond (DB), which may get ionized, providing an electron as the majority carrier and leaving a positively ionized donor behind. An acceptor yields one DB from one of its 1-nn host atoms which may get negatively ionized by capturing an electron of antiparallel spin at the acceptor, providing a hole as a majority carrier.

The impurity segregation to defect zones like interfaces or stacking faults (grain boundaries) increases with decreasing crystalline volume. This fundamental thermodynamical process is called self-purification [13]. As a result, E_{form} of an impurity increases with shrinking crystalline volume. Although the amount of increase depends on specific host material and impurity, self-purification occurs to some degree in most quantum systems, see top graph in Fig. 7.1 and references in [14].

Several forces contribute to self-purification. Mechanical stress originates from different bond lengths and angles of the impurity to its 1-nn host atoms, resulting in strained bonds. Latter have lower binding energies than relaxed bonds of the NC volume. This energetic difference fuels self-purification: the impurity is literally squeezed out so that the NC volume gains energy by taking on its most stable configuration.

Dopants introduce charge carriers to Si NCs, which leads to electrostatic stress. The electrostatic field forces a displacement upon host atoms depending on their charge. This displacement pushes atoms out of their most stable equilibrium position (which occurs at a maximum value of the integral over all binding energies between Si NC atoms) and therefore follows the same thermodynamical pattern described for mechanical stress. Codoping of a Si NC with one acceptor and one donor would cancel out electrostatic stress and therefore appears to be more likely as opposed to either acceptor or donor doping. Indeed, local density approximation (LDA) calculations have shown that E_{form} reaches a local minimum if a Si NC is codoped as opposed to doping with either an acceptor or a donor [67], see bottom graph in Fig. 7.1. However, codoping does not provide majority carriers as the electron from the positively ionized donor would be localized at the acceptor, whereby neither of the dopants could provide a majority carrier.

As an illustrative picture of self-purification, we consider a large transparent cube filled with billiard balls of same size and weight into which we put a ball of different size. If we shake the cube for some time (simulating thermal movement of atoms during an anneal), two things will happen: The billiard balls get arranged in the most dense possible packing fraction (hexagonally close-packed [HCP] "crystal"), and the ball with different size ends up at one of

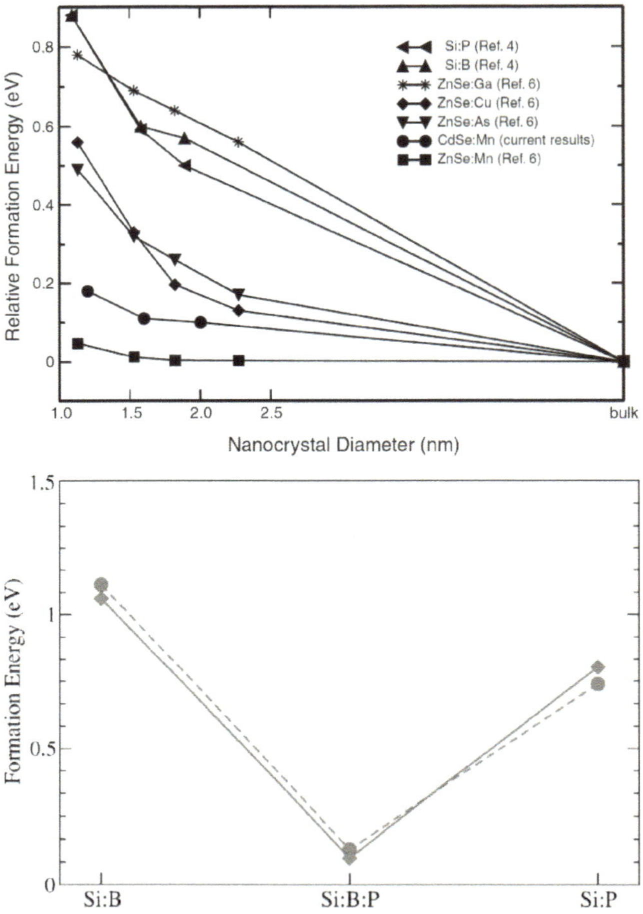

Figure 7.1 Relative values of E_{form} referring to bulk values for boron (B) and phosphorous (P) dopants in Si NCs, Mn, Cu, and Ga dopants in zinc selenide (ZnSe) NCs and Mn dopants in cadmium selenide (CdSe) NCs as a function of NC size (top) [14]. Copyright (2008) by the American Physical Society. Values of E_{form} of acceptor (B)- and phosphorous (P)-doped Si NCs relative to the codoped (B and P) case from LDA calculations of fully H-terminated Si NCs (bottom) [67]. Reprinted with permission from [Ossicini, S., Degoli, E., Iori, F., Luppi, E., Magri, R., Cantele, G., trani, F., and Ninno, D. (2005). Simultaneously B- and P-doped silicon nanoclusters: Formation energies and electronic properties, *Applied Physics Letters* **87**, 173120, 1–3]. Copyright [2005], AIP Publishing LLC. The full line with diamond symbols refers to 1.5 nm NCs ($Si_{87}H_{76}$) and the dashed line with full circles to 1.8 nm NCs ($Si_{147}H_{100}$).

the cube walls which corresponds to an interface or grain boundary in a crystalline volume. If the cube size is increased by an order of magnitude, we have to shake it a much longer time until the ball with the different size ends up at one of the cube walls. It is also more likely to become integrated into the HCP "lattice," although it will generate significant distortion and stress in its local lattice environment. This illustrates that we can anneal the sample longer to activate dopants without the vast majority of them ending up on an interface or grain boundary. Departing from this qualitative picture, we consider the number of bonds per atom in a face-centered cubic (fcc) lattice as a key parameter for self-purification.

The limit of bonds per lattice atom in a fcc crystal $N_{\text{bond}}^{\text{fcc}}$ with its extension r approaching infinity is $\lim_{r\to\infty}(N_{\text{bond}}^{\text{fcc}}) = 2$ if we count every bond just once. This can be easily verified by looking at a specific atom which has four bonds, each shared with exactly one 1-nn atom. We can now proceed to one of its 1-nn atoms and repeat the procedure, again ending up with four bonds which are shared with 1-nn atoms of the lattice site considered (one of them being the bond to the atom we had considered before). We can repeat this procedure until all bonds in an infinite crystal are accounted for, providing us with $4/2 = 2$ bonds per atom: $N_{\text{bond}}^{\text{fcc}} = 2$. For finite fcc crystals such as Si NCs we obtain $N_{\text{bond}}^{\text{fcc}} = N_{\text{NC}}(\text{Si–Si}) < 2$ due to interface bonds taking away more and more Si–Si bonds with shrinking NC size [49]. Octahedral Si NCs with eight $\{111\}$ interfaces are high symmetry Si NCs with the minimum number of interface bonds per unit volume. For these, recursive geometric series were derived [49] for the number of Si atoms forming the NC,

$$N_{\text{Si}}[i] = N_{\text{Si}}[i-1] + (2i+1)^2 \quad \forall i \geq 0 \tag{7.1}$$

and for the number of anions at the interface which is identical to the number of interface bonds:

$$\begin{aligned} N_{\text{X}}[i] &= N_{\text{X}}[i-1] + (8i+4) \quad \forall i \geq 0 \\ &= 4(i+1)^2, \end{aligned} \tag{7.2}$$

whereby X stands for the interface anion forming part of the embedding dielectric (N, O). As these Si NCs have a minimum number of interface bonds per unit volume, they serve as the lower size limit for a certain impact of the embedding dielectric. A second order recursive geometrical series describing the number of Si–Si bonds is given by

$$N_{\text{bond}}(\text{Si–Si})[i] = 16(i-1) + 12 + 2N_{\text{bond}}(\text{Si–Si})[i-1] - N_{\text{bond}}(\text{Si–Si})[i-2] \quad \forall i \geq 2\,, \tag{7.3}$$

with the starting terms $N_{\text{bond}}(\text{Si–Si})[0] = 0$ and $N_{\text{bond}}(\text{Si–Si})[1] = 12$. The quotient $N_{\text{bond}}(\text{Si–Si})[i]/N_{\text{Si}}[i]$ provides us with the number of Si–Si bonds as a function of NC size, $N_{\text{NC}}(\text{Si–Si},\ d_{\text{NC}}[i])$, whereby

$$d_{\text{NC}}[i] = \sqrt[3]{\frac{6}{\pi}\mathcal{V}_{\text{Si}} N_{\text{Si}}[i]} \quad \forall i \geq 0\,, \tag{7.4}$$

with $\mathcal{V}_{\text{Si}} = 2.005 \times 10^{-2}$ nm^3 as the unit volume per Si atom derived from its unit cell length of 0.54309 nm and a spherical shape of the NC. Latter is assumed here to work with one size parameter. The volume can be easily calculated for an octahedral shape[a]. Nevertheless, we shall use the spherical shape of the NC volume from Eq. 7.4 as it keeps the description simple and enables us to compare NC sizes with experimental values usually derived for NCs assuming a spherical shape. The interface impact can also be derived from the ratio of interface bonds to Si NC atoms $N_{\text{X}}[i]/N_{\text{Si}}[i]$ and from the ratio of interface bond to Si–Si bonds $N_{\text{X}}[i]/N_{\text{bond}}(\text{Si–Si})[i]$. These quotients are shown together with $N_{\text{bond}}(\text{Si–Si})[i]/N_{\text{Si}}[i]$ in Fig. 7.2, whereby the running parameter i was replaced by the respective $d_{\text{NC}}[i]$. Equations 7.1, 7.3 and 7.4 enable us to relate experimental results to the number of bonds per Si atoms per NC diameter d_{NC}, whereby d_{NC} serves as a lower size limit. We will use this in Section 7.3.2.

A simple structural reason why it does not make sense to increase doping densities into the alloy range ($\geq$ 0.1 mol-% or 5×10^{19} cm^{-3}) is the fact that very small Si NCs would cross the transition from very high dopant densities to an alloy. Under the assumption that every Si NC shall be doped with exactly one dopant, the doping density of 5×10^{19} cm^{-3} sets a lower limit to the size of Si NCs of $d_{\text{NC}} = 3.4$ nm (10^3 Si atoms). By going beyond this dopant density, the entire structure suffers from deteriorated

---

[a] $V = 1/6\, d_{\text{octa}}^3$, with d_{octa} as distance between two opposite peaks. d_{octa} only describes the maximum extension of the Si NC, requiring the distance of opposite planes $d_{\text{plane}} = 1/\sqrt{2}d_{\text{octa}}$ to be added for completeness.

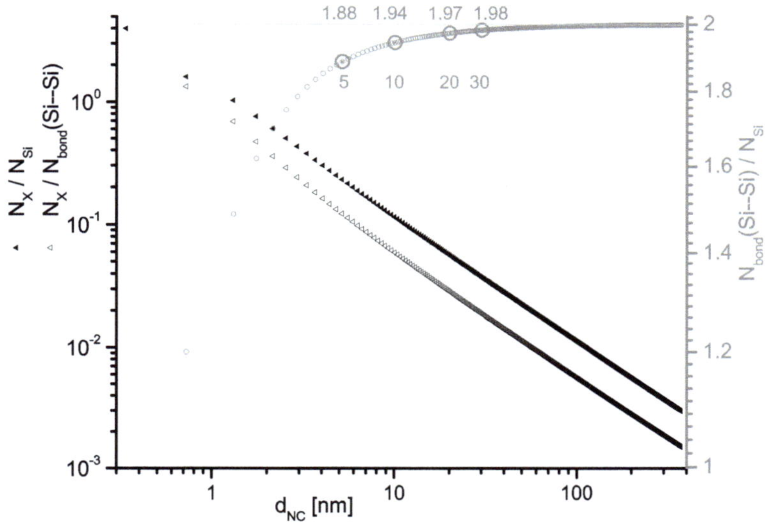

Figure 7.2 Quotients N_X/N_{Si} and N_X/N_{bond}(Si–Si) (empty and filled black symbols, left Y axis) for high-symmetry Si NCs of octahedral shape exclusively terminated by {111} surfaces shown as a function of $d_{NC}[i]$. The right Y axis and gray symbols show the average number of bonds per Si atom to its 1-nn Si neighbors N_{bond}(Si–Si)/N_{Si} as a function of $d_{NC}[i]$. The gray circles and values of N_{bond}(Si–Si)/N_{Si} (top row) with respective $d_{NC}[i]$ (bottom row) refer to Fig. 7.8 in Section 7.3.2. The dashed light gray line shows the asymptotic value of N_{bond}(Si–Si)/N_{Si} for an infinite Si crystal ($\lim_{N_{Si}\to\infty} = 2$).

quantum properties and starts to behave like an alloy, see Section 7.3.2. Another issue is the doping probability which was presumed to be 100% in the above consideration but is much smaller in reality.

The entropy of dopant activation in Si NCs was shown to lower the doping probability to extremely low values. Below we want to evaluate the question whether a short-time nonequilibrium situation provides the conditions for a much increased doping probability. Such a situation can be brought about by a rapid thermal anneal (RTA) or LASER spike anneal. Latter is state of the art in ULSI technology; typical temperatures and time windows are $T = 1240°C$ and a LASER pulse time of $t = 1$ ms [8]. Self-regulatory plasma doping is an alternative nonequilibrium approach [75], though a considerable dopant underdiffusion from the source or drain regions of a fin-field-effect transistor (FET) into the channel region occurs [37], see Section 7.3.1. As we shall see in Section 7.3.2, the situation is even worse for free-standing Si NCs, although very favorable high nonequilibrium conditions are used by in-situ plasma doping during their synthesis [42, 79]. When doping embedded Si NCs, dopant gettering (passivation of all its valence states by forming bonds to adjacent atoms) at the NC/dielectric interface and at point defects within the dielectric further diminishes the probability per dopant to provide a majority charge carrier, see Section 7.2.3. Judging from these experimental findings, nonequilibrium approaches to conventional doping of small Si nanovolumes present a major challenge, though to a smaller extent as compared to equilibrated dopant activation such as *in-situ* with Si NC formation in Si-rich dielectrics. Essentially, the challenge is to provide energy for dopant activation with minimum momentum as latter promotes diffusion. This is clearly pictured by the trend to provide activation energy by photons which have an extremely small momentum. Methods like tungsten halogen lamps in RTA or LASER pulses in spike anneals clearly show this trend. The advantage over conventional furnace anneal for Si NC formation in Si-rich dielectrics is a short burst of energy that heats the lattice only locally for a very short time, thereby minimizing diffusion. Unfortunately, this approach does not work well for Si NC formation in Si-rich dielectrics as the Si atoms themselves have to be promoted to diffuse to form Si NCs. However, even in ULSI integration, considerable blurring of doping profiles occurs. When dopants are provided with energy for their activation into the Si lattice, this energy will be used for both, diffusion and activation. While activation energies increase tremendously with shrinking NC volume, diffusion is usually not hampered much. The only leverage available to minimize diffusion is to provide the dopants with energy under the constraint that it brings along a minimum amount of momentum, that is, the lattice must either be kept possibly cold, or only heated for a very short time where its heat capacity starts to play a role in limiting momentum transfer to the dopant, lowering its ability to diffuse. The question is whether even shorter time slots in LASER spike anneals can improve dopant activation which would require to go for higher photon energies (UV to VUV range). Such photons would provide energy on a faster scale—assuming a constant photon flux—and are also helpful for shallow junctions as the absorption coefficients of semiconductors increase with photon energy. On the downside, high energy photons introduce radiation damage in both Si and gate dielectrics. Former plays a role where carrier lifetime or nonradiative recombination is important, for example, in solar cells, photo detectors or light emitting devices. The latter increases gate leakage in ULSI devices which deteriorates device performance. Self-regulatory plasma doping as alternative approach provides highly reactive dopant species which chemically interact with the Si lattice and form a compound. While a considerable partition is built into the Si lattice as active species, many dopant atoms stay inactive or even cluster due to the extremely high concentration required to achieve required active dopant densities, see Fig. 7.7.

From above discussion it became clear that Si NCs in dielectrics have to use other approaches and techniques than conventional doping. However, if the dopant density is in the range of 0.2 to 1 mol-% ($\approx$1 to 5 $\times$ 10^{20} cm^{-3}) without a clear evidence of high density Si NC doping, the question arises where the vast majority of these dopants end up, what their electronic nature is and how they contribute to the electronic structure of the entire NC/dielectric system. We shall consider the first part of this question and leave the electronic properties to Section 7.2.3. In general, an atom reaches its most stable position when its integral over all binding energies reaches a maximum. As the chemical bond contributes by far the most to the binding energy, the three key parameters which determine the maximum binding energy are the species the atom forms bonds with, the number of bonds established up to the full saturation of its valence states and the bond type to each 1-nn atom (single, double or triple). In Si, SiO_2 and Si_3N_4 virtually all interatomic bonds are single bonds, we therefore disregard the last criterion. The other two parameters have far-reaching consequences for dopant atoms, because donors (acceptors) have one more (one less) bond than Si. The occurrence of DBs like the P_{b1} or P_{b0} defect were shown to exist in the embedding dielectric and at the Si NC interfaces [34, 81]. Such DBs are very attractive for dopant gettering since the surplus or missing bond of a dopant allows for its full bond saturation and the saturation of the DB on one of its 1-nn atoms. The segregation anneal forming Si NCs does not use up all excess Si since this would require an extremely long annealing time. As a result, a small portion of excess Si is still present within the dielectric matrix whereby defects like interstitial sites, DBs, double bonds or strained bonds are formed. These will also contribute to dopant gettering. Regions which suffer from high strain like the NC/dielectric interface are more likely to getter dopants as these can decrease or disperse strain due to their different bond geometry.

To sum up, the conventional doping of separate Si NCs or nc-Si layers embedded in or sandwiched between SiO_2 or Si_3N_4 is not possible to a degree which would allow for nanodevice operation. Results presented in the literature as Si NC doping either use dopant concentrations of $\geq$ 0.1 mol-% ($\geq 5 \times 10^{19}$ cm^{-3}) what renders the entire material system to be an alloyed ternary compound or use excess Si concentrations so high that an interconnected Si NC network is created. While the doping probability is still too low for good device performance, such Si NC networks originating from $SiO_{x\leq 1}$ have to compete with a-Si as a cheap and established technology. We will corroborate these findings with experimental data in Section 7.3. Apart from very unfavorable thermodynamic boundary conditions there are also several reasons in the field of quantum electronics and physics as to why an incorporation of dopants into Si NCs does not yield to a majority carrier population. We discuss this in the next section.

7.2.2 *Electronic Properties: Quantum Structure vs. Point Defect*

Section 7.2.1 delivered a broad survey on the thermodynamic situation of conventional doping of small crystal nanovolumes. Here we get more specific regarding material properties, mainly elaborating on Si. Below we discuss the ability of thermal donors from bulk Si to provide electrons to Si NCs showing quantum confinement (quantum dots [QDs]). We assume that the donor atoms were built into the local Si lattice, presenting states with an unpaired electron. We will use phosphorous (P) which is a thermal donor in bulk Si because the literature provides much experimental data. Conclusions are also valid for acceptor states.

Quantum effects on which several novel electronic devices, such as all-Si tandem solar cells and ballistic transport MOSFETs are based upon, require Si NCs with diameters at or below twice the exciton radius a_{exc}. This class of Si NCs is therefore referred to as QDs within this chapter. For Si we obtain $a_{\mathrm{exc}} \approx 45$ Å which limits the range of interest to NC diameters of $d_{\mathrm{NC}} \leq 90$ Å. It is not helpful to provide an exact value of a_{exc} as it depends increasingly on expansive/compressive lattice stress (compressed/expanded Brillouine zone).

The ionization energy of the donor electron on P in bulk Si into the conduction band is $E^{\mathrm{D}}_{\mathrm{ion}}(\text{bulk Si}) = 0.049$ eV [18] and can be interpreted as the binding energy of the donor electron to the P atom. In addition, the donor electron experiences quantum confinement within a hyperbolic potential of a point defect in analogy to the ground level of the proton–electron system (H atom) [76],

$$
\begin{aligned}
E^{\mathrm{D}}_{\mathrm{QC}}(d_{\mathrm{NC}}) &= \frac{M m_0 e^3}{2(4\pi\varepsilon_0\varepsilon_{\mathrm{rel}}\hbar)^2}\,\frac{2a^{\mathrm{D}}_{\mathrm{exc}}}{d_{\mathrm{NC}}} \\
&= \frac{m_0 e^3}{2(4\pi\varepsilon_0\hbar)^2}\,\frac{M}{\varepsilon^2_{\mathrm{rel}}}\,\frac{2a^{\mathrm{D}}_{\mathrm{exc}}}{d_{\mathrm{NC}}} = \frac{1}{2}\mathrm{Ha}\,\frac{1}{\varepsilon_{\mathrm{rel}}}\,\frac{2a_{\mathrm{B},0}}{d_{\mathrm{NC}}}\,,
\end{aligned}
\tag{7.5}
$$

whereby $E^{\mathrm{D}}_{\mathrm{QC}}(d_{\mathrm{NC}})$ given in eV. The variable $\hbar = h/2\pi$ is the reduced Planck's constant, 1 Ha $= 27.2114$ eV, ε_0 is the dielectric constant in vacuum, m_0 is the electron rest mass, e is the elementary charge, $a_{\mathrm{B},0} = 52.9$ pm is the Bohr radius, $\varepsilon_{\mathrm{rel}} = 11.9$ is the relative dielectric constant of Si and $M = 1/2\,m_{\text{c-dos}} = 0.59$ is the reduced effective mass of the donor exciton in units of m_0 derived from the conduction-band density of states (DOS) effective mass of electrons $m_{\text{c-dos}}$. The exciton radius of the donor electron and the P atom in bulk Si is [4],

$$
a^{\mathrm{D}}_{\mathrm{exc}} = \frac{4\pi\varepsilon_0\hbar^2}{m_0 e^2}\,\frac{\varepsilon_{\mathrm{rel}}}{M} = a_{\mathrm{B},0}\,\frac{\varepsilon_{\mathrm{rel}}}{M}\,. \tag{7.6}
$$

For Si we obtain $a^{\mathrm{D}}_{\mathrm{exc}} = 1.07$ nm for a thermal donor in bulk Si. This is in accord with experimental values of the ionization cross section of thermal donor states in bulk Si, $\sigma^{\mathrm{D}}_{\mathrm{ion}}(\text{bulk Si}) = 10^{-14}$ to 10^{-13} cm^2 [4], yielding $a^{\mathrm{D}}_{\mathrm{exc}} = \sqrt{1/\pi\,\sigma^{\mathrm{D}}_{\mathrm{ion}}(\text{bulk Si})} = 0.56$ to 1.78 nm. The energy of the donor electron as $f(d_{\mathrm{NC}})$ is

$$
E^{\mathrm{D}}(d_{\mathrm{NC}}) = E^{\mathrm{D}}_{\mathrm{QC}}(d_{\mathrm{NC}}) + E^{\mathrm{D}}_{\mathrm{ion}}(\text{bulk Si})\,, \tag{7.7}
$$

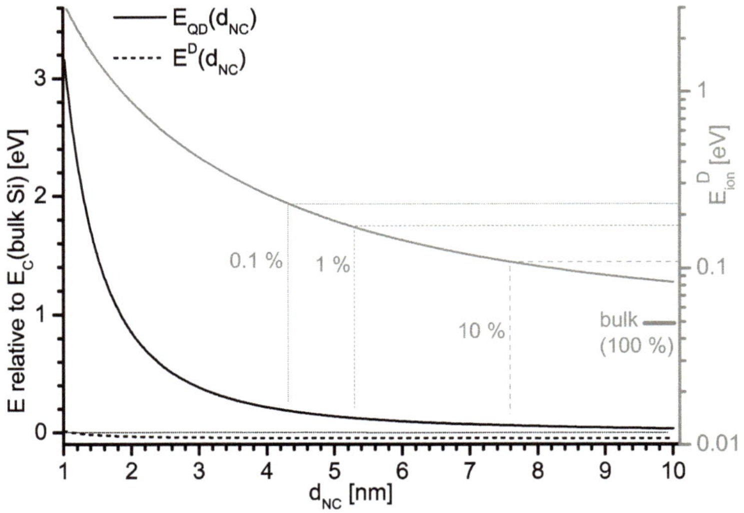

Figure 7.3 Energy of the NC LU state, E_{QD} of the donor state E^{D} (both left Y axis) and the donor ionization energy $E^{\mathrm{D}}_{\mathrm{ion}}$ (right Y axis) as a function of the NC diameter d_{NC}. The gray dashed, short dashed, and dotted lines show $E^{\mathrm{D}}_{\mathrm{ion}}$ and corresponding d_{NC} for the donor ionization probability of 10%, 1%, and 0.1% of the bulk value, respectively.

whereby $E^{\mathrm{D}}_{\mathrm{ion}}(\text{bulk Si})$ is the energetic boundary condition for the case of $d_{\mathrm{NC}} \gg a^{\mathrm{D}}_{\mathrm{exc}}$ (bulk Si case). The dashed curve in Fig. 7.3 shows $E^{\mathrm{D}}_{\mathrm{ion}}(d_{\mathrm{NC}})$.

The quantum confinement energy partition E_{QD} of the lowest unoccupied (LU) state in a spherical QD with finite potential V_{B} of rectangular shape and isotropic electron effective mass $m_{\mathrm{eff}}(n)$ can be described with the equations [76]

$$\xi \cot\xi = -\eta \quad \text{and} \quad \xi^2 + \eta^2 = \frac{m_0 m_{\mathrm{eff}}(n) V_0 d_{\mathrm{NC}}}{\hbar^2}, \tag{7.8}$$

whereby ξ, η, V_0 and E_{QD} are given by

$$\xi = \sum_{\nu=x,y,z} \xi_\nu = 3\xi_\nu;\ \xi_\nu = \frac{d_{\mathrm{NC}}}{2\hbar}\sqrt{2m_0 m_{\mathrm{eff}}(n) e E_{\mathrm{QD},\nu}}; \tag{7.9}$$

$$\eta = \sum_{\nu=x,y,z} \eta_\nu = 3\eta_\nu;\ \eta_\nu = \frac{d_{\mathrm{NC}}}{2\hbar}\sqrt{2m_0 m_{\mathrm{eff}}(n) e (V_{0,\nu} - E_{\mathrm{QD},\nu})};$$

$$V_0 = \sum_{\nu=x,y,z} V_{0,\nu} = 3V_{0,\nu};\ V_{0,\nu} = V_{\mathrm{B}} \quad \text{and} \quad E_{\mathrm{QD}} = \sum_{\nu=x,y,z} E_{\mathrm{QD},\nu}.$$

Equation 7.8 does not have a closed solution and must be solved by iteration. Equation 7.9 reflects the three-dimensional nature of the calculation to arrive at correct results. For Si, we use $V_B = 3.2$ eV as the conduction band offset from Si to SiO_2 [65] and $m_{eff}(n) = 1/3\,[m_{\parallel}^2(n) + m_{\parallel}^2(n) + m_{\perp}^2(n)]^{0.5} = 0.318\,m_0$ [4]. The black curve in Fig. 7.3 shows $E_{QD}(d_{NC})$ of the LU state. Figure 7.3 clearly shows the problem which arises for donor ionization in Si QDs. Confinement is strong for the LU state, while it is very weak for the donor electron. The ionization energy of the donor electron in the Si NC is given by

$$E_{ion}^{D}(d_{NC}) = E_{QD}(d_{NC}) - E^{D}(d_{NC}) \qquad (7.10)$$

and is shown by the gray curve in Fig. 7.3. As we know E_{ion}^{D}(bulk Si) $= 0.049$ eV, we can calculate $E_{ion}^{D}(d_{NC})$ for certain ionization probabilities $\mathcal{P}_{ion}^{D}$ referring to its bulk value $\left(\text{i.e., } \mathcal{P}_{ion}^{D}(d_{NC} \gg a_{exc}^{D}) \stackrel{\text{def}}{=} 1\right)$:

$$E_{ion}^{D}(\mathcal{P}_{ion}^{D}) = E_{ion}^{D}(\text{bulk Si}) \; - \; k_B T \; \ln\left(\mathcal{P}_{ion}^{D}\right). \qquad (7.11)$$

The Boltzmann constant is given by k_B. Since $E_{ion}^{D} = f(d_{NC})$, we can relate $\mathcal{P}_{ion}^{D}$ directly to d_{NC} which yields the NC size for a certain $\mathcal{P}_{ion}^{D}$. We made use of this in Fig. 7.3 where the values of $E_{ion}^{D}(d_{NC})$ are shown for $\mathcal{P}_{ion}^{D} = 10\%$, 1% and 0.1%. Ionization probabilities of 1 to 0.1% may not sound too bad given the maximum active donor density in Si is in the high 10^{19} cm^{-3} range. However, Fig. 7.3 shows that $\mathcal{P}_{ion}^{D}$ drops dramatically with decreasing d_{NC}. On the other hand, we expect more Si NCs in a superlattice (SL) system if d_{NC} decreases. This opposite trend renders the conventional doping of Si NCs with significant quantum confinement to be futile. If we have Si QDs with $d_{NC} = 4$ nm and 2 nm of dielectric between these spherical QDs so that their center distance is 6 nm, we obtain a QD density of 4.6×10^{18} cm^{-3}. We see from Fig. 7.3 that at this QD size the doping probability is already below 0.1% of the bulk saturation value so that the maximum density of active donors is ca. 10^{16} cm^{-3} range. We could only dope one QD out of a few hundred (presuming 100% of all dopants were activated into the local Si lattice), and the situation deteriorates rapidly for decreasing d_{NC}.

If Si QDs are attempted to be doped with very high densities of ≥ 0.2 mol-%, we have numerous QDs with no dopant, very few with one dopant and a minute partition of QDs with two dopants of the same type. The latter suffer from strong donor-level splitting due to much enhanced exchange interaction as the donor electrons are confined to the QD volume [69]. This results in a deep unionized donor level prone to SRH recombination and a flat donor level that may get ionized. However, our thermodynamical considerations have shown that the chance to build two dopants of the same type into one QD crystal lattice is extremely small, whereby this effect is not of eminent practical interest for Si QDs in embedded dielectrics.

The low donor ionization probabilities for Si QDs have to be multiplied with the low probability of dopant activation in Si NCs as discussed in the last section. The resulting probability product shows that Si NC doping below $d_{\mathrm{NC}} \approx 7.5$ nm is simply not feasible even for low quality applications with high recombination rates and low majority carrier densities. This is reflected in reliable experimental data discussed in Section 7.3.

For QD diameters of $d_{\mathrm{QD}} \leq 50$ Å, bulk values of effective carrier masses ($m_{\mathrm{eff}}(n)$, $m_{\mathrm{eff}}(p)$) and relative dielectric constant ($\varepsilon_{\mathrm{rel}}(\mathrm{Si})$) become increasingly inaccurate and meaningless as the underlying Bloch formalism of the effective mass approximation (EMA) [15] increasingly breaks down. When going deeper into the quantum regime, the assumption of an effective medium with average material values is no longer valid [45, 66, 85]. The next section investigates the electronic nature of P as donor and defect in SiO_2, $SiO_{0.9}$ and fully OH-terminated Si QDs within nonperiodic h-DFT using all-electron molecular orbital (MO) descriptions as an accurate ab initio method.

7.2.3 *Phosphorous as an Example: Hybrid Density Functional Theory Calculations*

In this section, we investigate the electronic nature of P as donor and defect in SiO_2, $SiO_{0.9}$ and fully OH-terminated Si QDs by h-DFT methods. Before we discuss results, we give a brief outline of the method.

We used the GAUSSIAN03 suite, rev. C.02, for nonperiodic (real space) h-DFT computations [22]. Approximants used the Hartree–Fock (HF) method for geometrical optimization and Gaussian type 3-21G(d) molecular orbital basis sets (MO-BSs) [3, 17].

Root mean square (RMS) and peak force convergence limits were 300 μHa/$a_{B,0}$ and 450 μHa/$a_{B,0}$, respectively. Electronic structures were computed with the B3LYP h-DF [1, 55] and the Gaussian type 6-31G(d) MO-BS [21, 32, 71]. Detailed accuracy evaluations for Si QDs referring to MO-BS and HF optimization can be found in [48, 49]. No symmetry constraints were applied to MOs and tight convergence criteria were set at the self-consistent field routine.

Figure 7.4 shows the α-quartz approximants of pure SiO_2 ($Si_{29}O_{76}H_{36}$, top left) and SiO_2 with one central Si atom substituted with P ($Si_{28}PO_{76}H_{36}$, top right), both seen along the $\langle 110 \rangle$ direction. The bond geometry of P is tetravalent and thus leaves an unpaired electron at the P atoms which is a requirement for a donor. The SiO_2-matrix surrounds P at least up to its 5-nn atom. The calculated band gap of SiO_2 is 7.49 eV which is ca. 85% of the experimental value of ca. 8.8 eV [65]. The highest occupied molecular orbital (HOMO) and the lowest unoccupied molecular orbital (LUMO) are associated with P as can be seen from the DOS plot. Due to the doublet nature of the unpaired electron introduced by P, the DOS is divided into two parts according to the two different spin orientations of the electronic states. We see that P cannot donate an electron in SiO_2 as its ionization energy is much too high for thermal excitation. We now compare these MOs with the HOMO and LUMO of the 1.5 nm Si QD completely terminated with hydroxyl (OH) groups ($Si_{84}(OH)_{64}$), cf. Fig. 7.5. The HOMO of the $Si_{28}PO_{76}H_{36}$ approximant is 0.13 eV above the HOMO of the $Si_{84}(OH)_{64}$ approximant so that it may work as a shallow recombination center for small Si QDs in SiO_2. The HOMO and LUMO associated with P in the $Si_{28}PO_{76}H_{36}$ approximant are in an excellent energetic position to promote carrier transport by hopping or trap-assisted tunneling. This finding shows an important aspect of the electronic nature of P in SiO_2: It can increase the conductivity of the SiO_2 matrix while *not working* as a donor. Several works build their evidence of Si QD doping on conductivities increasing with P doping densities, whereby P concentrations are $\geq$ 0.5 mol-% [9, 24, 30, 31, 41, 68, 70]. At such P concentrations, we can presume that SiO_2 has been turned into a ternary oxide, see Sections 7.2.1 and 7.3.3. The values obtained from the $Si_{28}PO_{76}H_{36}$ approximant are also valuable because the atomic ratio of $P/(Si + O + {}^1/_4\,H) = 0.009$ shows us the electronic structure of SiO_2 with

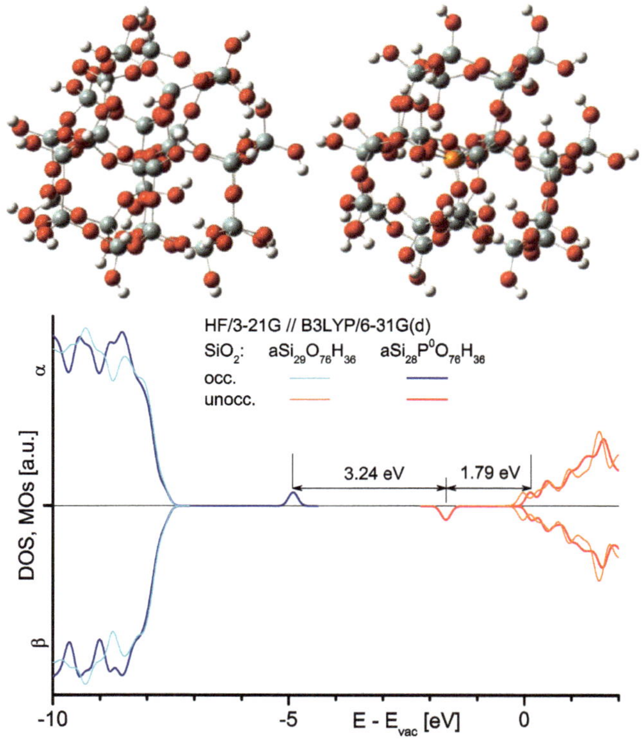

Figure 7.4 Optimized approximants of α-quartz, $Si_{29}O_{76}H_{36}$ presenting pure SiO_2 as reference (top left) and $Si_{28}PO_{76}H_{36}$ where one central Si atom was substituted for P, maintaining the tetravalent bond geometry (top right), both seen along the ⟨110⟩ direction. Si atoms are shown in gray, O atoms in red, H atoms in white, and the P atom in orange. The bottom graph shows the occupied (occ.) and unoccupied (unocc.) DOS of both approximants with the energy relative to the vacuum level E_{vac}, where α and β mark one of the two possible spin orientations of the MOs. MO energies were broadened by 0.1 eV. The α-HOMO and β-LUMO shown are due to P in SiO_2.

0.8 mol-% P, which is in the vicinity of the works just cited. We count H as 1/4 Si atom as it emulates one O–Si bond at the surface of the approximants.

Results for $SiO_{0.9}$ were obtained by an $Si_{73}PO_{55}(OH)_{35}H_{29}$ approximant where the 1-nn atoms of P are two O and two Si. This approximant used B3LYP h-DF with a 6-31G(d) MO-BS as a different optimization route throughout. Strictly speaking, these high accuracy data cannot be directly compared to results obtained by the default optimization using the HF method and a 3-21G(d) MO-BS, though deviations of the HOMO and LUMO energies for the pure approximants like $Si_{29}O_{76}H_{36}$ or $Si_{84}(OH)_{64}$ computed with both routes are less than 4% of the HOMO–LUMO gap. The gap of the $SiO_{0.9}$:P ($Si_{73}PO_{55}(OH)_{35}H_{29}$) approximant is 2.73 eV which overestimates the band gap from experiment (2.48 eV, [77]) by 10% which appears to be due to our perfect (crystalline) approximant devoid of defects. As in the SiO_2:P case ($Si_{28}PO_{76}H_{36}$), there is an α-HOMO and a β-LUMO associated with P. These are located 0.49 eV above the HOMO and 0.50 eV below the LUMO of the $SiO_{0.9}$ host lattice, respectively, leaving a gap of 1.74 eV between them. This energy (equivalent wavelength 710 nm) should be considered as an upper limit for photoluminescence (PL) of Si QD samples containing P since there is usually one or two monolayers (MLs) of $SiO_{1<x<2}$ around Si NCs formed by segregation anneal [87]. The HOMO and LUMO associated with P cannot be ionized as 0.49 to 0.5 eV is an energy of about 19 $k_B T$ (thermal energy, $T = 300$ K), the ionization probability is $\mathcal{P}_{\rm ion}^{\rm D} \approx 6 \times 10^{-9}$. We note that the composition of the $SiO_{0.9}$:P approximant is balanced, that is, there is no segregation of Si to a Si NC network as found in experiment, see Section 7.3.3. If the Si content is increased further, P is very likely to work as a conventional donor as in bulk Si due to an interconnected Si amorphous/NC network [56].

In Section 7.2.1 we pointed out that dopants are gettered at defect sites found at grain boundaries or interfaces. Here, we investigate P at a 1.5 nm Si QD with full OH termination including a complete bond saturation of the P atom. In addition to the reference approximant ($Si_{84}(OH)_{64}$, Fig. 7.5, top left), there are three different configurations: A bridge-bonded P atom replacing an outermost Si atom and backed by three OH groups ($Si_{83}P(OH)_3(OH)_{62}$, Fig. 7.5, top right); a P atom backed by four OH groups, forming an interface bond to the Si_{84} QD ($Si_{84}P(OH)_4(OH)_{63}$, Fig. 7.5, center left); a P atom attached to the Si_{84} QD via the bond configuration Si QD–O–P ($Si_{84}OP(OH)_4(OH)_{63}$, Fig. 7.5, center right). The bottom graph of Fig. 7.5 shows the electronic DOS. The four approximants considered here have no unpaired electron(s), hence they have a singlet configuration, that is, all MOs are either doubly occupied or

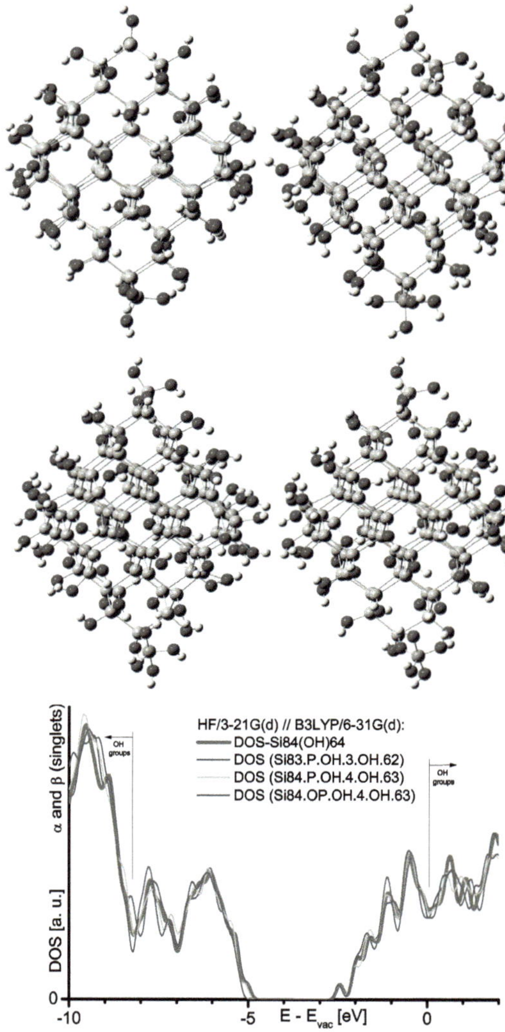

Figure 7.5 Optimized approximants of fully OH-terminated 1.5 nm Si QD, reference case ($Si_{84}(OH)_{64}$, top left), outermost bridge-bonded $Si(OH)_2$ replaced by $P(OH)_3$ ($Si_{83}P(OH)_3(OH)_{62}$, top right), OH group replaced by $P(OH)_4$ ($Si_{84}P(OH)_4(OH)_{63}$, center left), and H atom of OH group replaced by $P(OH)_4$ ($Si_{84}OP(OH)_4(OH)_{63}$, center left). Approximants seen along the ⟨110⟩ direction; for atom colors see Fig. 7.4. DOS of all four approximants (bottom). Their singlet configurations allow for one DOS. MO energies were broadened by 0.1 eV.

unoccupied. It is therefore not required to split the DOS into two different spin orientations as was the case of P with an unpaired electron in SiO_2 and $SiO_{0.9}$. The HOMO–LUMO gap of the 1.5 nm QD ($Si_{84}(OH)_{64}$) is 2.60 eV. The energy gap manifested by MOs associated with OH groups is 8.31 eV which is ca. 94% of the experimental band gap of SiO_2 (ca. 8.8 eV, [65]). The electronic DOS shows us that a fully saturated P atom does not introduce any state within the HOMO–LUMO gap of the reference case. If no states in the HOMO–LUMO gap of a 1.5 nm Si QD exist due to bond-saturated P, they are very unlikely to appear at bigger QDs as the HOMO–LUMO gap keeps shrinking with increasing QD size. So P gettered at the SiO_2/Si QD interface indeed can passivate Si DBs and does not introduce defect levels into the Si NC gap.

Next, we consider the electronic nature of P *within* a 1.5 nm fully OH-terminated Si QD, whereby the $Si_{84}(OH)_{64}$ approximant serves again as a reference. The two configurations of P inside the QD are: P as standard active dopant, replacing a central Si atom ($Si_{83}P(OH)_{64}$, Fig. 7.6, top left) and P at an interstitial site ($Si_{84}(OH)_{64}$) with interstitial P, Fig. 7.5, top right). The position of the interstitial P relative to its 1-nn Si atoms was fixed with experimental coordinates obtained by scanning tunneling microscope (STM) characterization [7]. Such interstitial P atoms form at very high P densities. The P atom on the lattice site of a central Si atom produces a HOMO at midgap 1.11 eV below the LUMO level. This is a very efficient deep recombination center for all Si NC sizes. Interstitial P introduces two states into the gap, a HOMO 0.7 eV above the HOMO of the 1.5 nm Si QD and a LUMO 0.41 eV below the LUMO of the QD. Both states cannot be thermally ionized at room temperature (RT) or indeed at elevated temperatures and provide efficient carrier recombination. The transition energy between the LUMO and HOMO of interstitial P is 1.49 eV which may spare large QDs from massive recombination. Care should be taken when interpreting PL spectra of QD species doped with P as this transition is optically active at a wavelength of ca. 830 nm. Both cases of P in $Si_{84}(OH)_{64}$ introduce deep or midgap defect levels within the HOMO–LUMO gap, whereby the deep defect levels of interstitial P are several orders of magnitudes more likely to occur. Diffusion of P through Si is proceeding at a high rate at annealing temperatures. Our thermodynamical considerations and

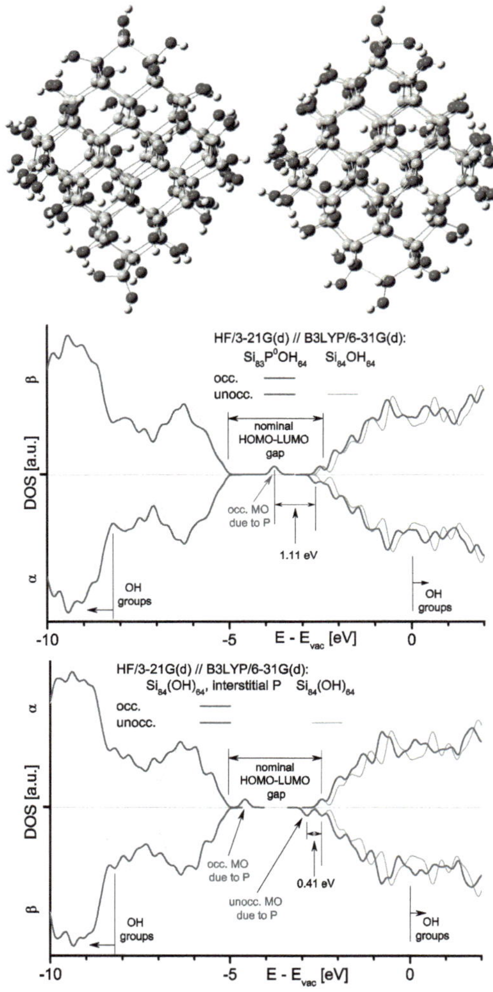

Figure 7.6 Optimized approximants of fully OH-terminated 1.5 nm Si QDs with a central Si atom substituted for P ($Si_{83}P(OH)_{64}$, top left), and with P on an interstitial site, as found in experiment [7]. Approximants seen along the ⟨110⟩ direction; for atom colors see Fig. 7.4. As in Fig. 7.5, the reference case is the $Si_{84}(OH)_{64}$ approximant. DOS of approximants with P on Si lattice site (center) and on interstitial site (bottom). MO energies were broadened by 0.1 eV. The α-HOMO and β-LUMO shown are due to P.

experimental data in Section 7.3.2 show us that—if P is incorporated into Si NCs—P is on an interstitial site with a probability of virtually 100%. The energy levels associated with P in Si QDs explain PL crunching for extremely high P doping densities as these are necessary to get P onto interstitial sites in most Si NCs [23, 24, 41].

To reiterate, h-DFT calculations of P in SiO_2, in $SiO_{0.9}$ at and in completely OH-terminated 1.5 nm Si QDs were carried out. P in SiO_2 introduces deep defect levels in the SiO_2 HOMO–LUMO gap that act as shallow recombination centers for Si QDs and facilitate transport throughout SiO_2 by defect-assisted hopping and tunneling while not working as a donor. Using increased conductivity at increased dopant densities as evidence for successful doping of Si NC structures is therefore misleading. Doping $SiO_{0.9}$ with P results in an occupied defect level 0.49 eV above the HOMO of $SiO_{0.9}$ and an unoccupied defect level 0.5 eV below the LUMO of $SiO_{0.9}$. Both of these defects cannot be thermally ionized, but should act as fast recombination centers for small QDs. Due to the likelihood of a ML of suboxide between the actual Si QD and the SiO_2 matrix, care must be taken when interpreting PL spectra of Si QD/SiO_2 samples containing P: The transition energy between defect levels puts an upper trust limit to PL at a wavelength of 710 nm (1.74 eV). Fully saturated P at a 1.5 nm Si QD with full OH termination does not introduce any defects into the HOMO–LUMO gap of the QD and thus passivates DBs. This statement holds for bridge-bonded (>$P(OH)_3$), single-bonded (−$P(OH)_4$), and P bonded via an oxygen bridge (−O−$P(OH)_4$) and appears to be the only beneficial effect of P in Si QD/SiO_2 samples. Regarding P inside the 1.5 nm QD, a midgap defect level occurred when one central Si atom in the QD was substituted for P. This defect is a very efficient recombination center. Interstitial P creates two deep defect levels 0.7 eV above the HOMO of the QD and 0.41 eV below its LUMO, with an interdefect transition energy of 1.49 eV. The corresponding wavelength of 830 nm can produce a PL signal, so once again this must be taken into account when characterizing Si QD/SiO_2 samples containing P. As for the Si QD with substitutional P, these defect levels should trigger massive recombination. This is corroborated by experimental results where PL intensities were found to decrease with increasing P concentration at high P densities. The h-DFT results obtained are of vital importance for interpreting experimental results as carried out in the next section.

7.3 Survey on Experimental Results of Conventional Si Nanovolume Doping

7.3.1 *Si Nanovolumes in Next-Generation Ultra-Large-Scale Integration*

With the 14 nm gate length technology node being rolled out by Intel in early 2014 [36, 64], the introduction of majority charge carriers is a major field of ongoing technological development and research. The drain and source doping areas require very high doping concentrations in order to achieve a certain minimum majority carrier density. As a consequence, clustering of inactive dopants is detrimental to performance due to increased inelastic carrier scattering which decreases carrier mobility (and thus clock frequency) and increases heat generation. The reduction of clock frequency is not crucial for the basic functionality of 14 nm FETs as short carrier lifetime is not an issue at GHz clock frequencies. Heat generation is a general concern which should be kept to a minimum. However, the major challenge is a sharp doping profile with the required very high dopant concentration. Dopants have to be activated by an anneal. Even with LASER spike anneals or self-regulatory plasma doping, substantial diffusion of dopants occurs as discussed in Section 7.2.1. We briefly look at arsenic (As) out-diffusion from drain and source regions of Si fin-FETs into the channel region underneath the gate dielectric [37]. The nominal As concentration is $\approx 7 \times 10^{20}$ cm^{-3} (1.4 mol-%). Fig. 7.7 shows the results of dopant positions in a fin-FET. The center graph of Fig. 7.7 shows the As distribution after a rapid spike annealing step at 1050°C for donor activation. The out-diffusion from the source and drain regions underneath the gate is 10 to 15 nm. Presumably, the current 14 nm technology node controls out-diffusion by dielectric spacers of 10 to 15 nm thickness between drain/source regions and the gate channel area which requires benign control of the diffusion mechanism as unwanted by-product of the spike anneal.

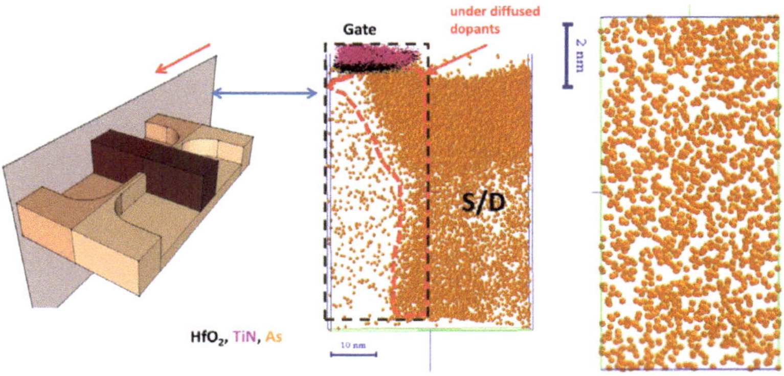

Figure 7.7 APT map of arsenic (As) doping and diffusion from the source/drain (S/D) region to the gate along the Si fin, as shown in schematic (left), following the gray plane through the Si fin (center of scheme where the two different brown colors meet) [37]. Incorporation of As was realized by self-regulatory plasma doping [75]. Silicon atoms are not shown, As shown in orange, the gate dielectric stack consists of HfO_2 shown in black, and TiN shown in pink. The red dotted line shows underdiffused dopant distribution into the channel under the gate. Detail of three-dimensional map showing inactive As atoms (right) [37] (reprinted with permission).

Clustering of As donors is present in the drain and source areas, with active donor densities reaching 3×10^{19} cm^{-3}, accounting for 4% of the total As density. These electrically inactive complexes cannot be observed by means of TEM, however they can reduce the dopant probability significantly [37]. Similar results regarding inactive dopant clustering and out-diffusion were obtained with boron (B) at concentrations of 10^{21} cm^{-3} (2 mol-%) [10]. An activation rate (doping probability) of merely 0.1 to 0.5% was reported for B concentrations of 0.5 to 0.03 mol-% (2.5×10^{20} to 1.5×10^{19} cm^{-3}) as detected by atom probe tomography (APT) and field ionization microscopy (FIM) after an anneal for 30 min at 800° C [43].

7.3.2 *Free-Standing Nanocrystals*

Free standing Si NCs were produced by microwave-induced decomposition of silane (SiH_4) in a low pressure microwave plasma reactor [42]. Attempts were made to incorporate phosphorous (P) as donor into these Si NCs by adding the dopant precursor gas phosphine (PH_3) to silane (SiH_4) during Si NC synthesis [79]. This is one of the most promising doping approaches due to *in-situ* incorporation of a dopant species to the host material, with both precursors in a highly reactive form provided by a plasma. The synthesis method produced Si NCs in a size range from 2 to 50 nm. These doped Si NCs were characterized meticulously by the Stegner group using electron paramagnetic resonance (EPR) as a reliable method to detect dopant atoms with an unpaired electron (neutral donor) in semiconductors. Shallow donors which are ionized at $T = 300$ K provide a shrinking EPR signal intensity with increasing T as they get increasingly ionized. Unfortunately, thermal noise restricts EPR measurements to very low T, hence does not allow for sampling donor ionization probabilities at RT. Low T EPR proves the existence of unpaired electrons in characteristic (element- and topology-specific) states by an electron spin–orbit electron resonance at a certain magnetic field strength and microwave frequency. Information about the 1-nn atoms of the dopant can be obtained from the fine structure of the EPR signal. Before EPR measurement, Si NCs were stripped of their 1.4 ± 0.4 nm thick native SiO_2 by a hydrofluoric acid (HF) dip, whereby 95% of the P content was removed. This is shown by the transition from $[P]_{nom}$ to $[P]_{core}$ in the left graph of Fig. 7.8 [80], whereby the nominal P density was obtained from secondary ion mass spectroscopy (SIMS) measurements. The vast majority of P donors at or in Si NCs does not contribute to the EPR signal. Of those which do, about 90 to 99% are charge compensated by DBs at the NC surface as shown by $[P]_{EPR}^{s.c.}/[P]_{nom}$ in the left graph of Fig. 7.8. The remaining EPR signal originates from P atoms with one unpaired electron shown by $[P]_{EPR}/[P]_{nom}$. Such P atoms are active donors under the condition that they can be ionized at RT. This cannot be derived from the EPR scans as these had to be carried out at $T = 20$ K in order to minimize signal blurring by thermal noise. To summarize the EPR data, we can conclude that Si NCs with $d_{NC} \leq 43$ nm have 95% of the P atoms located in their oxide shell. Starting from Si NCs of ca. 27 nm, the relative concentration of charge compensated P and P atoms with one unpaired electron drop dramatically. The relative concentration of P atoms with unpaired electrons which we described in this chapter

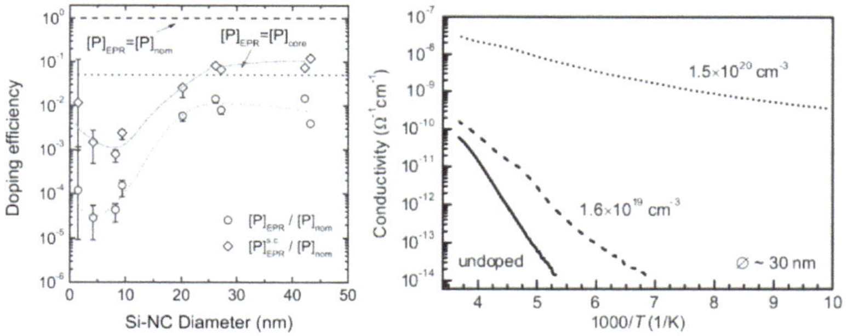

Figure 7.8 Relative doping concentrations measured by EPR referring to nominal density of P (left) [80]. Copyright (2009) by the American Physical Society. Relative concentration of donors incorporated into Si NCs, which were compensated by DBs, are shown by open diamond symbols, and the gray dot-dashed line is a guide to the eye. Relative concentrations of donors with one unpaired electron incorporated into Si NCs are shown by open circle symbols, and the gray dotted line is a guide to the eye. Conductivity of undoped and highly P doped 30 ± 2 nm Si NCs as a function of inverse temperature (right) [79]. Copyright (2009) by the American Physical Society.

as the doping probability drops to 3×10^{-5} for $d_{NC} \leq 8$ nm which is the range where quantum confinement occurs. Even with this tiny reminder of P atoms with an unpaired electron, the question remains whether these can donate an electron at RT. While our EMA calculations in Section 7.2.2 and our h-DFT calculations in Section 7.2.3 strongly suggest that this is not the case for NCs, the situation may be different for a-Si clusters. As a thermodynamical argument for self-purification, we had established the average number of Si–Si bonds per Si NC atom $N_{bond}(\text{Si–Si})/N_{Si}$ (see Eq. 7.3 and Fig. 7.2). With the data from the left graph in Fig. 7.8, we can relate self-purification to the decreased values of $N_{bond}(\text{Si–Si})/N_{Si}$ and estimate the dramatic drop in doping probability as $N_{bond}(\text{Si–Si})/N_{Si} \approx 1.9$. This value already takes into account that the values of Eq. 7.3 are an upper limit due to the high octagonal symmetry of the Si NCs considered. In other words, if the average number of bonds per Si atom of the NC deviates by $\geq 5\%$ from its asymptotic bulk value, the chances to conventionally dope a Si NC are virtually nil.

For conductivity measurements as a function of inverse temperature, Si NCs with $d_{NC} = 30 \pm 2$ nm were stripped of their native SiO_2 and deposited as densely packed films of 500 nm thickness onto Kapton poliimide substrates [79]. Electrical contacts were established by gold strips with 10 μm distance, conductivity measurements were carried out at 20 V bias voltage, rendering the average field strength between contacts to be 20 kV/cm. Conductivities under such conditions as a function of inverse temperature are shown in the right graph of Fig. 7.8 A defect activation energy of $\approx$0.5 eV was found for the undoped samples from the slope in Fig. 7.8, right graph. Apparently, the defect activation energy is reduced by a nominal P doping density of 1.6×10^{19} cm^{-3} (0.032 mol-%), while still maintaining the temperature dependence. The kink in the curve at 1000/4.75 K^{-1} = 210 K could be due to the freeze-out of donors [4], preventing their thermal ionization at lower temperatures. The conductivity at the maximum temperature of 1000/3.75 K^{-1} = 267 K only increases by roughly a factor of three although the donor density is within the range of the conduction-band DOS of c-Si. This corroborates the results of [80] and of Section 7.2 that conventional doping of Si NCs with $d_{NC} \leq 30$ nm does not work. As for the low values of absolute conductivity at RT, we have to take into account that ca. 400 Si NCs are required to bridge the gold contacts, which means that ca. 400 NC boundaries have to be penetrated. Increasing the P donor density by just 1 order of magnitude to 1.5×10^{20} cm^{-3} (0.3 mol-%) leads to a very low defect activation energy and temperature dependence of conductivity. This 9.4-fold increase in doping density increases conductivity at RT by a factor of ca. 170. Both, the very low temperature dependence of conductivity and its massive increase at room temperature corroborate the assumption that NCs behave very different, rendering them to be an alloy compound with metallic properties. This behavior is similar to the metallic behavior of the alloy $Al_{0.01}Si_{0.99}$ [59] which is used as metallic contact to p$^+$ doped Si. This behavior is exploited for Al-based p^+ back surface fields of Si solar cells [74]. The conductivity measurements of [79] therefore provide us with an estimate of ca. 0.1 mol-% where the transition from semiconducting Si NCs to a P-based metallic alloy occurs. This P concentration value will be of great importance in the next section.

7.3.3 *Embedded Nanocrystals Formed by Segregation Anneal*

There are many journal publications of doping attempts of the Si NC/SiO_2 system. We restrict our discussion to a few cornerstones from which we can derive the nature and behavior of dopants. However, it quickly emerges that most works do not distinguish between the absolute concentration of dopant atoms and the actual dopant density [9, 23, 24, 30, 31, 41, 68, 70]. These—as we have seen—are several orders of magnitude apart, even for free-standing Si NCs generated and doped under very favorable conditions.

Si NCs in SiO_2 were produced by cosputtering of Si and SiO_2 with either diphosphorous pentoxide (P_2O_5) or boron III oxide (B_2O_3) for incorporating P or B, respectively, and annealed in N_2 for 30 min in the range of 1100 to 1250°C [23, 24]. Si NC sizes were d_{NC} = 4 to 6 nm, undoped Si NCs showed substantial quantum confinement. Samples doped with alloy densities of either 1.24 mol-% B (6.2×10^{20} cm^{-3}) or 0.79 mol-% (4.0×10^{20} cm^{-3}) of P showed substantial PL quenching, which was ascribed to Auger recombination under the assumption that the dopants are located in the Si NCs, have thermal ionization energies and thus deliver majority carriers. The disappearance of free carriers was explained by codoping with B and P of the same density, ensuing a minimum sub-band gap absorption of Si NCs around a wavelength of 1.4 to 2.5 μm which was assigned to suppressed free carrier reabsorption within Si NCs [24]. From the maximum PL intensity for Si NCs with this balanced B-P codoping, the location of B and P was derived to be within Si NCs, cf. Fig. 7.9 for respective data. A strange feature in the absorption data is the *decreasing* wavelength of the absorption edge around 1 μm (increasing transition energy) with *increasing* codoping. Dopant levels are above the highest occupied (HO) state (acceptor) or below the LU state (donor) of the Si NCs and therefore should shift the absorption edge to longer wavelengths.

In the light of our recent h-DFT results of P in OH-terminated Si NCs (cf. section 7.2.3) and the experimental results of the Stegner group (cf. Section 7.3.2) the interpretation of PL and absorption data by [23, 24] appears to be incorrect. In the codoping case, dopants are already ionized as the B acceptor captures the electron from the

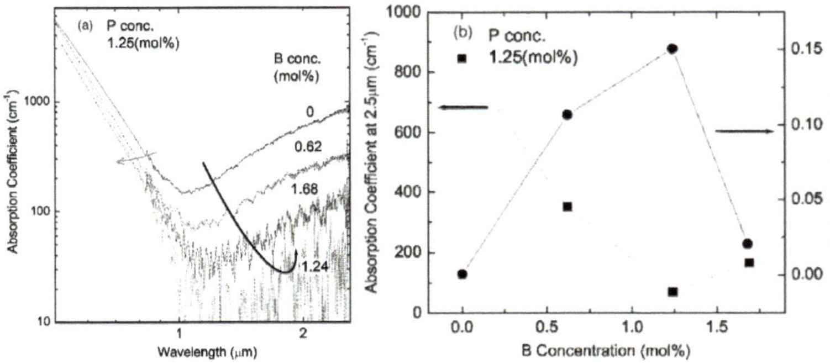

Figure 7.9 Absorption spectra measured on Si NCs in SiO_2 that were attempted to be doped with B and P (left). Note the *decreasing* wavelength of the absorption edge around 1 μm (increasing transition energy) with *increasing* codoping, as indicated by the gray arrow. PL intensities and absorption coefficients at 2.5 μm, showing maximum PL intensity and minimum absorption at balanced codoping (1.25 mol-% or 6.2×10^{20} cm^{-3}). All data from [24]. Reprinted with permission from [Fujii, M., Yamaguchi, Y., Takase, Y., Ninomiya, K., and Hayashi, S. (2005). Photoluminescence from impurity codoped and compensated SI nanocrystals, *Applied Physics Letters* **87**, 211919, 1–3.]. Copyright [2005], AIP Publishing LLC.

P donor ($h\nu$ = photon), $P^0{+}B^0 \rightarrow P^{+}{+}B^{-} + h\nu$, which deprives both dopants from the ability to deliver a free carrier. Codoped Si NC/SiO_2 samples showed a PL peak with photon energies below the bulk Si band gap (1.12 eV at 300 K) assigned to donor–acceptor transitions within NCs. However, PL does not provide us with the information where the dopants are located nor what transition generates the PL signal. All we know from PL is *that* the massive addition of donors and acceptors create a broad PL peak below the band gap of bulk Si. Higher dopant concentrations will result in more dopants near a Si NC. Then, the wave function overlap between free carriers created by optical generation and the respective ionized dopant in SiO_2 at the Si NC is big enough for carrier relaxation of the free carriers into the respective ionized dopant (e = electron, h = hole): $P^{+} + e(\mathrm{NC}) \rightarrow P^0$ and $B^{-} + h(\mathrm{NC}) \rightarrow B^0$. Radiative recombination proceeds immediately via $P^0{+}B^0 \rightarrow P^{+}{+}B^{-} + h\nu$ if the total number i of phonons ($\hbar\omega$) emitted by electron and hole is even: $\sum_i^{a=e,h}(\hbar\omega\ \text{emission})_i^a \in 2n$. This preserves the spin quantum number of electron and hole as a singlet whereby the

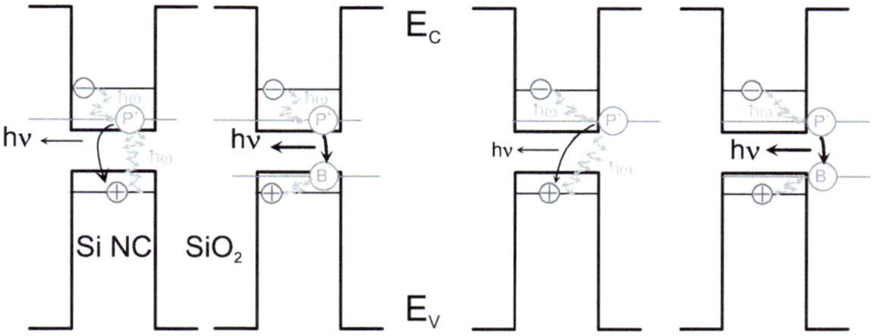

Figure 7.10 Two possible recombination mechanisms to obtain the experimental results advocated by [23, 24] shown in Fig. 7.9. The left two band diagrams show the mechanism as proposed by [24], requiring both dopants to be inside the Si NC for enhanced radiative recombination efficiency (PL signal intensity). The right two graphs show an alternative scenario with both dopants at the NC interface or in its immediate proximity, allowing for a carrier wavefunction overlap with the respective dopant and ensuing charge transfer. The latter is promoted by carriers dissipating energy by phonon emission (light gray arrows), whereby nonradiative recombination is suppressed by the strong optical transition provided by the donor–acceptor pair. Please note that in any scenario neither of the dopants can provide carriers due to quantum confinement of the HO and the LU state.

optical transition between P donor and B acceptor is allowed. Since the Si NC/SiO_2 interface provides a multitude of phonon states to scatter into, this restriction is no practical limitations for radiative recombination. Figure 7.10 illustrates two possible behaviors leading to the same PL spectrum. A similar process would also work for one of the dopants inside the NC and one of them in SiO_2 at or nearby the Si NC. At P alloy concentrations of 1.25 mol-%, the massive presence of P may lead to a minute portion of P being built into some Si NCs on a lattice site, while the huge majority of P creates defects within SiO_2 and $SiO_{x<2}$ as obtained with h-DFT, see Section 7.2.3. This corroborates the scenario with at least one of the dopants located outside the Si NC which is also in accord with experimental results [79, 80] in Section 7.3.2. PL is not an ideal tool for proving Si NC doping in a dielectric matrix. The very high dopant densities used in the literature create numerous radiative defects. Highly polar matrices like SiO_2 and to some extent Si_3N_4 provide an ideal environment for several types of strong radiative dipoles.

An assignment of a certain radiative dipole to a specific PL signal is somewhat problematic. Another rather trivial disadvantage of PL is the low spatial resolution as compared to the NC size, whereby it is virtually impossible to distinguish between radiation sources in the dielectric matrix, the Si NC/dielectric interface and the Si NCs themselves as well as the depth position of the radiative center.

SiO_2 of 200 nm thickness was implanted with Si (dose: 6×10^{16} cm^{-2}) and P or As (dose: 1 to 5×10^{15} cm^{-2}), whereby the energy of the implanted donors was chosen such that the dispersion profile matched the distribution of implanted Si [41]. The resulting excess Si rate in SiO_2 was 38% (excl. dopant content). Samples were annealed in N_2 for 4 h at 1100°C whereby Si clusters with a size of 3 ± 2 nm were grown. Below, we focus on the data obtained for P to maintain compatibility with results discussed previously. As the implantation profiles of Si and P are alike, we can derive the donor concentration by the ratio of implant doses—it is 1.6 to 8.3 mol-% and therefore at least 1 order of magnitude above the alloy threshold. Under such conditions, many donors are located within Si clusters as characterized by APT, *cf.* Fig. 7.11, left graph. From the distribution of P in this 2 nm thick material slice, we see that most P atoms are within Si clusters, roughly doubling the P concentration therein. This means that the P concentration in Si clusters is 3.3 to 16.7 mol-% or one P atom in 33 to 6 Si atoms. It is not clear from [41] whether the Si clusters are amorphous or crystalline. However, the extremely high P concentration suggests an amorphous structure due to symmetry breaking of the Si lattice by the massive presence of P atoms. It was also found in [41] that the average Si cluster size is bigger for P-doped samples as compared to undoped samples in accord with [30]. A massive presence of P in SiO_2 softens the glass matrix, enhancing diffusion which in return enables a faster growth of Si clusters or NCs. This softening effect is well known from reflow processing steps of spin-on glass (SOG) with 2 to 5 mol-% P as dopant source [60]. The implanted structures were deposited onto a p-Si substrate to form p/n junctions investigated by dark IV measurements, see Fig. 7.11, center graph. The linear scale of the current density does not provide detailed information in particular for the reverse bias range, though it is very clear that donor doping of Si clusters at extremely high P densities appears to

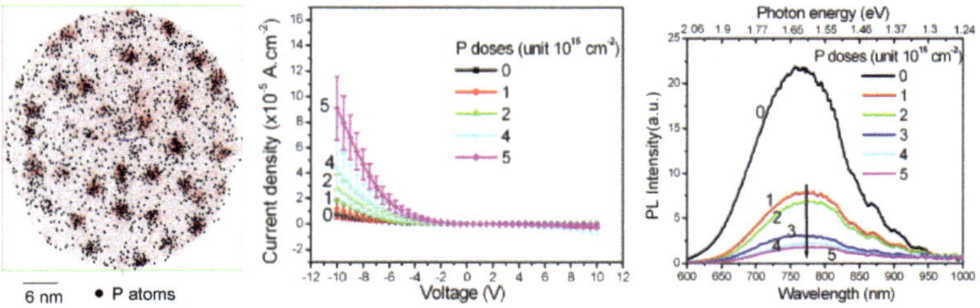

Figure 7.11 Detail of three-dimensional map over a slice thickness of 2 nm obtained by APT, showing P atoms in Si NC/SiO_2 sample (left). Red small dots show locations of Si atoms. Conductivity for P-doped Si NCs in SiO_2 on p-Si (center) increasing with P density; see text for details. Evolution of PL intensity with increasing P density, *cf.* h-DFT results on P in SiO_2 and $SiO_{0.9}$, Section 7.2.3. All data from [41]. Reprinted with permission from [Khelifi, R., Mathiot, D., Gupta, R., Muller, D., Roussel, M., and Duguay, S. (2013). Efficient n-type doping of Si nanocrystals embedded in SiO_2 by ion beam synthesis, *Applied Physics Letters* **102**, 013116, 1–4.]. Copyright [2013], AIP Publishing LLC.

work to some extent. The question is how good such a p/n junction rectifies with an extreme P density in SiO_2 with residual Si. From the current values at -10 V and $+10$ V in Fig. 7.11, center graph, we can derive a current ratio of ca. 9 and 35 for P implant doses of 4 and 5 $\times$ 10^{15} cm^{-2}, respectively. These implantation values correspond to concentrations of 6.7 and 8.3 mol-% P, respectively, which is equivalent to 3.4 and 4.2 $\times$ 10^{21} cm^{-3}. At such high P concentrations, many defects within SiO_2 with residual Si exist. Latter can be recognized by the Si signal in between Si clusters in the 2 nm thick slice obtained from APT, see Fig. 7.11, left graph. The Si signal originating from Si bonded to O was filtered out in this image [41]. The decreasing PL intensity with increasing P concentration also points to an increasing nonradiative defect density. In Section 7.2.3 we found out that the HOMO–LUMO transition of interstitial P in an OH-terminated 1.5 nm Si NC occurs at 1.49 eV. We can see this value as an upper limit as the minor quantum confinement a P donor experiences in a Si NC subsides with increasing NC size, see to $E^{\mathrm{D}}(d_{\mathrm{NC}})$, Fig. 7.3 For increasing P concentrations we would expect a shift of the PL peak to lower energies since the donor level of P is below the LU state of a Si NC. This shift shows up for As doped samples in [41]. Unfortunately, the authors do not address this peculiarity and h-DFT calculations on a whole different material system require a vast computation effort.

Successful doping of Si NCs with 0.42 mol-% (2.1×10^{20} cm^{-3}) P which were created by annealing $SiO_{0.7}/SiO_2$ (2 to 8 nm/ 1 to 2 nm) precursor stacks and solar cells made with this SL on top of a p-Si substrate were reported [9]. This work is illustrative for two reasons. Using $SiO_{0.7}$ as Si NC precursor material with very thin SiO_2 barrier layers results in a highly interconnected Si NC network [56] with SiO_2 islands. The intended SL is thereby destroyed as evident from the SIMS profile of the annealed sample [9] where the Si, SiO_2 and P signal does not show any oscillations indicating Si NC array layers and SiO_2 barriers. At such low O partitions, active doping should be feasible, though with much lower densities than the nominal P concentration. The Si solar cell produced with this structure on top is no real competitor to a c-Si or a-Si emitter doped with P on a p-doped c-Si wafer.

The capacitance–voltage (CV) characterization method was used by [57] to prove doping of a SL stack comprising Si NCs with $d_{NC} \approx 4$ nm in SiO_2 separated by 2 nm thick SiO_2 barriers. The usage of a quartz substrate removes any ambiguity about electronic properties influenced by semiconducting carriers such as doped Si wafers. The underlying 20 nm SiO_2 layer of the SL contains a substantial amount of either B or P incorporated by cosputtering of SiO_2 and the respective dopant oxide (B_2O_3 or P_2O_5), see Fig. 7.12, left graph. Samples were annealed for 40 min in N_2 at 1100°C for Si NC formation with *in-situ* doping by P out-diffusion from the bottom SiO_2 layer. As before, we will focus on the results obtained for P incorporation; results obtained for samples containing B were similar. The minimum dopant concentrations achievable by cosputtering of SiO_2 and P_2O_5 is ca. 0.5 mol-% (2.5×10^{20} cm^{-3}) in SiO_2. No quantitative data are provided by [57] on excess Si content or nominal P donor density. The evaluation of CV curves arrived at an active donor density of $\approx 10^{17}$ cm^{-3}. It is peculiar that the CV curves show the behavior of a conventional space charge region (SCR) which cannot form in a Si NC SL. In order to obtain a CV curve governed by a SCR, the SL layers have to be very Si rich (i.e., forming Si NC networks from $SiO_{x \leq 1}$) and interconnected. We can conclude

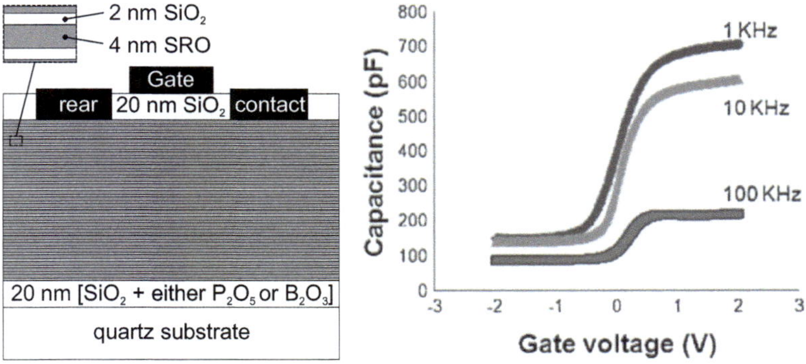

Figure 7.12 Schematic of samples with a 25-layer SL stack consisting of 2 nm SiO_2 barriers and 4 nm Si-rich oxide (SRO) layers (left, after [57]). The bottom SiO_2 layer contains a substantial amount of either B or P, which provides dopants to diffuse into the SL stack during Si segregation anneal to form Si NCs. The top SiO_2 layer serves as the gate dielectric; the lateral back contacts contact the SL underneath. Measured CV curves (right) from which an active P donor density of $\approx 10^{17}$ cm^{-3} was derived [57]. Copyright © 2011 WILEY-VCH Verlag GmbH & Co. KGaA, Weinheim.

on a qualitative base that the Si NC SL contains a substantial amount of excess Si which forms a Si NC network interconnected throughout the SL stack after the anneal. This is the only way of a bulk type semiconductor material capable of forming a SCR. The breakdown of the capacity with increasing frequency points to an extremely high defect density over the entire energy region with a very low DOS at the energy where the Urbach tails overlap. This is typical for amorphous semiconductors [4]. In addition, the missing recovery of the capacity under inversion conditions shows us that even at 1 KHz no inversion charge builds up. Again, this is typical of amorphous semiconductors where the mobility of minority carriers is too low to follow even low frequency signals. We appear to arrive at a similar material which was used by [9] and cannot be considered as a doped SL of isolated NCs or even quantum wells (QWs).

Our above discussions confirm our conclusions derived from thermodynamical, structural, electronical and quantum chemical theory. It is fairly certain that Si NCs embedded in dielectrics cannot obtain majority charge carriers from conventional doping with a probability high enough to allow for reasonable electronic device operation. Even for Si nanovolumes like drain and source areas of NG-ULSI FETs, which do not have interface and stoichiometry issues, efficient doping of Si for an abrupt junction is a real challenge. Experimental works on doping of Si NCs embedded in dielectrics suffered from two reoccurring issues: Dopant densities above the alloy limit and Si excess concentrations leading to interconnected Si NC networks as opposed to Si NCs embedded in (and separated by) dielectrics.

By any means, the precise and unambiguous characterization of the location and electronic nature of dopants in a dielectric matrix with embedded Si NC is a true challenge. In contrast to colloidal or free-standing NCs, it is paramount to distinguish between states in the dielectric matrix, states originating from the dielectric at the Si NC interface and states within the Si NCs. Elemental mapping by APT is a key tool, but its destructive nature together with field emission artefacts call for complementary techniques. In similar ways, PL is not a good complementary characterization technique to EPR. Below we give a brief overview on a characterization strategy for clarifying the location and in particular the electronic nature of dopants like P in Si NC/dielectric SLs.

The first task is to localize the excess Si within the annealed samples. This is important for two reasons: We must obtain information about the size, shape and distribution of *all* Si clusters, be they crystalline or amorphous, and we must analyse how much residual Si is left within the dielectric matrix (SiO_2, Si_3N_4). This can be achieved by 3D scans through APT. Unfortunately, APT suffers from different field evaporation rates per chemical element what makes it challenging to maintain atomic resolution on an absolute scale for all material constituents. However, energy electron loss spectroscopy (EELS) [11, 35] or high-angle annular dark-field scanning electron transmission (HAADF-STEM) spectroscopy [38] as nondestructive techniques do not suffer from artefacts originating from sample decomposition like SIMS [58] or APT and provide us with 2D information. They can be combined with transmission electron microscopy (TEM) to distinguish between amorphous and crystalline Si clusters for NCs with lattices aligned to the electron beam. Structure sizes require the scans to be carried out in the ultrahigh resolution (UHR) regime. Secondary neutral mass spectrometry (SNMS) is a destructive compositional scanning technique where the quantitative analysis does not depend as much on the sputter yield of the respective species as SIMS. This comes about by the missing ionization during the sputtering process. In SIMS, ionization increases the sputter rate for species with a lower ionization energy. The value of SNMS—like SIMS—is its ability to detect elemental concentrations down to the low 10^{16} cm^{-3} range which is very handy for detecting dopants at reasonable ($<$ 0.1 mol-% or 5×10^{19} cm^{-3}) concentrations. The next task is to clarify the immediate environment of the dopant atoms (1-nn and 2-nn). Thereby we learn about the electronic and structural boundary conditions of the dopant, for example, whether it is surrounded only by Si or O and Si or fully oxidized with 1-nn O atoms only. EPR has been proven to be an invaluable characterization technique, but has the disadvantage that the detection of unpaired electron states at RT is difficult due to thermal line broadening [53]. X-ray absorption near-edge spectroscopy (XANES) is a core level excitation technique which—apart from the element-specific absorption edge—contains information about the 1-nn atoms of the chemical element investigated [2]. As the core level (usually K shell electrons) is sensitive to the oxidation number of the specific element, we can also read out its ionization state which is very useful for dopant species.[a] Since no sample cooling is required, XANES can be carried out at RT. The ionization of a dopant changes its oxidation number which is detected by XANES. This already delivers important electronic information, though the detection limit of XANES requires a minimum doping density in the low 10^{19} cm^{-3} range. With the compositional, spatial and electronic information obtained, we can return to electronic characterization techniques of which CV and in particular deep level transient spectroscopy (DLTS) [54, 61] are of great value. Both techniques require single Si NC arrays in dielectrics adjacent to a well-defined semiconductor substrate such as a c-Si wafer. Undoped Si NC arrays in SiO_2 on Si wafers were characterized meticulously, delivering

[a]I am grateful to Sebastian Gutsch and Daniel Hiller, IMTEK, Albert Ludwigs University Freiburg, Germany, for fruitful discussions and their collaboration on XANES characterization work.

less spectacular but arguably correct results [19, 20]. Multiple NC arrays would be too difficult to interpret due to the superposition of charge capture and transport associated with traps/dopants as opposed to Si NCs, both occurring on multiple depth locations. As we already have compositional, structural, density and oxidation state information from preceding characterization, we can assign the detected densities, kinetic features, cross sections and energetic position within the dielectric band gap to these values. Thereby we should be able to obtain as much information as extractable which should bring us in the position to clarify the nature of conventional doping in the Si NC/dielectric system.

As already stated, it is fairly certain that Si NCs in dielectrics cannot be doped by conventional methods with a doping probability which allows for reasonable nanoelectronic device operation. Does this render any effort to introduce majority charge carriers into the Si NC/dielectric system to be a futile one? Certainly not. In the next section we explore alternative ways to achieve this task.

7.4 Alternatives to Conventional Doping

For alternative approaches it is useful to change the perspective on the problem and to broaden the view on material systems which may deliver clues to a solution. A survey of III–V semiconductor systems in particular for optoelectronic devices requiring smooth carrier transport and low nonradiative recombination rates points to modulation doping which is discussed in Section 7.4.1. Recent h-DFT calculations have shown that the interaction between Si NCs and the anions of the embedding dielectric has interesting features which may help to tune the preference of Si NCs for electrons or holes as majority carriers. We will briefly discuss preliminary results of this phenomenon in Section 7.4.2.

7.4.1 *Modulation Doping*

Modulation doping [16] is usually applied to interface electronics or SLs and has several advantages over conventional doping. In order to discuss these, we show its principle in Fig. 7.13 for donors (left

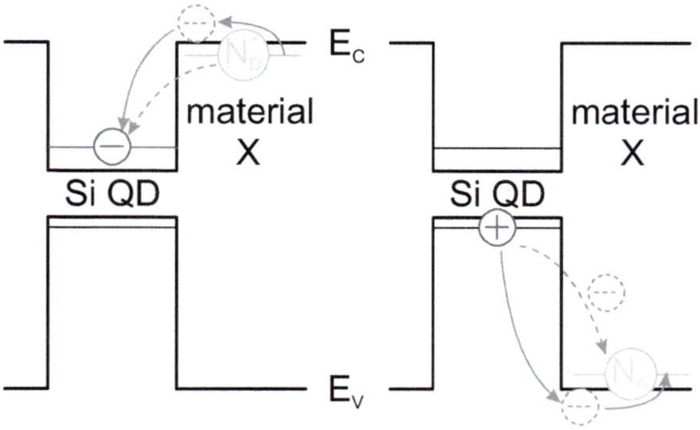

Figure 7.13 Principle of donor (left) and acceptor (right) modulation doping shown for Si QDs embedded in a barrier material. Conventional modulation doping via thermal ionization of dopant in the barrier material is shown by full arrows. Alternative ways to provide the respective charge carrier to the QD are shown by dashed arrows. These are direct carrier relaxation of the donor electron into the QD and the creation of a hole in the QD by direct relaxation of the electron in its HO state into the neutral acceptor. While modulation doping works over long distances by ionizing the respective band DOS of the barrier material, direct carrier relaxation only works in the proximity of the QD.

graph) and acceptors (right graph). The conventional modulation doping has dopants in the barrier material which can be thermally ionized to the respective band edge. From there, free carriers can diffuse into the quantum structures where they present the majority carriers. There are several advantages of this principle. With the dopant levels near the band edges of the barrier material, quantum and hopping transport through the barrier material occurs practically in the energy window between the dopants which minimizes carrier–dopant scattering. Since the dopant energies are outside the energy range of carrier transport, they do not contribute to carrier recombination. The dashed arrows in Fig. 7.13 show modulation doping by direct carrier relaxation. With a wide bandgap barrier material, the energy gain of the charge carriers can be 1 to 2 eV, providing an electric field of a few MV/cm through a few nm of barrier material to drive the carriers from the dopants into the QDs [47]. Another advantage of modulation doping is a flexible distribution of majority carriers as the carrier source is within the matrix material where there is no quantum confinement. This prevents the splitting of the dopant energy level within the QD due to exchange interaction [69]. In addition, it allows for an arbitrary NC size as dopants are ionized *outside* the Si NCs.

With all these favorable properties, we now have to find a material system of this kind which works for Si NCs in a barrier material. Ideally, we would use the Si-rich dielectric NC precursor layer as a dopant source for adjacent barriers. Modulation doping can proceed *in-situ* with the segregation anneal. As only a minute amount of Si is necessary, the growth of Si NCs would not be altered. Si is known to be a thermal donor in III–V materials. However, high donor Si densities face the problem of autocompensation [27] due to the amphoteric nature of Si regarding group III and V elements. If a sufficiently high density of Si donors on lattice sites of the group III element exist, additional Si atoms can obtain a higher binding energy by occupying the anion site (group V element) as acceptors which take up electrons from the Si donors. By using a possibly anionic group V element, the chemical nature of Si is shifted towards the group III element which increases the binding energy of Si on a group III element lattice cite relative to Si on an element V lattice site. This pushes the density limit of compensation to much higher values which is desirable for embedded Si NCs. The most anionic group V element is N which renders group III nitrides to be the most suitable barrier material. The next selection criteria are the band offsets between Si and group III nitrides to maintain quantum confinement. This rules out all group III nitrides apart from gallium nitride (GaN) and aluminum nitride (AlN) and their ternary compound $Al_xGa_{1-x}N$ [47]. Further investigations revealed that it is not straightforward to dope $Al_{>0.8}Ga_{<0.2}N$. The reaction enthalpy of Al with N is higher as opposed to Si which limits the incorporation of Si into the AlN lattice. There is also a limit to the type of the Si-rich dielectric. As the oxidation enthalpy of Al is higher than the oxidation enthalpy of Si, the only Si-rich dielectric we can use is Si_3N_4 (SiN_x). The nitridation enthalpy of Ga is smaller than the corresponding value for Si which sets an lower limit of $x = 0.4$ to $Al_xGa_{1-x}N$. Hence, Si donor modulation doping of $Al_xGa_{1-x}N$ works in the range $0.4 \leq x \leq 0.8$. The tunable stoichiometry allows for tailored band offsets to Si [47] which may be exploited for tuning

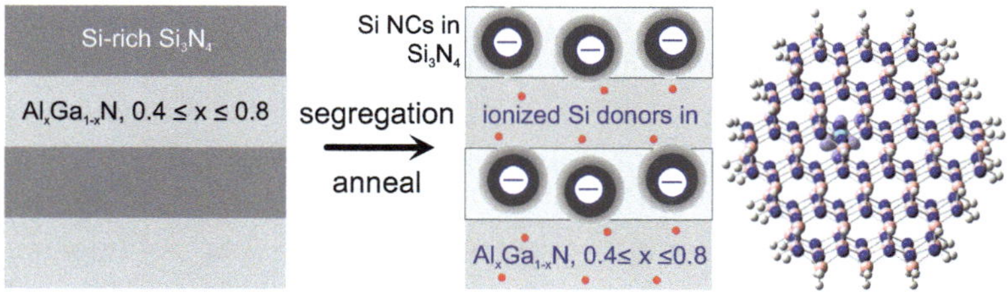

Figure 7.14 Principle of Si-rich $Si_3N_4/Al_xGa_{1-x}N$ SL segregation anneal with *in-situ* donor modulation doping (left and center graph). The same process works for Ge-rich Si_3N_4, see text. $Al_{107}GeN_{107}H_{126}$ approximant with the donor MO density of Ge shown as iso-plot of $0.008\ e/a_{B,0} = 0.19\ e/V_{Ge}^{atom}$, see text for details of h-DFT calculations. Ge atom shown in cyan, Al atoms in pink, N atoms in blue, and H atoms terminating surface bonds in white.

SL properties. The principle is illustrated in the left and center graph of Fig. 7.14. Both AlN [83] and GaN [63] can be deposited by sputtering which is also used for SiN_x/Si_3N_4 layer deposition [78]. Donor activation anneals of Si in $Al_xGa_{1-x}N$ were carried out at 980 to 1040°C [73, 82] which is covered by temperatures used for SiN_x/Si_3N_4 segregation anneals carried out at 1100°C [78]. The same process also works for Ge-rich Si_3N_4 which was annealed at 900°C [62], whereby higher temperatures can be used. Thermal ionization of the Ge donor in $Al_xGa_{1-x}N$ occurs up to $x = 0.3$ as found by local density approximation (LDA) calculations [6]. For AlN, the Ge donor ionization energy was calculated by h-DFT and found to be 0.16 eV, whereby its donor ionization probability is only 3.6×10^{-3}, which likely explains the absence of published experimental data of Ge in AlN [47]. The right graph of Fig. 7.14 shows a H-terminated AlN approximant with one Al atom substituted for Si ($Al_{107}GeN_{108}H_{126}$). The density of the donor MO $\langle \Psi_{MO} | \Psi_{MO}^* \rangle$ is shown as an iso-plot. The approximant was calculated using h-DFT with the route HF/3-21G(d)//B3LYP/6-31G(d) as described in Section 7.2.3. With direct modulation doping of AlN, Ge NCs in Si_3N_4 donor electrons can relax into Ge NCs through 3 to 4.5 nm of AlN or Si_3N_4.

The Si/SiO_2 system is one of the best-known and manageable electronic material combinations with a wealth of research and technology data collected over decades. It would be of great benefit if modulation doping could be realized for Si NCs in SiO_2. We cannot use $Al_xGa_{1-x}N$ barriers as O from SiO_x would immediately oxidize Al and Ga at high temperatures, disintegrating the entire material system. Preliminary h-DFT calculations showed that scandium (Sc) in SiO_2 forms an acceptor state which is able to capture an electron from a nearby Si NC, the relaxation energy is ca. 1 eV [50]. No modulation donor candidates were found since O is too anionic to allow for an unpaired electron state with a reasonably low ionization energy. The integration of Sc as a new material into Si technology requires substantial precursor and process development [28].

7.4.2 *Exploiting Interface Energetics: Nanoscopic Field Effect*

For Si QDs, the anions of the dielectric have a strong influence on their electronic structure [49]. This interface impact shifts HOMO and LUMO of the Si QDs, in particular for O and N representing SiO_2 and Si_3N_4. Thereby, it may be possible to create Si QDs with a built-in preference for electrons or holes [44]. While this is not a doping method, it could help to define the majority carrier type in Si QDs. The charge transfer over the interface can be interpreted as a nanoscopic field effect [46].

7.5 Conclusion and Outlook

The conventional doping of separate Si NCs or nc-Si layers embedded in or sandwiched between SiO_2 or Si_3N_4 is not possible to a degree which would allow for nanodevice operation. Results presented in the literature as Si NC doping either use dopant concentrations of ≥ 0.1 mol-% ($\geq 5 \times 10^{19}$ cm^{-3}) what renders the entire material system to be an alloyed ternary compound or use excess Si concentrations so high that an interconnected Si NC network is created. Another problem arises from the fact that dopants are point defects whereby they respond much less to quantum confinement as compared to the Si QD in which they reside. As a consequence, the dopant ionization energy increases which dramatically reduces the ability of the dopants to create free carriers. Results of h-DFT calculations investigating P as typical conventional donor in the materials SiO_2, $SiO_{0.9}$, at and in Si QDs completely terminated with OH groups showed that P does not work as donor but introduces defect levels within QDs or within SiO_2 and $SiO_{0.9}$. No defect levels are created within the HOMO–LUMO gap of the Si QD if P getters DBs at the Si NC/SiO_2 interface. Groundbreaking works of Stegner et al. on free-standing Si NCs doped *in-situ* with P went through the meticulous characterization work to pin down the electronic nature of P fused into Si NCs. Even for such a favorable plasma process, the probability of an active P donor was in the few 10^{-5} range relative to the nominal P doping concentration. As the huge partition of nonactive P donors creates defects in pure SiO_2 and oxide with residual Si, the electronic impact of conventional doping onto Si NCs is virtually nil. Conduction behavior over temperature typical for metallic materials were measured for nominal P doping densities of ca. $\geq 5 \times 10^{19}$ cm^{-3} (≥ 0.1 mol-%), which proves the point about defect-assisted carrier transport. We laid out a characterization strategy for embedded Si NCs with dopants to overcome the uncertainties and ambiguities of experimental data presented in the literature. Dopant activation and out-diffusion is also a major challenge for NG-ULSI FET devices. The clustering of inactive dopants provides additional scattering centers which slow down carrier transport and generate additional heat. Out-diffusion from the D and S areas into the gate channel region is another concern which severely deteriorates device performance.

Turning to alternative methods to introduce majority charge carriers into Si NCs, we showed that modulation doping adapted from III–V SL structures holds great promise for Si NCs in Si_3N_4. For the Si NC/SiO_2 material system, ongoing theoretical research revealed so far that donor modulation doping is not possible due to the strong anionic nature of O. Acceptor modulation doping of SiO_2 is likely to work with Sc, though the ionization of Si NCs has to proceed directly, setting the maximum distance of the Sc acceptors to a few nm from the Si NCs. This can be achieved in a Si NC SL structure where the SiO_2 barrier thickness is on the order of 2 nm with full NC segregation control [29]. The interface impact of the embedding dielectric onto the electronic structure of small Si NCs may turn out to be useful for setting a preference to either electrons or holes as majority charge carriers of Si NCs. While this is not a doping process, it provides Si NCs with a nanoscopic field effect which can separate charges in analogy to a macroscopic field effect junction.

We are at the beginning of a new era with a great variety of quantum and ultrasmall nanoelectronic devices. To make these work, different principles of majority carrier introduction or band structure manipulation are necessary. This calls for a concerted research effort in the fundamental science of quantum chemistry, quantum electronics, and quantum physics, covering the whole ground from theoretical research and material simulations via meticulous and detailed characterization to careful and precise sample preparation. When talking of samples in this context, I do not have in mind complete devices. We need to build and fortify the scientific foundation for accurate, reliable and repeatable results and underpin these with a sound theory. Only after this is done to a degree which delivers enough insight, we shall go forward and engage in the science and engineering of embedded Si NC devices.

Acknowledgments

I am very thankful to many colleagues I worked with on silicon in any shape and size, in particular M. Rennau, M. Henker, N. Zichner, and G. Ebest (Center for Microtechnologies and Professorship of Electronic Devices, Chemnitz University of Technology), J. Rudd (Australian Institute of Advanced Photovoltaics, University of New South Wales, Sydney), and C. Flynn (Silanna Semiconductors, Sydney), as well as D. Hiller, S. Gutsch and M. Zacharias (IMTEK, Albert Ludwigs University, Freiburg).

I am grateful for substantial compute power provided by the Leonardi compute cluster of the Engineering Faculty, UNSW.

This work has been supported by the Australian government through the Australian Renewable Energy Agency (ARENA). Responsibility for the information expressed herein is not accepted by the Australian government. Financial support from the Australian Research Council (ARC) Centre of Excellence funding scheme, by the Global Energy Climate Project (GECP), by the Australian Centre of Advanced Photovoltaics, and by the Go8-DAAD joint research cooperation scheme is gratefully acknowledged.

References

1. Becke, D. (1988). Density-functional exchange-energy approximation with correct asymptotic behaviour, *Physical Review A* **38**, 3098–3100.
2. Bianconi, A., Dell'Ariccia, M., Durham, P., and Pendry, J. (1982). Multiple-scattering resonances and structural effects in the x-ray absorption near-edge spectra of Fe II and Fe III hexacyanide complexes, *Physical Review B* **26**, 6502–6508.
3. Binkley, J., Pople, J., and Hehre, W. (1980). Self-Consistent Molecular Orbital Methods. 21. Small Split-Valence Basis Sets for First-Row Elements, *Journal of the American Chemical Society* **102**, 939–947.
4. Boer, K. (1990). *Survey of Semiconductor Physics, Vol. I: Electrons and Other Particles in Semiconductors* (Van Nostrand Reinhold, New York).
5. Boer, K. (1992). *Survey of Semiconductor Physics, Vol. II: Barrier, Junctions, Surfaces and Devices* (Van Nostrand Reinhold, New York).
6. Bogusławski, P., and Bernholc, J. (1997). Doping properties of C, Si, and Ge impurities in GaN and AlN, *Physical Review B* **56**, 9496–9505.
7. Brown, G., Uberuaga, B., Grube, H., Hawley, M., Schofield, S., Curson, N., Simmons, M., and Clark, R. (2005). Observation of substitutional and interstitial phosphorus on clean Si(100)-(2×1) with scanning tunneling microscopy, *Physical Review B* **72**, 195323, 1–5.
8. Chen, S., Wang, X., Thompson, M., Wang, Y., Lu, C., and McWhirtere, J. (2010). Characterization of Dopant Activation, Mobility and Diffusion in Advanced Millisecond Laser Spike Annealing, in *18th IEEE Internat. Conference on Advanced Thermal Processing of Semiconductors* (Gainesville, Florida), 66–70, http://ieeexplore.ieee.org/stamp/stamp.jsp?tp=&arnumber=5623601.
9. Cho, E.-C., Park, S., Hao, X., Song, D., Park, S.-G., Conibeer, G., and Green, M. (2008). Silicon quantum dot/crystalline silicon solar cells, *Nanotechnology* **19**, 245201, 1–5.
10. Cojocaru-Mirèdin, O., Cristiano, F., Fazzini, P.-F., Mangelinck, D., and Blavette, D. (2013). Extended defects and precipitation in heavily B-doped silicon, *Thin Solid Films* **534**, 62–66.
11. Colliex, C., Trencé, M., Levévre, E., Mory, C., Gu, H., Bouchet, D., and Jeanguillaume, C. (1994). Electron Energy Loss Spectroscopy Mapping, *Mikrochimica Acta* **114–115**, 71–87.
12. Current, M., Lopes, D., Foad, M., England, J., Jones, C., and Su, D. (1998). 200 eV - 10 keV boron implantation and rapid thermal annealing: Secondary ion mass spectroscopy and transmission electron

microscopy study, *Journal of Vacuum Science and Technology B* **16**, 327–333.

13. Dalpian, G. M., and Chelikowsky, J. R. (2006). Self-purification in semiconductor nanocrystals, *Physical Review Letters* **96**, 226802, 1–4.
14. Dalpian, G. M., and Chelikowsky, J. R. (2008). Dalpian and Chelikowsky reply:, *Physical Review Letters* **100**, 179703, 1.
15. Davies, J. (1998.). *The Physics of Low-Dimensional Semiconductors* (Cambridge University Press, Cambridge, UK).
16. Dingle, R., Störmer, H., Gossard, A., and Wiegmann, W. (1978). Electron mobilities in modulation-doped semiconductor heterojunction superlattices, *Applied Physics Letters* **33**, 665–667.
17. Dobbs, K., and Hehre, W. (1987). Molecular-orbital theory of the properties of inorganic and organometallic compounds. 5. Extended basis-sets for 1st-row transition-metals, *Journal of Computational Chemistry* **8**, 861–879.
18. Faulkner, R. (1969). Higher donor excited states for prolate-spheroid conduction bands: A re-evaluation of silicon and germanium, *Physical Review* **184**, 713–721.
19. Flynn, C., König, D., Perez-Wurfl, I., Conibeer, G., and Green, M. (2009). Capacitance and conductance characteristics of silicon nanocrystal metal-insulator-semiconductor devices, *Solid State Electronics* **53**, 530–539.
20. Flynn, C., König, D., Perez-Wurfl, I., Conibeer, G., and Green, M. (2010). Correlation between fixed charge and capacitance peaks in silicon nanocrystal metal-insulator-semiconductor devices, *Semiconductor Science and Technology* **25**, 045011, 1–8.
21. Francl, M., Pietro, W., Hehre, W., Binkley, J., DeFrees, D., Pople, J., and Gordon, M. (1982). Self-Consistent Molecular Orbital Methods. 23. A polarization-type basis set for 2nd-row elements, *Journal of Chemical Physics* **77**, 3654–3665.
22. Frisch, M., Trucks, G., Schlegel, H., Scuseria, G., Rob, M., Cheeseman, J., Jr., J. M., Vreven, T., Kudin, K., Burant, J., Millam, J., Iyengar, S., Tomasi, J., Barone, V., Mennucci, B., Cossi, M., Scalmani, G., Rega, N., Petersson, G., Nakatsuji, H., Hada, M., Ehara, M., Toyota, K., Fukuda, R., Hasegawa, J., Ishida, M., Nakajima, T., Honda, Y., Kitao, O., Nakai, H., Klene, M., Li, X., Knox, J., Hratchian, H., Cross, J., Bakken, V., Adamo, C., Jaramillo, J., Gomperts, R., Stratmann, R., Yazyev, O., Austin, A., Cammi, R., Pomelli, C., Ochterski, J., Ayala, P., Morokuma, K., Voth, G., Salvador, P., Dannenberg, J., Zakrzewski, V., Dapprich, S., Daniels, A., Strain, M., Farkas, O., Malick,

D., Rabuck, A., Raghavachari, K., Foresman, J., Ortiz, J., Cui, Q., Baboul, A., Clifford, S., Cioslowski, J., Stefanov, B., Liu, G., Liashenko, A., Piskorz, P., Komaromi, I., Martin, R., Fox, D., Keith, T., Al-Laham, M., Peng, C., Nanayakkara, A., Challacombe, M., Gill, P., Johnson, B., Chen, W., Wong, M., Gonzalez, C., and Pople, J. (2006). Gaussian 03 Revision C.02, *Gaussian, Wallingford, CT.*

23. Fujii, M., Yamaguchi, Y., Takase, Y., Ninomiya, K., and Hayashi, S. (2004). Control of photoluminescence properties of Si nanocrystals by simultaneously doping n- and p-type impurities, *Applied Physics Letters* **85**, 1158–1160.
24. Fujii, M., Yamaguchi, Y., Takase, Y., Ninomiya, K., and Hayashi, S. (2005). Photoluminescence from impurity codoped and compensated Si nanocrystals, *Applied Physics Letters* **87**, 211919, 1–3.
25. Gelpey, J., McCoy, S., Lerch, W., Paul, S., Cristiano, J. N. F., and Bolze, D. (2005). Advanced activation and stability of ultra-shallow junctions using flash-assisted RTP, in *13th IEEE Internattional Conference on Advanced Thermal Processing of Semiconductors* (Santa Barbara, CA), 73–76, http://ieeexplore.ieee.org/stamp/stamp.jsp?tp=&arnumber=1613700.
26. Green, M., Emery, K., Hishikawa, Y., and Warta, W. (2011). Solar cell efficiency tables (version 37), *Progress in Photovoltaics* **19**, 84–92.
27. Greiner, M. E., and Gibbons, J. F. (1984). Diffusion of silicon in gallium arsenide using rapid thermal processing: Experiment and model, *Applied Physics Letters* **44**, 750–752.
28. Gutsch, S., and Hiller, D. (2013). Private discussion.
29. Gutsch, S., Laube, J., Hartel, A., Hiller, D., Zakharov, N., Werner, P., and Zacharias, M. (2013). Charge transport in Si nanocrystal/SiO_2 superlattices, *Journal of Applied Physics* **113**, 133703, 1–10.
30. Hao, X., Cho, E.-C., Scardera, G., Bellet-Amalric, E., Bellet, D., Shen, Y., Huang, S., Huang, Y., Conibeer, G., and Green, M. (2009). Effects of phosphorus doping on structural and optical properties of silicon nanocrystals in a SiO_2 matrix, *Thin Solid Films* **517**, 5646–5652.
31. Hao, X., Cho, E.-C., Scardera, G., Shen, Y., Bellet-Amalric, E., Bellet, D., Conibeer, G., and Green, M. (2009). Phosphorus-doped silicon quantum dots for all-silicon quantum dot tandem solar cells, *Solar Energy Materials & Solar Cells* **93**, 1524–1530.
32. Hehre, W., Ditchfield, R., and Pople, J. (1972). Self-Consistent Molecular Orbital Methods. 12. Further extensions of Gaussian-type basis sets for use in molecular-orbital studies of organic-molecules, *Journal of Chemical Physics* **56**, 2257–2261.

33. Hiller, D., Goetze, S., and Zacharias, M. (2011). Rapid thermal annealing of size-controlled Si nanocrystals: Dependence of interface defect density on thermal budget, *Journal of Applied Physics* **109**, 054308, 1–5.

34. Hiller, D., Jivanescu, M., Stesmans, A., and Zacharias, M. (2010). P_{b0} centers at the Si-nanocrystal/SiO_2 interface as the dominant photoluminescence quenching defect, *Journal of Applied Physics* **107**, 084309, 1–4.

35. Hillier, J., and Baker, R. (1944). Microanalysis by Means of Electrons, *Journal of Applied Physics* **15**, 663–675.

36. Jelec, D. (16 Oct. 2013). Intel's Broadwell Manufacturing Delayed Until 1Q 2014 Due to a Defect, http://www. brightsideofnews.com/news/ 2013/10/16/intele28099s-broadwell-delayed-until-2014-due-to-a-defect.aspx, retrieved on March 12, 2015.

37. Kambham, A. K., Kumar, A., Florakis, A., and Vandervorst, W. (2013). Three-dimensional doping and diffusion in nano scaled devices as studied by atom probe tomography, *Nanotechnology* **24**, 275705, 1–8.

38. Kaneko, K., Kawasaki, M., Nagano, T., Tamari, N., and Tsurekawa, S. (2000). Determination of the chemical width of grain boundaries of boron- and carbon-doped hot-pressed β-SiC by HAADF imaging and ELNES line-profile, *Acta Materialia* **48**, 903–910.

39. Kazmerski, L., Nelson, A., Dhere, R., Yahia, A., and Abou-Elfotouh, F. (1988). Neutralization and bonding mechanisms of shallow acceptors at grain boundaries in polycristalline silicon, *Journal of Vacuum Science and Technology A* **6**, 1007–1011.

40. Keck, P., and Golay, M. (1953). Crystallization of Silicon from a Floating Liquid Zone, *Physical Review* **89**, p. 1297.

41. Khelifi, R., Mathiot, D., Gupta, R., Muller, D., Roussel, M., and Duguay, S. (2013). Efficient n-type doping of Si nanocrystals embedded in SiO_2 by ion beam synthesis, *Applied Physics Letters* **102**, 013116, 1–4.

42. Knipping, J., Wiggers, H., Rellinghaus, B., Roth, P., Konjhodzic, D., and Meier, C. (2004). Synthesis of high purity silicon nanoparticles in a low pressure microwave reactor, *Journal of Nanoscience and Nanotechnology* **4**, 1039–1044.

43. Koelling, S., Richard, O., Bender, H., Uematsu, M., Schulze, A., Zschaetzsch, G., Gilbert, M., and W. Vandervorst, W. (2013). Direct imaging of 3D atomic-scale dopant-defect clustering processes in ion-implanted silicon, *Nano Letters* **13**, 2458–2462.

44. König, D. (2007). Research seminar (confidential), ARC Photovoltaics Centre of Excellence, University of New South Wales.

45. König, D., Green, M., Conibeer, G., Takeda, Y., Ito, T., and Motohiro, T. (2006). DF-HF computation of Si QDs: Intrinsic stress, iso-valent impurities and validity of the effective mass approximation, in *21st European Photovoltaic Science and Engineering Conference (21 E-PVSEC)* (Dresden, Germany), 164–167 [presentation 1CO.6.4].

46. König, D., Hiller, D., Gutsch, S., and Zacharias, M. (2014). Energy offset between silicon quantum structures: interface impact of embedding dielectrics as doping alternative, *Advanced Materials Interfaces* **1**, 1400359.

47. König, D., and Rudd, J. (2013). Formation of Si or Ge nanodots in Si_3N_4 with in-situ donor modulation doping of adjacent barrier material, *AIP Advances* **3**, 012109, 1–8, open access.

48. König, D., Rudd, J., Green, M., and Conibeer, G. (2008). Critical Discussion of Computation Accuracy, EPAPS Document E-PRBMDO-78-101827, http://journals.aps.org/prb/supplemental/10.1103/PhysRevB.78.035339/EPAPS_Rev3.pdf.

49. König, D., Rudd, J., Green, M. A., and Conibeer, G. (2008). Role of the interface for the electronic structure of Si quantum dots, *Physical Review B* **78**, 035339, 1–9.

50. König, D., Rudd, J., Shrestha, S., Conibeer, G., and Green, M. (2009). Acceptor modulation doping of SiO_2 for providing holes to Si quantum dots, in *24th European Photovoltaic Science and Engineering Conference (24 E-PVSEC)* (Hamburg, Germany), 575–578 [presentation 1CV.4.53].

51. Lallement, F., and Lenoble, D. (2005). Combination of plasma doping and flash anneal RTP for 45 nm CMOS node, in *13th IEEE International Conference on Advanced Thermal Processing of Semiconductors* (Santa Barbara, California), 153–155, http://ieeexplore.ieee.org/stamp/stamp.jsp?tp=&arnumber=1613700.

52. Lammers, D. (2010). CMOS transitions to 22 and 15 nm, http://scsong.wordpress.com/2010/01/04/cmos-transitions-to-22-and-15-nm-2010-01-01-070000-semiconductor-international/; retrieved on December 24, 2014.

53. Lancaster, G., Wyk, J. V., and Schneider, E. (1964). Spin-lattice relaxation of conduction electrons in silicon, *Proceeding of the Physical Society* **84**, 19–24.

54. Lang, D. (1974). Deep level transient spectroscopy: A new method to characterize traps in semiconductors, *Journal of Applied Physics* **45**, 3023–3032.

55. Lee, C., Yang, W., and Parr, R. (1988). Development of the Colle-Salvetti correlation-energy formula into a functional of the electron density, *Physical Review B* **37**, 785–789.
56. Liedke, B., Heinig, K.-H., Mücklich, A., and Schmidt, B. (2013). Formation and coarsening of sponge-like Si-SiO_2 nanocomposites, *Applied Physics Letters* **103**, 133106, 1–4.
57. Lin, D., Ma, L., Conibeer, G., and Perez-Wurfl, I. (2011). Study on electrical properties of Si quantum dots based materials, *Physica Status Solidi B* **248**, 472–476.
58. Magee, C., Harrington, W., and Honig, R. (1977). Secondary quadrupole ion mass spectroscopy for depth profiling: Design and performance evaluation, *Review of Scientific Instruments* **49**, 477–485.
59. McCaldin, J., and Sankur, H. (1971). Diffusivity and solubility of Si in the Al metallization of integrated circuits, *Applied Physics Letters* **19**, 524–527.
60. Mercier, J., Calder, I., Beerkens, R., and Naguib, H. (1985). Rapid isothermal diffusion of PSG films, *Journal of the Electrochemical Society* **132**, 2432–2435.
61. Miller, G., Lang, D., and Kimerling, L. (1977). Capacitance transient spectroscopy, *Annual Review of Material Science* **7**, 377–448.
62. Mirabella, S., Cosentino, S., Gentile, A., Nicotra, G., Piluso, N., Mercaldo, L. V., Simone, F., Spinella, C., and Terrasi, A. (2012). Matrix role in Ge nanoclusters embedded in Si_3N_4 or SiO_2, *Applied Physics Letters* **101**, 011911, 1–5.
63. Miyazaki, T., Takada, K., Adachi, S., and Ohtsuka, K. (2005). Properties of radio-frequency-sputter-deposited GaN films in a nitrogen/hydrogen mixed gas, *Journal of Applied Physics* **97**, 093516, 1–7.
64. Myslewski, R. (10. Sept. 2013). Intel reveals 14 nm PC, declares Moore's Law'alive and well', http://www.theregister.co.uk/2013/09/10/intel_reveals_14nm_pc_declares_moores_law_alive_and_well/, retrieved on March 12, 2015.
65. Nicollian, E. H., and Brews, J. R. (1982). *MOS (Metal Oxide Semiconductor) Physics and Technology* (Wiley & Sons, New York).
66. Niquet, Y., Delerue, C., Allan, G., and Lannoo, M. (2000). Method for tight-binding parametrization: Application to silicon nanostructures, *Physical Review B* **62**, 5109–5116.
67. Ossicini, S., Degoli, E., Iori, F., Luppi, E., Magri, R., Cantele, G., trani, F., and Ninno, D. (2005). Simultaneously B- and P-doped silicon nanoclusters: Formation energies and electronic properties, *Applied Physics Letters* **87**, 173120, 1–3.

68. Perego, M., Bonafos, C., and Fanciulli, M. (2010). Phosphorous doping of ultra-small silicon nanocrystals, *Nanotechnology* **21**, 025602, 1–6.

69. Pereira, R., Almeida, A. J., Stegner, A., Brandt, M., and Wiggers, H. (2012). Exchange-coupled donor dimers in nanocrystal quantum dots, *Physical Review Letters* **108**, 126806, 1–4.

70. Pi, X. D., Gresback, R., Liptak, R. W., Campbell, S. A., and Kortshagen, U. (2008). Doping efficiency, dopant location, and oxidation of Si nanocrystals, *Applied Physics Letters* **92**, 123102, 1–3.

71. Rassolov, V., Ratner, M., Pople, J., Redfern, P., and Curtiss, L. (1982). Basis set for third-row atoms, *Journal of Computational Chemistry* **22**, 976–984.

72. Rath, J. (2003). Low temperature polycrystalline silion: a review on deposition, physical properties and solar cell applications, *Solar Energy Materials and Solar Cells* **76**, 431–487.

73. Rowland, L., Doverspike, K., and Gaskill, D. (1995). Silicon doping of GaN using disilane, *Applied Physics Letters* **66**, 1495–1497.

74. Sana, P., Salami, J., and Rohatgi, A. (1993). Fabrication and Analyisis of High-Efficiency Polycrystalline Silicon Solar Cells, *IEEE Transactions on Electron Devices* **40**, 1461–1468.

75. Sasaki, Y., Okashita, K., Mizuo, B., Kubota, M., Ogura, M., and Nishijima, O. (2012). Conformal doping mechanism for fin field effect transistors by self-regulatory plasma doping with AsH_3 plasma diluted with He, *Journal of Applied Physics* **111**, 013712, 1–6.

76. Schiff, L. I. (1968). *Quantum Mechanics (3. Ed.)* (McGaw-Hill, Singapore).

77. Singh, A., and Davis, E. (1990). The SiO_x: H_y thin film system II. Optical bandgap behavior, *Journal of Non-Crystalline Solids* **122**, 233–240.

78. So, Y.-H., Gentle, A., Huang, S., Conibeer, G., and Green, M. (2011). Size dependent optical properties of Si quantum dots in Si-rich nitride/Si_3N_4 superlattice synthesized by magnetron sputtering, *Journal of Applied Physics* **109**, 064302, 1–5.

79. Stegner, A. R., Pereira, R., Klein, K., Lechner, R., Dietmueller, R., Brandt, M. S., Stutzmann, M., and Wiggers, H. (2008). Electronic transport in phosphorus-doped silicon nanocrystal networks, *Physical Review Letters* **100**, 026803, 1–4.

80. Stegner, A. R., Pereira, R. N., Lechner, R., Klein, K., Wiggers, H., Stutzmann, M., and Brandt, M. S. (2009). Doping efficiency in freestanding silicon nanocrystals from the gas phase: Phosphorus incorporation and defect-induced compensation, *Physical Review B* **80**, 165326, 1–10.

81. Stesmans, A. (1993). Structural relaxation of P_{b0} defects at the (111)Si/SiO_2 interface as a function of oxidation temperature: The P_{b0}-generation-stress relationship, *Physical Review B* **48**, 2418–2435.

82. Taniyasu, Y., Kasu, M., and Kobayashi, N. (2002). Intentional control of n-type conduction for Si-doped AlN and $Al_xGa_{1-x}N$ ($0.42 \leq x \leq 1$), *Applied Physics Letters* **81**, 1255–1257.

83. Vashaei, Z., Aikawa, T., Ohtsuka, M., Kobatake, H., Fukuyama, H., Ikeda, S., and Takada, K. (2009). Influence of sputtering parameters on the crystallinity and crystal orientation of AlN layers deposited by RF sputtering using the AlN target, *Journal of Crystal Growth* **311**, 459–462.

84. Voutsas, A., Hatalis, M., Boyce, J., and Chiang, A. (1995). Raman spectroscopy of amorphous and microcrystalline silicon films deposited by low-pressure chemical vapor deposition, *Journal of Applied Physics* **78**, 6999–7006.

85. Wood, D., and Zunger, A. (1996). Successes and failures of the $k \cdot p$ method: A direct assessment for GaAs/AlAs quantum structures, *Physical Review B* **53**, 7949–7963.

86. Zacharias, M., Heitmann, J., Scholz, R., Kahler, U., Schmidt, M., and Bläsing, J. (2002). Size-controlled highly luminescent silicon nanocrystals: A SiO/SiO_2 superlattice approach, *Applied Physics Letters* **80**, 661–663.

87. Zimina, A., Eisebitt, S., Eberhardt, W., Heitmann, J., and Zacharias, M. (2006). Electronic structure and chemical environment of silicon nanoclusters embedded in a silicon dioxide matrix, *Applied Physics Letters* **88**, 163103, 1–3.

Chapter 8

Electrical Transport in Si-Based Nanostructured Superlattices

Blas Garrido, Sergi Hernández, Yonder Berencén, Julian López-Vidrier, Joan Manel Ramírez, Oriol Blázquez, and Bernat Mundet

MIND-IN[2] UB, Departament d'Electrònica, Universitat de Barcelona, Carrer Martí i Franquès 1, 08028 Barcelona, Spain

blas.garrido@ub.edu

8.1 Introduction and Scope

A *superlattice* (SL) is a multilayered structure with a large number of alternating layers of two semiconductors or insulators with different band-gap energies. The materials are deposited or grown with sublayer thicknesses up to 10 nm. Usually, a multilayered structure of alternate materials is called a *multiquantum well* (MQW) when sublayers are thicker than this. In practice, this thickness division is somewhat arbitrary and a particular definition should be done for each material system. In an ideal SL all the quantum wells (QWs) are coupled and thus carriers are delocalized. Hence, in a SL the individual levels of each QW turn into a *miniband* of the whole structure. In contrast, the width of the barriers in an MQW

Nanotechnology and Photovoltaic Devices: Light Energy Harvesting with Group IV Nanostructures
Edited by Jan Valenta and Salvo Mirabella

ISBN 978-981-4463-63-8 (Hardcover), 978-981-4463-64-5 (eBook)
www.panstanford.com

is large enough to prevent carrier tunneling between wells. Esaki and Tsu were the first to propose a semiconductor 1D SL [1]. In addition to the compositional SLs they also envisioned doping SLs, in which both sublayers are of the same material but have different doping type. SLs can be fabricated from crystalline, amorphous, or nanocrystalline materials. Amorphous and nanocrystalline materials provide great flexibility in composition and have relaxed requirements of lattice matching and interfacial strain. Nevertheless, understanding transport in these materials is much more difficult than in crystalline ones due to the presence of *defects, gap states, disorder,* and *hopping transport.*

This chapter is devoted to introduce the conduction of amorphous and nanocrystalline SLs in a comprehensive way. The selection of literature and the general approach adapted to present theory and models are quite generic. Nevertheless, the last sections point to particular results of the authors and collaborators mainly focused on the suitable applications in third generation photovoltaics. Section 8.2 of this chapter is devoted to develop the basic concepts and theory of SLs and minibands. Section 8.3 reviews accomplishments and main results of amorphous and nanocrystal (NC) SLs and introduces a simple model of electronic structure that will be used to study the transport. Section 8.4 introduces the basic theory and some experimental results of conduction in SLs. Finally, Sections 8.5 and 8.6 present experimental results of vertical and horizontal transport in SLs that are formed of nanocrystalline silicon alternated with SiO_2 or a-SiC layers. These experimental results are oriented toward applications in tandem solar cells. We shall demonstrate that nanocrystalline SLs allow designing layers in which (i) NCs are very close in the horizontal plane, while (ii) in the vertical growth direction, separation by barriers allows limiting its size, size distribution, and vertical transport properties.

8.2 Superlattices and Minibands

The most well-known example of a semiconductor SL is the alternating system GaAs/AlGaAs. The lattice constant mismatch between GaAs and $Al_{0.5}Ga_{0.5}As$ is only 0.08% and therefore eases

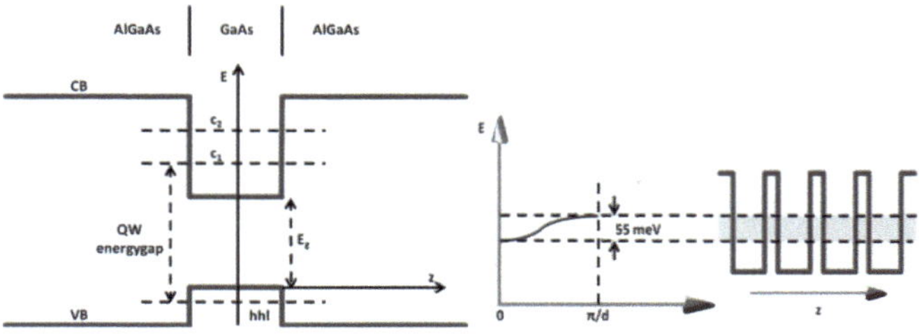

Figure 8.1 Energy band diagram versus position of an AlGaAs/GaAs/AlGaAs heterostructure. Electrons and holes are confined in the GaAs quantum well (type I heterostructure) (left). Sketch of the conduction band of a superlattice versus wavevector and position (center and right). The barriers are thin enough, so energy states into the wells become delocalized and extend through the multilayer, forming a miniband.

the growth of strain-free and low-dislocation-density multilayered structures. The GaAs semiconductor has smaller band-gap energy that the ternary alloy AlGaAs and the band offsets between conduction and valence bands are collocated in a way that both electrons and holes are confined in the GaAs layer. Figure 8.1 shows conduction and valence bands of an AlGaAs/GaAs/AlGaAs heterostructure, whose band alignment is referred as "type I" heterostructure. In case that the energy confinement scheme is such that electrons are confined in one sublayer and holes are confined in the other sublayer, its band alignment is referred as "type II" heterostructure (e.g., an InAs/GaSb SL). Moving from QWs to SLs consists in preparing multiple piled-up heterostructures in which AlGaAs barriers are so thin that the quantum states of individual wells are strongly delocalized, that is, with wavefunctions extending throughout the multilayered structure.

The formation of a SL requires something more than structural periodicity and thin "transparent" barriers. Interfaces must be ideally abrupt, without roughness and imperfections and with a low density of trapping or interfacial states. This usually calls for an equal crystal structure of both sublayers and very close lattice constants which is necessary for pseudomorphic or commensurate strain-free growth. These requirements are essential so that the electronic envelope wavefunction becomes coherent across the entire SL. Deviation from these conditions reduces coherence and may produce localization of the wavefunction within a few QWs. As we will see, the application of an electric field is a way of inducing localization.

A SL constant (i.e., period d) is usually between 10 and 100 times the lattice parameter (a_0). So, we expect to have a SL first Brillouin zone edge (π/d) which is reduced in the same factor (10–100) compared to the first Brillouin zone edge of the crystalline material (π/a_0). SLs are thus artificial "supercrystals" in which one can devise and tailor a reciprocal space depending on the actual needs (this has been called sometimes band-gap engineering). The electronic structure of a SL can be calculated in a number of ways: from Krönig–Penney models to self-consistent ab initio methods. Smith and Mailhiot have reviewed the methods for calculating the electronic structure of semiconductor SLs [2]. The systems of interest in this chapter are mostly amorphous or NC sublayers with relatively wide band-gap energy, closely matched (i.e., small strain) and with band alignments of type I. As a result, there is no mixing of the different symmetry sublayer wavefunctions to form the SL wavefunction (i.e., s–p–d mixing or mixing between the electron and hole bands). Hence, the Krönig–Penney models scaled by the appropriate effective masses and with barrier potentials given by the band offsets provide a good and intuitive description of the energy levels. Additionally, for transport behavior we are interested in states located close to the band edges in which those approximations better hold [2].

We refer to the schematic structure of a SL shown in Fig. 8.2. A and B are two different materials with sublayer thicknesses a and b and bulk band-gap energy E_{A} and E_{B}. The period of the SL is $d = a + b$ and hence the potential is periodic in d, that is, $V(z) = V(z + d) = V(z+2d) = \ldots$ It is quite instructive to remind the solutions for one particular well of material A; this will be also of interest when speaking about NCs. The solution of the 1D Schrödinger equation gives oscillating solutions for the classically allowed region (well) and exponentially decaying solutions for the classically forbidden regions (barriers):

$$\frac{\hbar^2}{2m^*(z)}\frac{d^2\psi(z)}{dx^2} + V(z)\psi(z) = E\psi(z) \qquad (8.1)$$

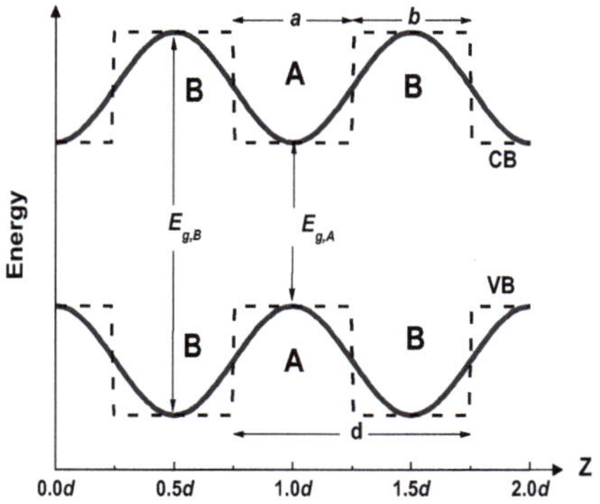

Figure 8.2 Band energy versus wavevector plot (solid line) for the conduction and valence bands of a superlattice with sublayers A and B (dashed lines); a is the thickness of sublayer A, b is the thickness of sublayer B, and $d = a + b$ is the period of the superlattice. $E_{g,A}$ and $E_{g,B}$ are the band-gap energies of materials A and B, respectively.

where $\hbar$ is the reduced Planck constant, ψ is the wavefunction, and n is an integer number:

$$\psi(z) = \sqrt{\frac{2}{a}} \sin\left(k\left(z+\frac{a}{2}\right)\right), \quad ka = n\pi, \; k = \sqrt{\frac{2m_A^* E}{\hbar^2}} \tag{8.2}$$

Different effective mass $m^*(x)$ values are allowed for materials A and B. This method of calculating the electronic structure is called the "effective mass approximation" or the "envelope function approach." It has been well justified for slowly varying potentials but it is more difficult to justify for heterostructures where there is an abrupt jump of potential due to band offsets. Nevertheless, Burt [3] has identified that the envelope approximation still works for abrupt heterojunctions provided that the envelope wavefunction varies slowly on the scale of the lattice period.

The bound-state solutions are quantized, that is, only certain energy levels are allowed. The approximation of infinite height of the barriers is quite good for large energy band offset between A and B materials. Hence, at the walls of the well A the potential is infinite and solutions decay immediately to zero. Additionally, the potential energy origin is taken at the bottom of the well, that is, $V(x) = 0$. Thus, the zero solutions for the energy at the left boundary are given by:

$$E_n^{inf} = \frac{\hbar^2 k_n^2}{2m_A^*} = \frac{\hbar^2 \pi^2 n^2}{2m_A^* a^2}, \qquad (8.3)$$

where k is the allowed values of the wavevector.

For the finite square well with barriers V_0 the energy levels can be written as:

$$E_n = \frac{\hbar^2 k_n^2}{2m_A^*} V_0 - E_n^{inf} = \frac{\hbar^2 \kappa_n^2}{2m_B^*}, \qquad (8.4)$$

where k_n are the solutions of

$$\tan\left(\frac{ka}{2}\right) = \begin{cases} -\dfrac{m_A^* \kappa}{m_B^* k}, & \text{symmetric solutions} \\ -\dfrac{m_B^* k}{m_A^* \kappa}, & \text{antisymmetric solutions} \end{cases} \qquad (8.5)$$

We see from Eq. 8.3 that the energy levels scale inversely with a^2. As the well gets narrower, each energy level and the gaps between them become larger. It is remarkable that the band gap of the QW is increased with respect to the bulk material due to electron confinement as $\frac{\hbar^2 \pi^2}{2m^* a^2}$, and a corresponding quantity for hole confinement (same expression but with the hole effective mass). It is also worth to mention that the confinement energy of the particle is a consequence of the Heisenberg uncertainty principle. If we consider that the particle is confined within the well, the uncertainty of its momentum increases by an amount of the order of $\frac{\hbar}{a}$, which corresponds to a kinetic energy of $\frac{\Delta p^2}{2m^*} = \frac{\hbar^2 \pi^2}{2m^* a^2}$, being equal to the minimum energy of the particle.

For the periodic potential of the SL it is reasonable to look for functions in the form of Bloch waves where $u(z)$ has the periodicity of the SL.

$$\psi(z) = u(z)\, e^{ikz} = u(z + nd) e^{ikz} \qquad (8.6)$$

By inserting this expression into the Schrodinger equation we obtain the solutions for $u(z)$ in the well and the barriers. Following the Krönig–Penney model, as developed in Ref. [4], we solve the equations separately as in the single QW and then impose continuity and periodicity for $u(z)$ and its derivative. In this way, each of the bound states in the single well (n state) evolves into a band of states (n band), whose states can be classified according to their

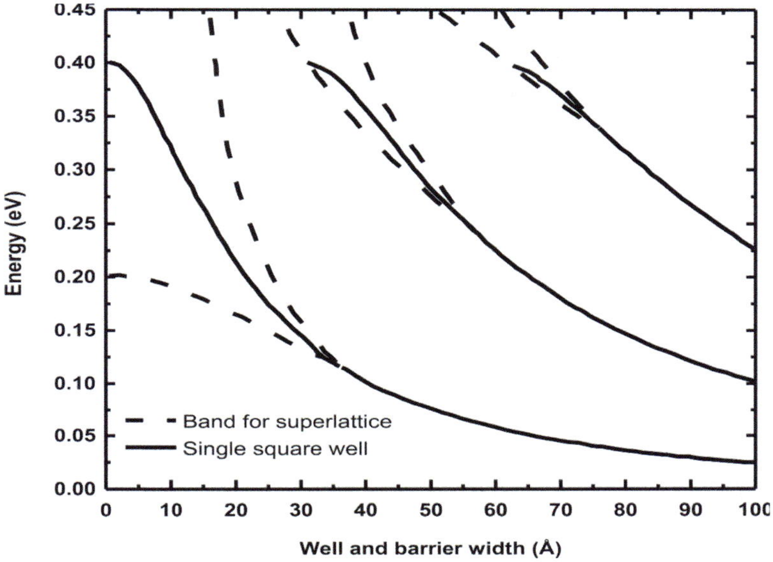

Figure 8.3 Evolution of the energy levels from quantum wells to minibands, depending on the well and barrier widths. The results are obtained by applying the Krönig–Penney model, as worked out in the text. Notice that the band average and band width increase as an effect of quantum confinement when the well and barrier width decrease. Adapted from Ref. [5].

wavevector **k** (see Fig. 8.3). The energy of each state depends upon k, leading to a periodic function $E_n(k)$ over $2\pi/d$ (see Fig. 8.2) that is termed dispersion relation in the n band. For the particular case of large barrier heights and thick walls and denoting by $2\Delta_n$ the energy width of the band, the complete result in Ref. [4] can be simplified to (plotted in Fig. 8.2):

$$E_n(k) = \Delta_n[1 \mp \cos(kd)] \tag{8.7}$$

The energy origin is taken at the bottom of the miniband. The minus sign $(-)$ is for the symmetric solutions and the plus sign $(+)$ is for the antisymmetric solutions of the square well. It is striking that the velocity of the Bloch states $v = \hbar^1 \mathrm{dE/dk}$ changes its sign when $kd = \pi$ (zone edge). This means that when the electron is gaining energy, as for example from a constant electric field, we expect that the electron reflects back when k gets to the limit of the first Brillouin zone. Thus, the electron oscillates between the extreme values of the energy minibands (if we rule out interband transport).

In real space this comes out as an oscillation back and forth which is called *Bloch oscillation* (see Appendix B for more details of this effect).

8.3 Amorphous and Nanocrystal Superlattices

SLs with crystalline sublayer materials were the first to be envisioned and produced [1, 5]. We have already mentioned in Section 8.2 the importance of having lattice matching, abrupt interfaces and a low density of interfacial states and defects. Additionally, it is now possible to grow crystalline SLs which are not perfectly lattice matched, that is, they grow strained. There is a certain limit to the thickness of a strained layer (critical layer thickness) before it breaks releasing mechanical strain energy via creation of misfit dislocations. By preserving the SL sublayers thinner than the critical layer thickness and by careful growth (by molecular beam epitaxy, MBE, or metalorganic chemical vapor deposition, MOCVD), it is possible to grow defect-free strained-layer SLs. An important example is the Si/Ge SL in which lattice mismatch is 4% and many others that are important for optoelectronics [6].

The next step is growing amorphous SLs, with sublayer materials in which long-range periodic lattice no longer exist. This idea came about with the work by Abeles and Tiedje in 1983 [7]. They demonstrated SLs from amorphous hydrogenated semiconductors such as the pairs a-Si:H/a-Ge:H, a-Si:H/a-SiN$_x$, and a-SiN$_x$/a-Si$_{1-x}$C$_x$:H. With these amorphous silicon containing materials the growth is neither lattice matched nor epitaxial. The authors obtained interfaces basically free from defects and nearly atomically abrupt. Thus, in these material systems there is no need of matching the composition versus lattice constant. On the contrary, it can be arbitrarily changed without limiting only to the lattice matched composition, as it happens in the crystalline SLs. Do we expect some sort of quantum confinement effects (QCEs) in amorphous SLs (a-SLs)? Actually, similar effects have been observed in a-SLs to those that have been seen in QWs. Abeles and Tiedje [7] demonstrated QCE experimentally in their a-SLs for varying sublayer thickness. As an illustrative example, QCE in low-dimensional a-Si (QWs, QDs, and

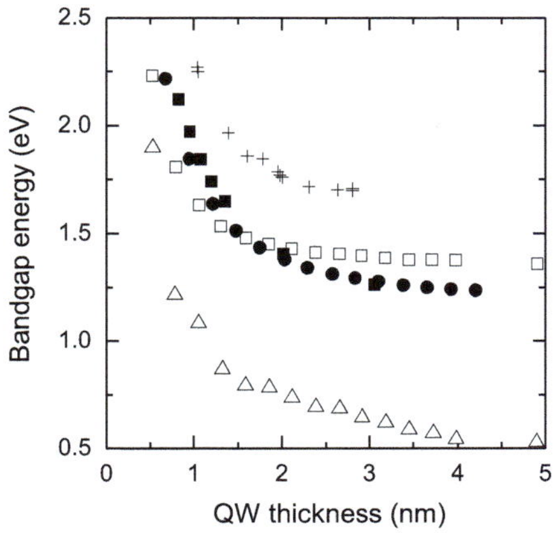

Figure 8.4 Compilation of theoretical results of band-gap energy calculations for quantum wells from tight-binding methods for c-Si (black dots), first principle for c-Si (black squares), a-Si (triangles), and a-Si:H (empty squares), as a function of the quantum well thickness (defined as *a* in this chapter). Crosses are the experimental photoluminescence peak energy for c-Si. The authors state that the tight-binding method underestimates confinement energy. From Ref. [8]. Reprinted with permission from [G. Allan, C. Delerue, M. Lannoo, *Appl. Phys. Lett.* **71**, 1189 (1997)]. Copyright [1997], AIP Publishing LLC.

SLs) have been demonstrated both theoretically and experimentally [7–9]. We can see in Fig. 8.4 (from Ref. [8]) the band energy evolution of a-Si and a-Si:H QWs as a function of the wall thickness, in a way similar to that of c-Si QWs. The results show an increase of the band-gap energy for decreasing thickness with a dependence close to a^{-2} (see Eq. 8.3). Clearly, the quantum confinement energy for a QW (2D) is much smaller than for a quantum dot (0D) [8] and this has been taken into account for recently developed a-SLs, as we discuss below.

In addition to the relaxation of lattice matching in a-SLs, we have to contemplate that the methods used for their growth are much less complex (and much more cheap) than those used for the crystalline SLs. The synthesis methods include chemical vapor deposition (CVD)—low-pressure chemical vapor deposition (LPCVD) or plasma-enhanced chemical vapor deposition (PECVD)—thermal or e-beam evaporation, and/or sputtering. The fact that most a-SLs can be grown with a low number of interfacial defects is a consequence of passivation, a circumstance that is well known for a-Si:H and other amorphous semiconductors in bulk form. Hydrogen plays an important role by passivating bulk and interfacial coordination defects (dangling bonds). On the contrary, making abrupt interfaces in amorphous SLs is a tough issue because it is difficult to sharply change the growing conditions in the reaction chamber to get atomically thin transition layers. Nevertheless, by carefully lowering and controlling the growth rate it is possible to avoid fabrication of transition layers when the deposition conditions change. It is also possible to include interfacial smooth transition layers leading to sinusoidal potential profiles instead of the step-like discussed so far [10].

Since the first demonstration in 1983 [7], numerous publications have appeared on a-SLs of silicon-compatible materials, especially during the decade of the 1990s. Roxlo and Abeles reported on the growth and electronic properties of a-Si:H/a-SiO_x SLs [11], Hattori et al. on a-Si:H/a-SiC:H [12], Williams on a-Ge/a-SiO_x [13], and Silva et al. on a-C/a-C:H SLs [14].

In the last eight years there has been a renewed interest on a-SLs and/or MQWs for optoelectronic and photovoltaic applications. A further reduction of dimensionality has been described for those new a-SLs. It has been achieved by keeping one of the sublayers—the barrier—amorphous: usually stoichiometric SiO_2, Si_3N_4, or SiC. Meanwhile, in the other sublayer—the well—some nanotexture is introduced in the form of *nanoprecipitates*, *nanocrystals*, or *nanoclusters* inside an amorphous matrix (usually Si NCs, but also Ge, SiC, or C nanoparticles, to be abbreviated in the following by Si NC, etc.). The barrier layers have been successfully used to limit the growth of the NC layers, and thus they afford a mechanism for NC size selection. It has also been used to improve the quality of NC/matrix interfaces. This scheme leads to complex sublayer/SL structures sometimes called Si NC SLs (nc-SLs), such as (i) Si NC/SiO_x wells combined with SiO_2 barriers, (ii) Si NC/SiC_x wells with SiC barriers, and (iii) Si NC/SiO_xN_y wells with a-Si or SiO_2 barriers and other combinations; results on those structures have been published recently [15–30]. For detailed information about the growing methods and morphology of Si NC SLs, please read the chapters by D. Hiller and by C. Summonte in this book.

A simple model of Si NC SLs considers them as 3D cubic potential wells (following Jiang and Green [31]) that can be used to get some insight into the electronic structure and transport properties as a function of NC size or inter-NC distance. For cubic NC 3D wells the Schrödinger equation is separable into the three dimensions and results in three formally identical equations as for the 1D QW. The expressions and some examples are fully developed in Appendix A. The resulting dispersion relation for the NC SL will be useful for studying transport properties and is:

$$E(\vec{k}) = \frac{2\hbar^2}{m^*}\left(\frac{v_{n_x}^2}{L_x^2} + \frac{v_{n_y}^2}{L_y^2} + \frac{v_{n_z}^2}{L_z^2}\right) - \left(\beta_{n_x} + \beta_{n_y} + \beta_{n_z}\right) - \left(\Delta_{n_x}\cos(k_x d_x) + \Delta_{n_y}\cos(k_y d_y) + \Delta_{n_z}\cos(k_z d_z)\right) \quad (8.8)$$

where symbols have the same meaning as defined before: $d_{x,y,z}$ are the periods in the x, y, and z directions, $L_{x,y,z}$ are the well widths in the x, y, and z directions, and β *and* Δ are related to the overlap integrals in the tight binding approximation as shown in Appendix A.

We finish this section by considering amorphous wells or amorphous nanoparticles embedded in the wells. This is expected to have a tremendous influence on the transport properties of the multilayers. For a more thorough introduction to the electronic structure of amorphous solids, the reader is referred to some excellent books on the subject [32–35]. The characteristic feature of amorphous materials is the element of disorder, which turns into a complete lack of long-range order in their structure. Nevertheless, there is a certain degree of short- and medium-range order which makes that bonding between nearest neighbors becomes quite similar to that of crystalline materials (as for example a-Si which is by far the best known amorphous semiconductor). Similar bonding does not mean "the same" and in fact instead of having fixed bond lengths and bond angles like in a crystalline material, amorphous materials are characterized by a distribution of possible bond lengths and bond angles and this is what introduces disorder. The other important "ingredient" of amorphous materials is the existence of a large number of "unsatisfied" bonds or "dangling bonds" that are a direct consequence of the degree of disorder and/or the method of growth.

We consider now the electronic properties of amorphous materials. What are the conceptual modifications to be done in the band theory of crystalline materials to understand the electronic structure of amorphous materials? First of all, amorphous materials have a band-gap energy which is quite similar to that of the crystalline equivalent (although it is rather fabrication dependent). If we put a-Si as an example, it shows a band-gap energy which is between 1.2 and 1.8 eV which has to be compared with 1.12 eV for crystalline silicon. The existence of a band-gap energy (E_g) becomes evident by considering their optical and electrical properties: (i) we have an absorption threshold at a photon energy equal to that of E_g, and (ii) we can speak about conduction electron and holes in extended states much like carriers in the conduction and valence bands of crystalline materials. Nevertheless, in amorphous materials we cannot classify the extended states by the wavevector **k** because **k** and the related crystal momentum are a direct consequence of the translation symmetry of the crystal (recall Bloch theorem). Thus, the distinction between direct and indirect transitions between conduction and valence bands is meaningless. Moreover, the band gap is not completely empty of electron and/or hole states. The limits of the conduction and valence bands (extended states) are not abrupt and there is a long tail of states that go deep into the band gap (see Fig. 8.5). This can be seen by optical absorption which starts for energies much lower than the expected band-gap energy.

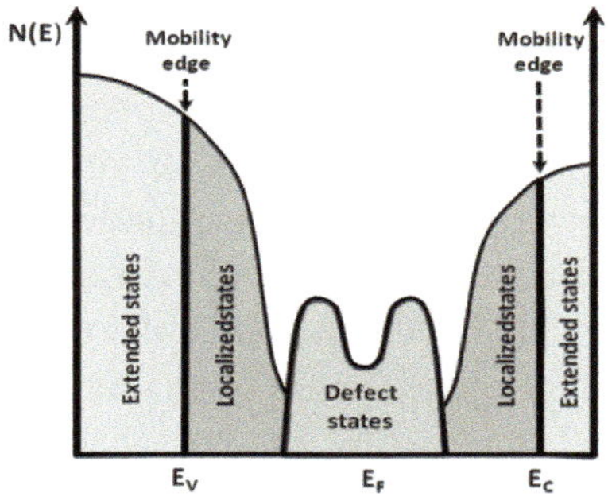

Figure 8.5 Scheme of the dependence of the density of states for an amorphous semiconductor showing the delimitation between the extended and localized states (mobility gap), tail states, and defect states.

Though a "continuation" of the conduction and valence band states, the properties of the tail states are rather different from the carrier transport point of view. Tail states are characterized by a much lower mobility than extended states and can be regarded as localized states (disorder-induced localized states, that is, the wavefunction is localized among a few bond lengths). Thus, the "true band gap" of amorphous solids can be understood as the delimitation between the extended and localized states and can be regarded as a "mobility gap." Thus, we expect that the optical and the mobility gaps are somewhat different in amorphous solids.

Not only that, but also there is a high density of defect states with energy deep in the band gap and mostly located close to the central energy of the band gap (Fig. 8.5). Those defect states (also present in crystalline materials but in much lower quantities) have their origin in the dangling bonds and the impurities, that is, any deviation of the perfect chemical bonding and perfect stoichiometry. Those defects states can be present in such a very high quantity that the Fermi level remains pinned near the middle of the gap allowing for concentration of electrons and holes only in the localized states. In that case, conductivity remains very low and insensitive to doping (hence insensitive to our control). In fact, Spear and LeComber [36] discovered that conductivity increased orders of magnitude when a significant amount of H was introduced when depositing a-Si under certain conditions (by either PECVD or reactive sputtering), passivating most of the defects states. This fact opened the door to use amorphous materials in many applications in the field of thin film electronics and photovoltaics. Apart from high mobility states contribution (extended states) like in crystalline materials, we expect to have in amorphous materials a much more complex transport picture in which localized and defect states participate also in the conduction either actively (hopping conduction) or passively (charge traps and recombination centers).

8.4 Transport in Nanocrystal Superlattices

We shall consider separately vertical and horizontal transport in NC SLs and we shall assume that both of them are uncoupled in a first approximation. By vertical transport we refer to the mechanisms that govern carrier transport in the direction of growth of the SL (axis z). We contemplate that vertical transport takes place only in 1D. On the contrary, we refer to horizontal transport to the mechanisms that govern carrier transport in the plane perpendicular to the growth of the SL (plane x-y). We consider that horizontal transport takes place in the 2D wells. Issues related to vertical transport are important for the behavior of NC SLs applied to solar cells and optoelectronic devices. Issues related to horizontal transport are important for taking into account leakage currents. Horizontal transport is also essential for applications in electronic devices such as memories and MOS transistors.

Transport can be first classified by the sign of the carriers producing the current. It can be either bipolar (both electrons and holes) or unipolar (only one type, usually electrons). Bipolar currents are common in crystalline materials in which electron and hole effective masses and mobilities are comparable, injecting contacts are ohmic and/or conduction and valence band offsets are similar. For a-SLs and nc-SLs these conditions are usually not met, specially the last one, and thus transport is usually unipolar and carried by electrons. However, a particular case that will be addressed is silicon nitride in which the hole current is significant for Si substrate and/or polysilicon injecting electrodes due to the low valence band offset.

In general, in SLs we expect to have a macroscopic transport signature of quantum confinement (QC) if the wavefunction remains coherent at least within a few SL periods. The theoretical framework to account for this effect is called coherent transport. This kind of transport is the dominant one when the inelastic scattering length of carriers in the SL is much longer than the well width a. Crystalline III–V semiconductor SLs with low lattice mismatch ($<1\%$) measured at low temperature have shown some interesting features related to coherent resonant tunneling and negative differential conductance (NDC) for well widths of several nanometers [37–42]. In contrast, coherent transport has been barely seen for amorphous and NC SLs, in part due to the shorter scattering lengths expected for disordered materials.

Perpendicular transport has been studied in a-Si:H, a-SiN_x, a-SiO_x, and chalcogenide SLs. Tsu et al. [38] have shown that QC would be still observable in amorphous layers with well width up to 4 nm and for inelastic scattering lengths lower than 1 nm; in connection to this, they have demonstrated resonant tunneling in a-Si/a-SiO$_2$ SLs. Miyazaki et al. [40] have associated current bumps in the $I(V)$ curves at low temperatures of amorphous multiple QWs of a-Si:H/a-SiN$_x$:H to resonant tunneling between quantized levels in the wells. Evidence of minibands and resonant tunneling has also been reported for a-Si:H/a-SiC$_x$:H [43]. Current bumps and regions of NDC have been reported for Se/Se-Te structures [44]. However, for most of the applications of amorphous and NC SLs working at room temperature, we expect that the dominant mechanisms of conduction are related to incoherent transport that will be treated semiclassically except for sequential barrier tunneling.

8.4.1 *Semiclassical Miniband and Band Transport*

If the barrier width is thin enough (<4 nm) and coherence length is long enough (several nanometers) the transport in the miniband extended states can be significant.

Close to the minimum of the miniband the transport can be approximated by that of a parabolic band with effective mass m^* (see also Eqs. 8A.12–8A.16 in Appendix A):

$$q\vec{F} = m^*\frac{d\vec{v}}{dt} + \frac{m^*\vec{v}}{\tau} \tag{8.9}$$

where $\vec{F}$ is the electric field, $q = -e$ for electrons (a similar equation holds for holes) and τ is the average time between scattering events which allows introducing an ad hoc phenomenological internal friction force proportional to the velocity. An elemental treatment from the Boltzmann equation in the relaxation time approximation gives the same result upon redefinition of an energy-dependent time between collisions. The stationary solution of previous equation is the one given in standard text books for the drift transport at low fields:

$$\vec{v} = \frac{q\tau}{m^*}\vec{F} = \mu_n\vec{F}$$

$$\vec{j}_{\text{drift}} = qn\mu_n\vec{F} = \sigma_n\vec{F} \tag{8.10}$$

where μ_n and σ_n are the electron mobility and conductivity, respectively.

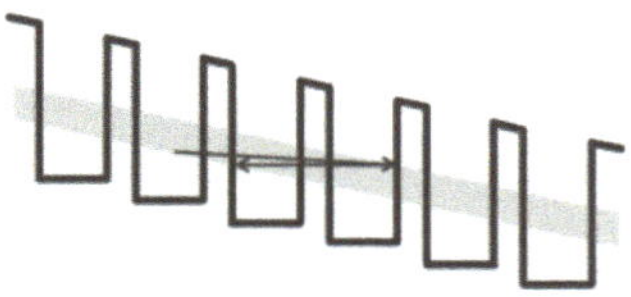

Figure 8.6 Superlattice with applied electric field showing miniband transport and reflection of Bloch electron upon arriving at the zone boundary.

For n-type (same treatment for p-type) crystalline semiconductors with a N_D doping concentration of shallow carrier that are ionized at room temperature, $n = N_D$. For increasing electric fields, the parabolic approximation of the miniband dispersion relation is no longer valid and we must use the whole dispersion expression:

$$E_n = \frac{\hbar^2 k_x^2}{2m_A^*} + \frac{\hbar^2 k_y^2}{2m_A^*} + \frac{n^2 \hbar^2 \pi^2}{2m_A^* a^2} - \beta_n - \Delta_n \cos(k_z d) \qquad (8.11)$$

The group velocity of a Bloch electron in the z direction is given semiclassically [45] by:

$$v_z = \frac{1}{\hbar}\frac{\partial E}{\partial k_z} = \frac{\Delta_n d}{\hbar} \sin(k_z d) \qquad (8.12)$$

which means that the velocity increases up to $k_z d = \pi/2$ and then decreases. Without scattering, the velocity oscillates and results in null current, as electrons are oscillating back and forth with energies between the limits of the miniband producing Bloch oscillations (see Fig. 8.6). If we add the collision time, the net drift velocity would be proportional to the electric field. At a certain electric field, the average drift velocity would reach the maximum near $k_z d = \pi/2$ (the exact point would be depending on average k_z) and then would decrease with increasing electric field. Bear in mind that velocity decreases because the electron has negative acceleration for positive electric field and this means negative effective mass! Because the drift velocity is directly proportional to the current, the miniband transport shows NDC. It can be observed by simply measuring the $I(V)$ curves of the SL. The first reports of NDC in crystalline and amorphous MQWs and SLs appeared in the late 1980s for epitaxial, defect-, and strain-free deposition techniques. Appendix B develops further the semiclassical transport in the extended states of a SL, including specific issues that can be observed only at very low temperatures such as the *Wannier–Stark ladders*.

Bloch oscillations have never been reported for amorphous and NC SLs so far. Mean free paths of carriers in SiO_x, SiN_x, a-Si:H, etc., are of the order of few nanometers and thus scattering induced by disorder and defects kills the oscillating behavior. Nevertheless, there are interesting proposals for Bloch oscillators in crystal SLs that can be taken into account for NC SLs. For example, Daniel et al. [46] proposed a SL in which the electrical field instabilities are suppressed by a direct lateral attachment of a parallel transport channel. Electron trapping which lead to electric field instabilities can be eliminated in these systems since the electrons can flow off into the parallel transport channel.

Additional considerations arise for amorphous and NC SLs. In ternary and quaternary crystalline III–V compound SLs, the disorder introduced by the random lattice mixing is treated as a perturbation and is taken into account as a scattering phenomenon. On the contrary, for amorphous and NC SLs the lattice periodicity is so strongly disrupted that we cannot treat it as a perturbation. We expect some sort of carrier localization and thus the *breakdown of the semiclassical treatment* in the extended states of the miniband. Thus, miniband transport as introduced before is no longer a valid approach in these systems, as indicated by some of the results presented in the literature. As stated by Sibille in Refs. [41, 42] there will be a correlation between state localization and energy. We thus expect the low energy states to be localized and the existence of a mobility gap in the miniband. Thermal and field assisted carrier de-trapping from the low mobility localized states will favor miniband transport and the equations for describing this behavior must account for carrier exchange between localized and extended states. Existence of minibands in a-SLs has been experimentally established as for example in a-Se/a-SeTe a-SLs [47].

In semiconductors and nondefective bulk insulators like SiO_2, transport is mostly due to drift in the conduction or valence bands (Ohmic type at low fields). High-field effects are observed when drift velocity becomes significant in comparison with thermal velocity and when energy exchange between the carrier and the lattice cannot dissipate the energy acquired by the carrier from the field.

High-field effects can be seen frequently as a saturation of the carrier velocity (at around 10^7 cm/s for relevant semiconductors). In the case of SiO_2, high-field effects (>5 MV/cm) induce electron heating that is usually described by a much higher temperature than that of the lattice (hot electrons). For even higher fields other effects appear, such as impact ionization [48].

8.4.2 *Transport with Field-Assisted Carrier Exchange between Localized and Extended States*

In contrast with the shallow carriers in crystalline semiconductors, amorphous and nanocrystalline materials usually have deeper levels that exchange carriers with the extended states:

$$\vec{j}_{\text{drift}} = qn\mu_n\vec{F} = \sigma_n\vec{F}, \quad n = N_t e^{-\frac{\varphi_0}{k_B T}} \tag{8.13}$$

where k_B is the Boltzmann constant and N_t is the density of donors and φ_0 is the energy distance of the level to the miniband or conduction band (extended states). The Fermi level is supposed much deeper into the gap than the donor level, as for nondegenerated semiconductors, that is, Maxwell–Boltzmann statistics holds. This carrier exchange with extended states gives two types of possible transport:

(i) **Miniband transport or transport in the conduction band of the well** if carriers in the wells are delocalized and the donor or localized levels are located in the minigap and exchange carriers with them. The energy distance is usually small and can be thought as an activation energy for the conduction.

(ii) **Band transport or transport in the conduction band of the barrier** material if carriers are strongly localized in the wells or remain in localized states because of disorder. Thus, the donor levels and/or the wells exchange carriers with the conduction band of the barrier material and the energetic carriers are transported above the barriers. This case is frequently found in nanocrystalline materials and NC SLs of SiO_x and SiN_x and can be referred as a high-field *transport in the matrix extended states* [26, 30].

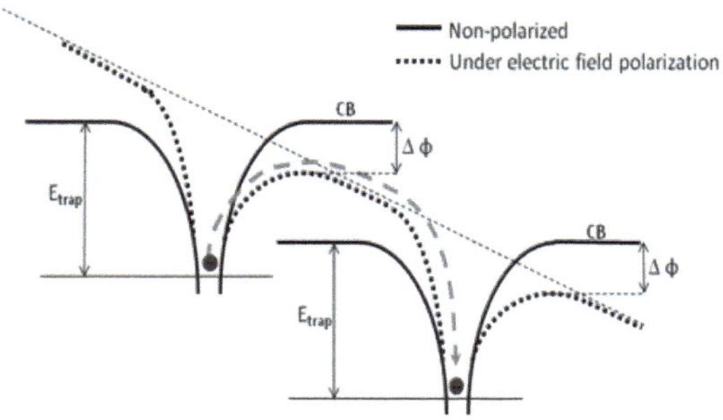

Figure 8.7 Sketch of the mechanism of barrier lowering by the applied electric field for a Coulombic trap known as the Poole–Frenkel effect.

When applying the drift equation to defective solids such as amorphous and insulators for local or tail states one has to take care of a field correction when electric fields are high. The *Poole–Frenkel effect* (P-F) describes how, in a large electric field, the electron ionization from the local or tail state to the extended states of the conduction band is assisted by the electric field. The current detaches from the pure ohmic behavior and becomes field dependent. The potential barrier of the trap is lowered by the electric field with the requirement that the level is positively charged when empty and uncharged when filled. The interaction between the positively charged trap and the electron gives rise to a Coulombic barrier (see Fig. 8.7) [37]. Essentially, the electric field lowers the barrier and pulls the electron to the conduction band. A straightforward calculation of the barrier lowering corrects drift equations to give the well-known P-F law [37]:

$$\vec{j}_{\mathrm{PF}} = qN_t\mu_n\vec{F}e^{-\frac{\left(\varphi_0-\sqrt{\frac{\left|q^3F\right|}{\alpha\pi\epsilon}}\right)}{k_{\mathrm{B}}T}} \tag{8.14}$$

where α is a factor that accounts for deviations of the 1D Coulombic field and dispersion of the distance between traps, and ε is the dielectric constant of the material. A representation of $\ln(J/E)$ versus $E^{1/2}$ should give a straight line and from the slope one can obtain the dielectric constant for consistency check. A study of current versus T allows obtaining the energy distance of the trap to the conduction band φ_0. The P-F effect is the dominant mechanism of bulk conduction in silicon nitride materials and defective silicon oxides (SiN_x, SiO_x). This is not the case for nondefective SiO_2, where the absence of relatively shallow traps leaves the conduction only for the extended states.

8.4.3 *Conduction through Localized States (Hopping by Tunneling)*

We cannot speak of miniband conduction or conduction in the extended states when the thickness of the barriers is relatively large. In this case, coupling between adjacent wells is weak and/or scattering length is short in comparison with the SL period. This happens most often in amorphous and NC SLs. In those cases conduction proceed by *sequential hopping* between adjacent wells. If barriers are high enough (like in SiO_x/SiO_2 SLs) then we can rule out thermionic currents and concentrate only in tunnel currents. For lower barriers (like in SiC_x/SiC SLs) *thermionic classical hopping* can be important and shall be treated with the P-F equations shown in the previous section. We shall concentrate in this section only in the quantum mechanical hopping, that is, hopping by tunneling.

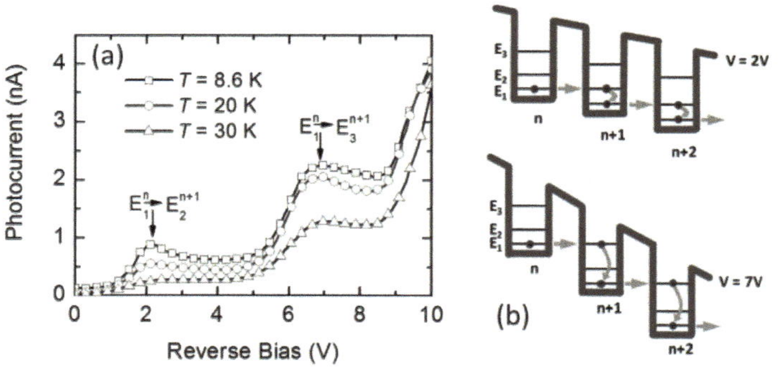

Figure 8.8 (a) Resonant sequential tunneling in photocurrent–voltage characteristics for a 35-period superlattice AlInAs/GaInAs when the detuning introduced by the electric field is equal to the energy difference between two levels of the well [49]. Curves obtained at different temperatures are displayed. (b) Corresponding schemes where sequential tunneling is shown for different voltage values. Reprinted with permission from [F. Capasso, K. Mohammed, A.Y. Cho, *Appl. Phys. Lett.* **48**, 478 (1986)]. Copyright [1986], AIP Publishing LLC.

This situation of hopping by sequential tunneling is sketched in Fig. 8.8 (right). If the states in the QWs are localized (no miniband) and if the applied electric field makes that the E_1 in the QW n matches the E_2 energy level in the QW $n+1$, then the situation is similar with all pairs of adjacent QWs in the SL. Thus, we expect a peak in the current for that particular electric field (voltage) that is related to the effect called resonant sequential tunneling. If the applied electric field is increased further so that E_3 matches E_1 then tunneling reaches another maximum and so on. This effect can be identified as a series of NDC regions between maxima in the $I(V)$ or photocurrent of SLs. The first report on sequential resonant tunneling in SLs dates back to 1986 by F. Capasso et al. (Fig. 8.8, left) [49]. A derivation of resonant sequential tunneling plus relaxation to the bottom level of the potential well by photon emission is the basic principle of the quantum cascade laser [50]. This transition is usually called intersubband transition and the carriers are not recombined and disappear like in interband recombination, on the contrary they still reside in the QW and can be further transported.

It is important to state that we shall not refer to coherent tunneling in the MQW; for it, the wavefunction preserves the phase (i.e., there is not inelastic scattering) between reflections at the barriers. In this situation one must add the amplitudes of reflected and transmitted waves for allowing interference effects and afterward calculate the current. On the contrary, for sequential tunneling we calculate first the probabilities from the amplitudes and afterward we add the probabilities for the current calculation. Coherent resonant tunneling is important for the treatment of two or three QWs and recently several resonant tunneling devices with excellent NDC properties have been proposed [51].

Consequently, we shall deal only with incoherent resonant tunneling and the conduction would proceed as shown in Fig. 8.8, that is, through hopping between states created by the detuning of the band by the electric field. For quantifying the current we need to calculate the transparency of the barriers. Let us try first finding a compact model to present in a simple way the results. For two incoherent wells the transmission can be expressed as:

$$T_{12} = \frac{T_1 T_2}{1 - R_1 R_2} \frac{R_{12}}{T_{12}} = \frac{R_1}{T_1} + \frac{R_2}{T_2} \tag{8.15}$$

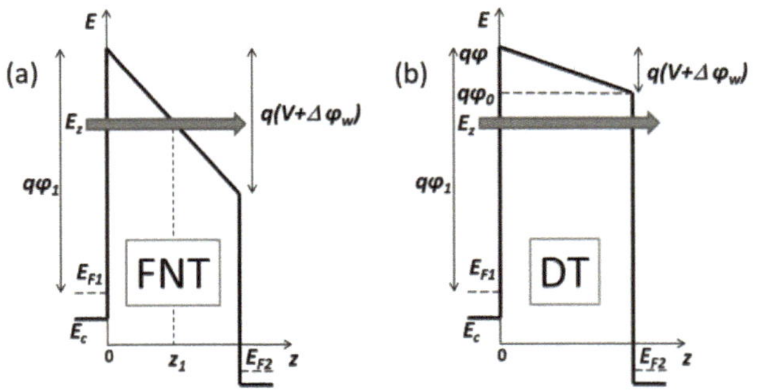

Figure 8.9 Sketch showing the barrier shapes for Fowler–Nordheim and direct tunneling mechanisms.

where T_i is the transmission of the single barrier i and R_i is the reflection of barrier i. Generalizing to λ barriers per unit length, that is, a total of λL barriers, where L is the total thickness of the SL it reads, one can express the total transmission T_t as:

$$T_t = \frac{T_i}{\lambda L(1 - T_i) + T_i} \tag{8.16}$$

Thus, we have reduced the problem to calculating the transparency of a single barrier. The electron (hole) essentially sees a trapezoidal energy barrier and has two ways of crossing it depending of its energy: through the triangular part (Fowler–Nordheim tunneling, FNT) or the rectangular part (direct tunneling, DT) (see Fig. 8.9). Usually, when studying the transport through an insulator—that is, the barrier layers—one has to deal with such a problem. It is shown elsewhere that the Fowler–Nordheim and the direct tunnel currents will limit conduction in capacitors of insulators like SiO_2 depending on the thickness and the applied electric field [52, 53]. In those capacitors, the supply of carriers from the electrodes is restricted due to the barrier they have to overcome to access the extended states of the conduction band.

Using the well-known Wentzel–Kramers–Brillouin (WKB) approximation we obtain the transmission coefficients:

$$T_{\mathrm{FN}}(\varepsilon) = \exp\left(-\frac{8\pi\sqrt{2m^*}}{3hqF}(\phi - \varepsilon)^{\frac{3}{2}}\right)$$

$$T_{\mathrm{DT}}(\varepsilon) = \exp\left(-\frac{8\pi\sqrt{2m^*}}{3hqF}\left((\phi - \varepsilon)^{3/2} - (\phi_0 - \varepsilon)^{3/2}\right)\right) \tag{8.17}$$

where b is the thickness of the barrier. We assume that the electrodes are degenerated (metals) and we shall use the Fermi–Dirac distribution (a step at 0 K). Electrons can be regarded as quasifree with effective mass m^*. Finally, we can use the Tsu–Esaki formula to calculate the current through a single barrier in both cases:

$$J = \frac{4\pi m_e^* q}{h^3} \int_{E_{\min}}^{E_{\max}} T(E)\, N(E)\, dE$$

$$N(E) = \int_0^{\infty} (f(E - E_{F1}) - f(E - E_{F2}))dE_\rho,\ dE_\rho = 4\pi k_\rho^2 dk_\rho \tag{8.18}$$

where E_{F1}, E_{F2} are the Fermi levels at the left and at the right of the barrier, respectively, with $E_{F1}-E_{F2} = \text{eV}$, V being the applied voltage. $N(E)$ is the supply function which stands from integrating the density of states of the metal along the x and y directions. A further integration along the energy in the z direction results in the following useful expressions:

$$J_{FN} = \frac{q^3}{8\pi h\phi_0} F^2 e^{-\frac{8\pi\sqrt{2m^*}}{3hqF}\phi_0^{3/2}}$$

$$J_{DT} = \frac{q^3}{8\pi h\left(\phi_0^{1/2} - (\phi_0 - qFb)^{1/2}\right)^2} F^2 e^{-\left[\frac{8\pi\sqrt{2m^*}}{3hqF}\left(\phi_0^{3/2} - (\phi_0 - qFb)^{3/2}\right)\right]} \tag{8.19}$$

The so called Fowler–Nordheim plot consists in representing $\ln(J/F^2)$ versus $1/E$ (or versus $1/V$) and should yield a straight line for a FNT current. From the slope of this plot we can extract the barrier height. This current is characteristic of silicon oxides and other dielectrics with a small number of traps. Particularly, the barrier between Al (or polysilicon) and SiO_2 is very large ($\Phi_0 =$ 3.2 eV) and negligible currents are obtained for fields of less than 5 MV/cm. If the dielectric barrier has traps, the tunnel injection can be assisted by the traps and consequently an effective reduction of the injection barrier height occurs. Also, the FNT expression may have two (or more) barrier heights that will appear subsequently. This particular behavior can be easily detected by a slope change in the $I(V)$ characteristic, once plotted in the FN representation

[52, 53]. Fowler–Nordheim *trap-assisted tunneling (TAT)* is basically a two-step sequential process in which injected electrons tunnel first to the traps or wells inside the material and afterward to the conduction band of the matrix. TAT is important in materials with nancocrystals and although conceptually very similar to Fowler–Nordheim, there are some significant differences that are developed in Appendix C.

Finally, in the case that the density of traps is too large, the current can follow a preferential path of leakage through defects (percolative filamentary paths) by hopping from one trap to the next one with assistance by phonons; this type of conduction is called Mott variable range hopping [32] and is typical of amorphous semiconductors at low temperatures (no carriers in extended states) and at low voltages (no tunnel injection from the electrodes and/or ohmic contacts).

8.4.4 *Injection and Space Charge–Limited Currents*

Fowler–Nordheim and tunnel injection in single layers are usually regarded as an *injection-limited conduction* type in insulators as the only interfaces are those between the electrodes and the layer. The barriers formed depend on the work function difference (work function is the distance between the Fermi level and the vacuum level). FNT and DT as injection processes occur when the barriers for injection are high and there are not thermal carriers (thermionic) promoted to energies high enough to jump over the barrier. The expressions for FNT and DT currents are those of the previous paragraph for a single layer or interface with the appropriate barrier.

Nevertheless, for low energetic barriers, such as in a *p–n* junction, in a Schottky diode or in a MOSFET channel, injection can occur for carriers with energies above that of the barrier. For low fields this is just the typical thermionic or Richardson current:

$$J_{\mathrm{TI}} = \frac{4\pi q m^*}{h^3} (kT)^2 \, e^{-\frac{\phi_0}{kT}} \tag{8.20}$$

And for medium and high fields there is a barrier lowering (similar to that of P-F effect) due to the electric field and the image effect. The current is called the Schottky current or field enhanced thermionic current:

$$J_{\mathrm{TI}} = \frac{4\pi q m^*}{h^3} (kT)^2 e^{-\frac{\left(\phi_0 - \sqrt{\frac{|q^3 F|}{\alpha\pi\epsilon}}\right)}{kT}} \tag{8.21}$$

Charging effects and other space charge effects are important in insulator materials used as barriers for nc-SLs. Also, not only defects in the insulator are responsible for the charge trapping, but also the NCs themselves (or their surrounding media). The shift of the $I(V)$ and $C(V)$ characteristics due to the trapped charge can be used to quantify its amount and profile. The charge starts to be trapped close to the injecting interface and a trap current will be shown as a displacement current with an *RC* time constant. The so called tunnel front separates spatially the filled from the nonfilled traps and advances logarithmically with time due to the fact that the tunnel probability depends exponentially with distance [54]. Trapped charge is deleterious mainly because of (i) instabilities of operation and (ii) creation of defects that can eventually percolate and breakdown the device. Charge trapping at the many heterointerfaces of a-SLs makes that large number of bias-induced metastable conductance effects may appear especially when dipole layers and built-in fields are formed, and even including switching phenomena (like memristor-type effects) [55]. Other space charge effects are related to doping and depletion of carriers due to diffusion and creation of internal electric fields such as in $p-n$ junctions. These effects are not important for insulators in which doping is negligible. Otherwise, space charge effects may also show off for low doping and low background conductivity, and also when injecting high currents due to local electron–electron interaction. The space charge–limited current for a solid without traps has the well-known *Mott–Gurney quadratic law* which is modulated by a Frenkel-type expression when considering a single set of traps with density N_{t} at a distance E_{A} from the conduction band:

$$J = \frac{9}{8}\mu\epsilon\epsilon_0 \frac{V^2}{L^3}\theta_0 \exp\left\{\frac{0.891}{k_{\mathrm{B}}T}\left(\frac{q^3 V}{\pi\epsilon\epsilon_0 L}\right)^{1/2}\right\}\frac{\rho_{\mathrm{f}}}{\rho_{\mathrm{f}}+\rho_{\mathrm{t}}}$$

$$\theta_0 = \frac{N_{\mathrm{c}}}{N_{\mathrm{t}}}\exp\left(-\frac{E_A}{k_{\mathrm{B}}T}\right) \tag{8.22}$$

where ρ_{f} and ρ_{t} are the free and trapped charge density, respectively, and N_{c} is the amount of carriers in the conduction band.

8.4.5 *Horizontal Transport*

It was soon recognized that in crystalline SLs it is possible to spatially separate free carriers and impurity doping atoms, such as in modulation doped GaAs–GaAlAs SLs, for increased carrier mobility by orders of magnitude in the undoped materials [56]. This is extremely important for new devices such as the high-electron-mobility transistors (HEMTs) used in high-frequency applications, for which carrier transport proceeds in-plane parallel to the heterointerface. These effects also stimulated the research in a-SLs. Thus, the in-plane transport in a-SLs has also been studied for doping schemes similar to those of modulation doping. However, there are only few reports on the rise of mobility in the intrinsic layers of *n–i–p–i*-type a-Si:H SLs at low temperature [57]. Furthermore, it was found that the reduction in the density of states (2D), the increase of the band-gap energy (i.e., quantum confinement) and carrier scattering at the interfaces are even stronger than the effect of modulation doping providing for the reduced conductivity if compared with the bulk layer (BL) [58]. Nevertheless, other groups have reported an increase in the lateral conductivity of a-Si:H/a-SiN_x SLs with decreasing a-Si:H layer thickness which has been ascribed to electron transfer (a kind of modulation doping) from a-SiN_x to the a-Si:H wells [59].

For the particular case of nc-SLs there are few reports in the literature for the in-plane conductivity. For example, CdSe nanocrystalline layers separated by a-SiO_x barriers have been studied in [60]. While carrier confinement along the SL axis takes place in the CdSe layers, carriers do not undergo 3D confinement in the CdSe NCs because of the low potential barrier between them in the plane. Thus, horizontal transport in those nc-SLs turns out to be similar to that of polycrystalline materials with reduced dimensionality. Interfacial potential barriers are formed at the grain boundaries due to the large number of defects and a decrease of conductivity with thickness of the layer is ascribed to a reduction of CdSe NC size [61].

For nc-SLs one expects that the NCs in the horizontal plane are not ordered and consequently the interdot distance has a certain distribution (between some limits). For the particular case of SiO_x

and SiN_x materials the average interdot distance value depends on the silicon excess added to the compound and thus on the density of NCs. When the volume fraction of the Si excess precipitated is large enough, there start to appear some percolation paths. These percolation paths are formed for Si volume fraction larger than 30% for 3D layers [62], even though theoretical value for random structures is 15% [63]. Those percolations are essentially highly conductive paths of Si NCs in contact. A sharp rise in conductivity is expected right at the percolating threshold of volume fraction. Above threshold it shows an Ohmic type of conduction with intergrain potential activation energy as a consequence of the semiconducting nanocrystalline precipitated phase. In contrast, the in-plane conduction for precipitated fractions below the percolation limit is expected to follow a behavior closer to an amorphous semiconductor and/or a dielectric. Thus, we presume that some kind of hopping can occur between Si NC and/or defect states, either by thermally activated hopping (P-F) for energies of donor or localized states not far from the extended states, or by tunneling hopping for large barrier offsets (Fowler–Nordheim for high-field and direct tunneling for low fields). There are many works devoted to conduction in BLs with Si NCs and to review this subject it is outside of the scope of this chapter [52, 54]. We only introduce the transport following a simple model by Ron and DiMaria [64]; refer to Fig. 8.10 for details of geometry.

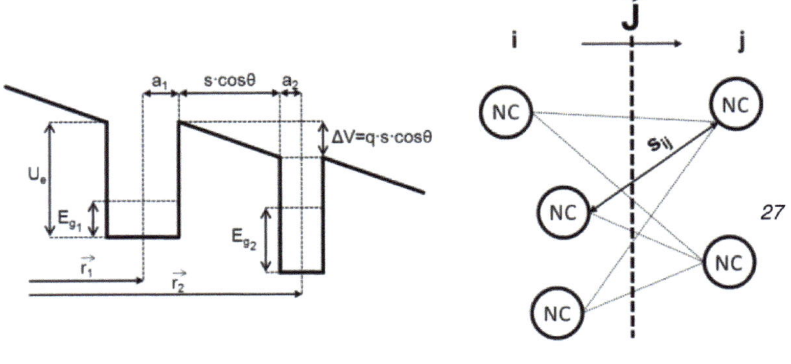

Figure 8.10 Sketch showing the barrier shapes and parameters for relative dot position and current [64].

The current density J is equal to the sum of electronic charges which cross the plane per unit time (Fig. 8.10, right). The basic assertion of the percolation treatment is that the effective current link T_{ij} (tunnel probability) between NCs can be replaced by $T_c \approx e^{-\zeta}$, with ζ an effective link-normalized distance, and exponents $\ll 1$ are extremely rare, and those exponents $\gg 1$ are not effective for conduction. The essential proposition of the percolation method is that the value of the critical percolation exponent, ζ, can be estimated by geometrical arguments. Ron and DiMaria [64] arrive to the following expressions:

$$J = qN_{\omega}l^{*}\frac{E_g}{\hbar}e^{-\zeta}$$
$$\zeta = 2\left[\frac{2m_e^*}{\hbar^2}\right]\frac{\left[\left(U_e - E_{g_1}\right)^{3/2} - \left(U_e - E_{g_1} - qFs_c\right)^{3/2}\right]}{\frac{3}{2}qF} \qquad (8.23)$$

where U_e is the depth of the potential well, l^* is considered a "typical length" of the order of the interwell distance and N_{ω} is the average number of islands per unit volume.

Result that is similar to the Fowler–Nordheim and direct tunneling results. The s_c is the critical length:

$$s_c = \left(\frac{3v_c}{4\pi N_{\omega}}\right)^{1/3} \qquad (8.24)$$

For low fields and low temperature, the model yields the famous Mott result for variable range hopping conduction in amorphous semiconductors. For high fields, the model gives a pure Fowler–Nordheim current which transverses the sample through the weakest links (percolation model):

$$J = CqN_{\omega}s_c\frac{E_g}{\hbar}e^{-\bar{F}/F} \qquad (8.25)$$

C is a numerical factor with value of the order of unity.

The next sections are devoted to the presentation of experimental results on amorphous and NC SLs. The results will be analyzed in the framework of the theory and models introduced in the present and previous sections.

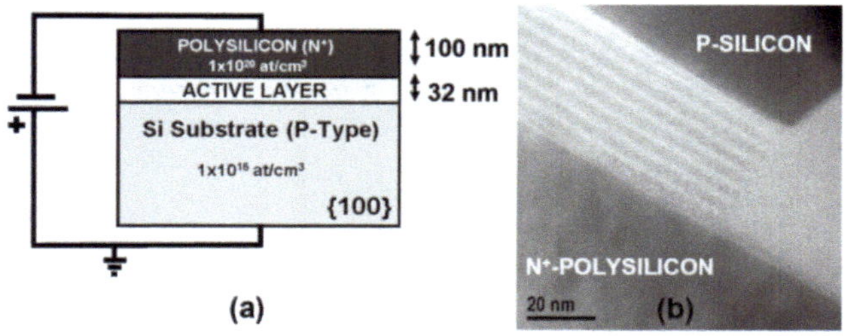

Figure 8.11 (a) Device geometry and preferred polarization and (b) TEM image of the SL embedded in a capacitor device.

8.5 Vertical Transport in SRO/SiO_2 Superlattices

This section is dedicated to illustrate the conduction in nanocrystalline SLs with some specific samples developed for electroluminescent (EL) purposes. In particular, Er^{3+} ions are used here as luminescent centers, taking advantage of the convenient EL emission at 1.5 μm provided when such rare earth ions de-excite from the first excited level [65]. Active layers have been produced by alternating silicon oxide/silicon-rich silicon oxide (SiO_2/SRO) layers sandwiched between a *p*-type silicon substrate and a highly *n*-doped polysilicon electrode ($\sim 10^{20}$ at/cm^3) forming an NMOS-like capacitor, as can be seen in Fig. 8.11a. A device with a bulk SRO monolayer (BL) is also used as a reference to distinguish between bulk and SL conduction. Both layers have been fabricated by means of plasma enhanced chemical vapor deposition (PECVD) in a standard CMOS line (at CEA-LETI in Grenoble, within the framework of EU project HELIOS). The bulk SRO layer was deposited with an average 12 at.% silicon excess. For the case of the SL, the Si excess may be referred to each single SRO layer of the SL (20 at.%) or to an average value of the whole SL, that is, taking into account the SiO_2 barriers also (12 at.% Si excess in that case). In particular, the SL gate stack is as follows: $\{[2\ \text{nm}(SiO_2, 0\%) + 3\ \text{nm}(SRO, 20\%)] \times 6\} + 2\ \text{nm}(SiO_2)$, resulting in a nominal thickness of 32 nm. Figure 8.11b shows a transmission electron microscopy (TEM) image of one of these devices, revealing the SL gate stack. In addition, both the BL and the SL structure have been implanted with a flat erbium profile with a concentration of 5×10^{20} at./cm^3.

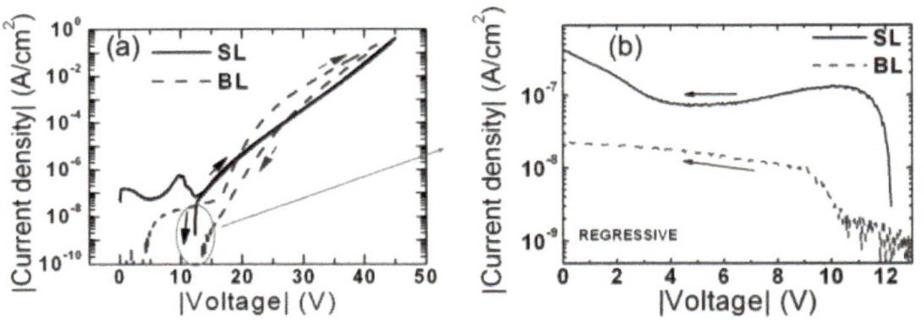

Figure 8.12 (a) Comparison between the current density (J) versus applied voltage (V) for superlattice (solid line) and the bulk layer (dashed line) in NMOS-like capacitors. (b) Current–voltage characteristic at low voltages when the sign of the current has changed.

Subsequently, a post annealing treatment was performed in order to mitigate implantation-induced defects. Further fabrication details can be found elsewhere [53].

Figure 8.12a depicts the quasistatic $J(V)$ characteristic of both the SL and the BL structures. Two different regimes can be identified: one at low voltages, where there is an important contribution of the displacement current (almost constant with V), and another one at voltages in excess of a threshold voltage, V_{th}, where the real transport of current across the capacitor predominates. Therefore, it has to be noticed that the V_{th} in the SL is much lower than the BL (i.e., roughly 5 V lower). Thus, the onset for conduction in the SL proceeds at a lower electric field. In addition, the current slope of the SL is smaller than that of the BL, suggesting different transport mechanisms. Moreover, the BL displays a significant shift between progressive and regressive sweeps which indicates charge trapping. On the other hand, charge trapping is nearly absent in the SL and we believe this is the reason for much higher operation lifetimes in comparison with the BLs. In fact, the BL showed a catastrophic failure only after a few minutes of operation.

To further illustrate the charging effect, in Fig. 8.12b is shown the $J(V)$ curve at low V in the regressive sweep where current is reversed to discharge the sample. The discharging current is much higher in the SL indicating that trapped charge is evacuated much faster. So, the cumulative effects of the trapped charge are less

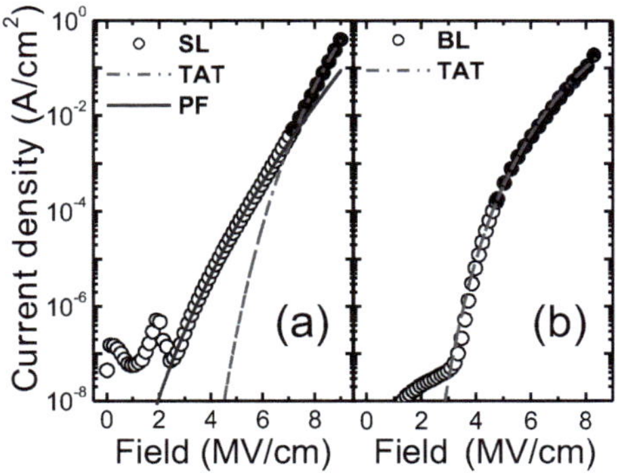

Figure 8.13 (a) PF and TAT conduction mechanism at low and high voltages, respectively, for SL and BL devices. (b) Trap-assisted tunneling mechanism for the BL. Black solid circles indicate the range where the EL emission takes place in both samples [66]. Reprinted with permission from [J.M. Ramírez, Y. Berencén, L. López-Conesa, J.M. Rebled, F. Peiró, B. Garrido, *Appl. Phys. Lett.* **103**, 081102 (2013)]. Copyright [2013], AIP Publishing LLC.

deleterious for the SL, especially if the device works under pulsed voltage conditions.

A comprehensive study on the conduction mechanisms in both the BL and the SL has been carried out. The fittings of Fig. 8.13a show that the P-F bulk limited mechanism governs the transport in the SL at low voltages, whereas at high voltages, a Fowler–Nordheim TAT predominates. This mechanism change takes place when at high voltages the electrode is not able to supply all the carriers needed for the bulk conduction and thus the conduction is limited by the rate at which carriers can tunnel from the electrode. Only TAT type of conduction can be seen in BLs (Fig. 8.13b). All this is clearly the signature that the SL is much more conductive due to intrinsic carriers than the BL; it might be the interfacial states that are relatively close to the conduction band of the SiO_2 matrix.

For further insight, the region with significant electroluminescence (EL) from the Er doping atoms is indicated by the filled black circles. Notice that the EL appears solely in the TAT regime and not during the P-F conduction. Although we will come back to this point later, this feature also provides a clear fingerprint of the main

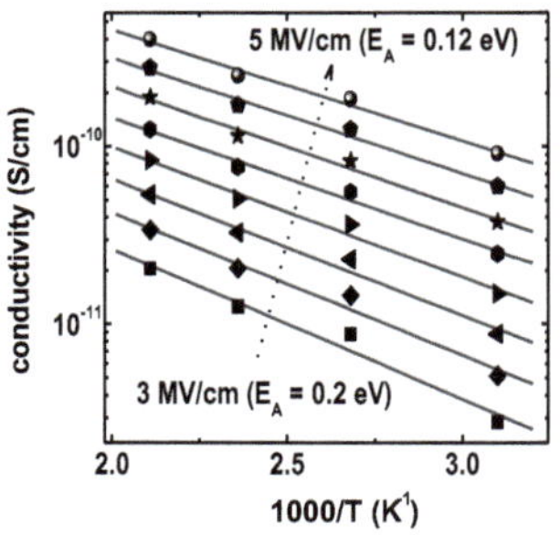

Figure 8.14 Arrhenius plot of the conductivity of the superlattice in the Poole–Frenkel conduction region.

excitation mechanism of Er ions, which is expected to be impact ionization by the hot electrons accelerated in the extended states.

The relative dielectric permittivity (ε_r) can be extracted from the P-F region (see Eq. 8.14) and is found to be $\varepsilon_r = 11$ (SiO_2 has $\varepsilon_r = 4$ and silicon has $\varepsilon_r = 12$). This is a value quite high for a mixture with 20 at.% Si excess but it is reasonable to expect this in our SLs, as the Si excess is concentrated only in the SRO wells which locally have much more than 12 at.% Si excess. A barrier height of $\phi_B = 2.1$ eV can be extracted from the TAT region (see Eqs. 8C.1–8C.10 in Appendix C). This barrier value is in further agreement with previous studies of SRO conduction. Therefore, the difference with the polysilicon-SiO_2 barrier (3.2 eV) is due to the trap assisted tunneling provided by defects and/or Si NCs to injected carriers from the electrode. Detailed analysis on the transport mechanisms of this particular SL structure can be found in Ref. [66].

Temperature dependence studies were performed to ascertain whether conduction can be assigned to PF conduction mechanism. In Fig. 8.14 we display the conductivity (σ) of the SLs in the P-F region (low to medium voltages) at different temperatures, which presents an exponential behavior with the inverse of temperature that can be fitted by means of the Arrhenius law

$$\sigma(T) = \sigma_0 e^{-\left(\frac{E_A}{k_B T}\right)}, \tag{8.26}$$

where k_B is the Boltzmann constant, σ_0 a pre-exponential factor and E_A the activation energy of the conduction mechanism. We obtain 0.2 eV for the distance of the trap site to the conduction band of

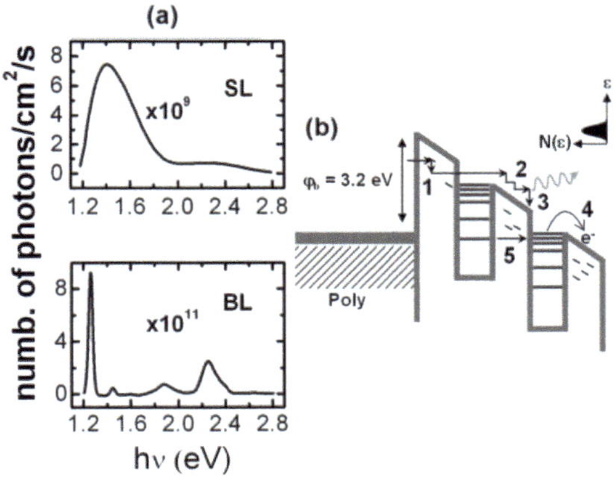

Figure 8.15 (a) Visible Er^{3+} spectra in the bulk layer and the superlattice. (b) Energy band diagram of the SL showing the Er excitation by hot electrons generated in the TAT regime. The approximate electron kinetic energy distribution is schematically shown at superlattice–substrate interface [66]. Reprinted with permission from [J.M. Ramírez, Y. Berencén, L. López-Conesa, J.M. Rebled, F. Peiró, B. Garrido, *Appl. Phys. Lett.* **103**, 081102 (2013)]. Copyright [2013], AIP Publishing LLC.

the SL, which is the fitting value at low voltages. At higher voltages, as can be seen on the plot, the energetic distance is lower, which is expected due to the field lowering of the barrier in the P-F effect [67]. This result is in agreement with the fact that the Si NCs introduce relatively shallow trap levels in the SiO_2 band gap.

The carriers in the P-F conduction regime remain at energies close to $k_B T$, while in the TAT (i.e., Fowler–Nordheim assisted by traps) the carriers become hot after being accelerated in the conduction band. This can be distinguished if we take care of the visible EL spectra. As can be seen in Fig. 8.15, the visible emission of the BL is composed of peaks coming from higher Er excited levels, while in the SL these peaks are absent and only the emission from Si NCs arises. The explanation is simple: in the SL the carriers remain at $\sim k_B T$ and have not enough energy to excite the Er^{3+} ions to the highly energetic levels responsible of the visible emission. For further details about the emission properties of Er-doped SLs, please refer to Ref. [66]. A further insight into the conduction properties is expressed in Fig. 8.16, where the infrared EL emission is represented

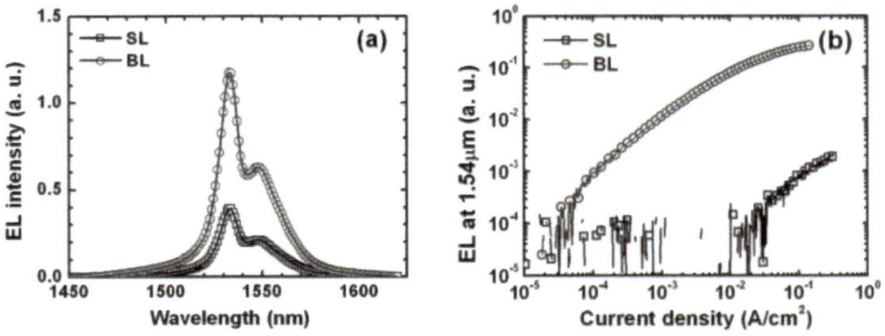

Figure 8.16 (a) Typical Er spectrum at 1.54 μm of both the bulk layer and the superlattice, taken at 10^{-4} A/cm^2 and 0.1 A/cm^2, respectively. Reprinted with permission from Ref. [68]. (b) EL at 1.5 μm as a function of current density [66]. Reprinted with permission from [J.M. Ramírez, Y. Berencén, L. López-Conesa, J.M. Rebled, F. Peiró, B. Garrido, *Appl. Phys. Lett.* **103**, 081102 (2013)]. Copyright [2013], AIP Publishing LLC.

as a function of the injected current for both the BL and the SL. The BL is much more efficient, by some orders of magnitude, than the SL. This is in agreement with the fact that carriers can accelerate much easier in the bulk system, hence obtaining the required energy for Er excitation (0.8 eV) in few nanometers. This fact provides for higher probability of Er^{3+} impact excitation for each single injected electron. On the contrary, the SL would act as an efficient "thermal driver" for injected electrons in the conduction band, diminishing the average kinetic energy below 1.26 eV (see Fig. 8.15a) and, hence, the thermal release of electrons within the structure. Then, bearing in mind that poorly energetic electrons are being injected in the SL, we postulate that another excitation mechanism different from direct impact excitation provides the Si NC emission observed at the top of Fig. 8.15a [30]. Although the origin of such an emission has not been deeply investigated for the present case, we speculate on the possibility of an electron–hole radiative recombination within a previously ionized Si NC (or defects related to its surrounding media). Thus, devices containing a SL structure are not good systems for accelerating injected carriers, probably due to the SL geometry that promotes additional scattering centers. In any case, a wide variety of applications can be foreseen, especially those where tight control of kinetic energy of injected carriers in the conduction band is required.

8.6 Transport in SRON/SiO_2 and SRC/SiC Superlattices

8.6.1 *Horizontal Transport in SRC/SiC Superlattices*

This section aims at further illustrating the transport in NC SLs by presenting some experimental results concerning the electrical and electro-optical properties of NC SLs targeted at photovoltaic applications. In particular, we focus on the lateral and vertical transport in SRON/SiO_2 and SRC/SiC SLs. SRON stands for silicon rich silicon oxynitride, SiO_xN_y, and SRC stands for silicon rich silicon carbide, SiC_x. Both silicon rich materials, upon annealing at temperatures exceeding 1000°C, undergo a phase separation in which the silicon excess precipitates in the form of NCs (see chapter by Hiller). For annealing temperatures lower than this, the precipitates remain amorphous or partially crystallized [23]. All the samples introduced here have been produced at IMTEK University of Freiburg (SRON SLs, see Ref. [23]) and at CNR-IMM Bologna (SRC SLs, see Ref. [29]) in the framework of the EU project NASCEnT.

We first introduce the horizontal transport in the SRON/SiO_2 SLs. The SLs have been deposited by PECVD on silicon substrates thermally oxidized with bilayers consisting of 3.5 nm SRON layers and 2 nm SiO_2 barriers. The oxygen-to-silicon ratio (x in SiO_xN_y) was varied between 0.64 and 0.93, and the N concentration remained constant in the range $y = 0.23$–0.25 [69]. The stoichiometries employed here lead to a silicon excess in the SRON regions of 27.1 and 16.7 at.% for $x = 0.64$ and 0.93, respectively (we have considered that $[Si]_{excess} = (1-0.5 \cdot x-0.75 \cdot y)/(1 + x + y)$, x and y being the [O]/[Si] and [N]/[Si] ratios, respectively). A bulk SRON layer was deposited for reference purposes. The SLs were annealed for 1 hour at 1150°C to precipitate and crystallize the silicon in excess in the form of NCs. After annealing, the samples were laterally contacted with a TiPdAg metal stack, as shown in Fig. 8.17a, the nominal separation between contacts being 50 μm. The thick thermal oxide deposited on the Si substrate and the particular geometry of the contacts ensure that all the current flows laterally through the SL, avoiding any leakage currents through the substrate.

The $J(V)$ curves of the horizontal devices are shown in Fig. 8.17b. It is clear that the conduction is ohmic in this range

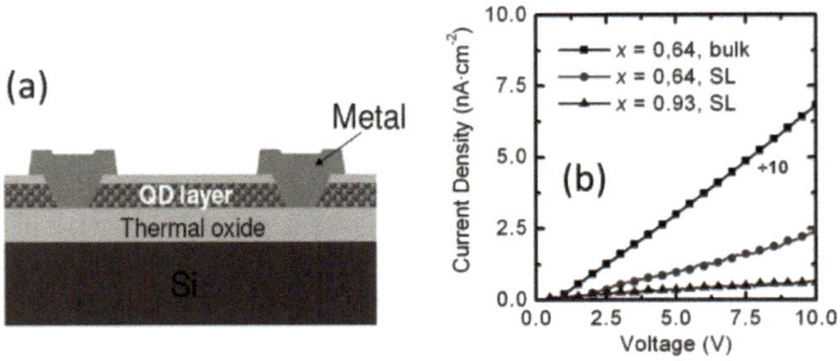

Figure 8.17 (a) Cross-sectional scheme of the devices employed for the horizontal characterization of SRC/SiC layers. (b) J (V) curves corresponding to the studied devices, containing different SiO_x stoichiometry (x).

of applied voltage. In addition, the regressive sweep presented no hysteresis. Therefore, the limiting factor in these devices is the employed contact material for injecting and collecting the charge. The conductivity values extracted from the characteristics are shown in Table 8.1. It has to be mentioned that, at very high applied voltage (not shown in the graph), we observed an exponential trend, which is an indicator of a Schottky current. This, again, corroborates the current dependence on the selected contacts.

In addition, the observed increase in current density with the Si excess, as well as its dependence with the sample structure (the BL is far more conductive than the equivalent one with SLs), leads to a series of hypothesis for the carrier transport taking place in the horizontal direction. As expected, and as can be seen in the

Table 8.1 Conductivity values obtained from the ohmic curves of devices containing different active-layer structures (bulk or superlattice) and different [O] to [Si] ratios (x in SiO_xN_y). The Si excess is also presented, whose values, in the case of SL samples, were calculated considering both the whole SL structure and only the SRON layers

[O]/[Si] (x in SiO_xN_y)	Average [Si] content (at.%)	σ ($S\cdot cm^{-1}$)
0.64 (bulk)	27.1	3.4×10^{-11}
0.64 (superlattice)	17.2	1.2×10^{-12}
0.93 (superlattice)	16.8	3.2×10^{-13}

figures and the conductivity values, there is a strong dependence of conductivity on the SRON composition or silicon excess, obtaining higher conductivity values for a larger amount of Si. The conductivity of bulk sample (with the highest Si excess) is close to amorphous or intrinsic silicon, stating that nanograins coalesce and that the conduction proceeds via the extended states of the nanocrystalline semiconducting material. Consequently, the conduction through these systems can be interpreted considering an in-plane transport through a single QW instead of hopping between quantum dots. Thus, the lateral conduction can be thought to proceed through percolative paths of silicon NCs within this range of composition. Hopping would dominate conduction only for very small silicon excess when dots are far apart. The results of the bulk sample agree with this interpretation. This device displays a larger current density than the observed one for the equivalent SL structure. We believe that the SL is a partially ordered nanocrystalline structure and the bulk sample is a totally disordered one. The random location of the NCs in the bulk sample makes it more probable that the electrodes are connected with numerous paths of high conductivity due to very near or almost touching silicon NCs (weakest barriers for the current).

The case of lateral transport in SRC/SiC SLs is quite analogous. The main difference lies in the fact that the SiC matrix (with a band-gap energy between 2.5 and 3 eV) is far more conductive than SiO_2. This implies a larger current density through this material than for SRON SLs. For this study we used devices with varying SRC layer thickness (from 2 to 4 nm) while holding the SiC barrier thickness at 6 nm. The stoichiometry of the SRC layers was fixed, with $x = 0.85$ (in Si_xC_{1-x}). The Si NC precipitation was achieved by annealing the SLs at 1100°C. Finally, a metallization of Ti/Pd/Ag was carried out, the intercontact distance being 1 mm. A sketch of the employed structure is presented in Fig. 8.18a.

In analogy with the results from horizontal SRON/SiO_2 devices, SRC/SiC devices present a linear $J(V)$ characteristic (shown in Fig. 8.18b), both in direct and reverse bias polarization, that clearly corresponds to an ohmic behavior. Moreover, it was found that the obtained conductivity values range from microcrystalline SiC to microcrystalline Si, scaling accordingly to the amount of Si presented

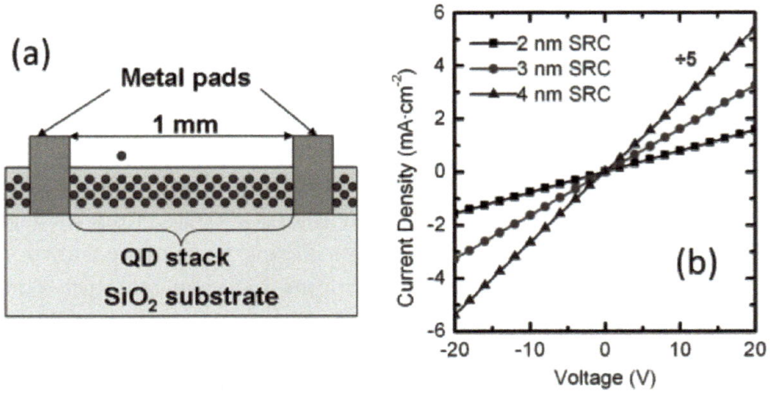

Figure 8.18 (a) Cross-sectional scheme of the devices employed for the horizontal characterization of SRC/SiC layers. (b) $J(V)$ curves corresponding to the devices containing SRC/SiC superlattices, with different SRC layer thickness.

in the SRC/SiC active layers (i.e., the SRC layer thickness). In Table 8.2 we present a summary of the conductivity and resistivity values obtained for the SRC/SiC SLs, compared to the ones in *i*-Si and *i*-SiC. It is evident from the conductivity values that SRC/SiC SLs are much more conductive (by many orders of magnitude) than SRO/SiO_2 ones, due to the fact that the SiC matrix has a nonnegligible contribution to the conduction and the potential barriers are much smaller. This ohmic behaviour in the SRC states, as in the SRON samples, that the lateral conduction proceeds via the extended states and the wells behave like a semiconducting material. Thus, the silicon excess is high enough so that conduction proceeds via percolative paths and not via hopping between nanocrystallites.

As aforementioned in Section 8.5, further useful information can be extracted from a study of the conductivity as a function of the temperature. An Arrhenius plot is shown in Fig. 8.19 for the conductivity of all studied devices. The activation energy can be directly obtained from the experimental data, giving practically identical values around 0.26 eV, independently of the thickness. This behavior does match with the hypothesis of conduction in the extended states and shallow trap levels, that for high enough voltages should present a P-F type of conduction.

Table 8.2 Conductivity values obtained from the ohmic curves of devices containing different active-layer structures (bulk or superlattice) and different SRC thickness

SRC thickness	σ ($S\cdot cm^{-1}$)
2 nm	6.0×10^{-6}
3 nm	8.2×10^{-6}
4 nm	9.2×10^{-6}
i-Si	1.6×10^{-5}
i-SiC	2.3×10^{-7}

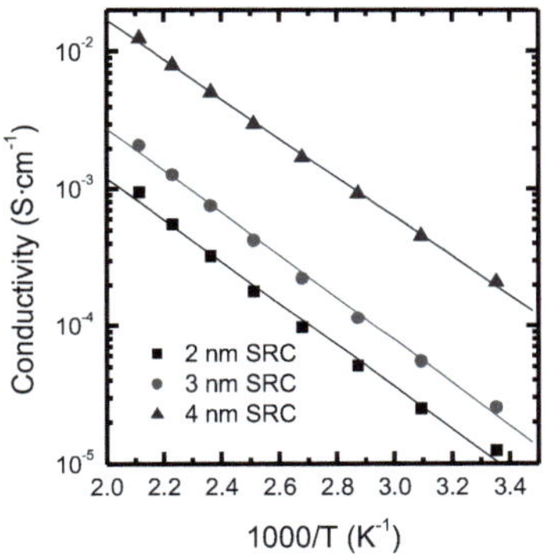

Figure 8.19 Conductivity dependence on the inverse of temperature for devices containing SRC/SiC superlattices, with different SRC layer thickness. The solid lines indicate the Arrhenius fits performed on the experimental data.

8.6.2 *Vertical Transport in SRON/SiO_2 Superlattices*

The 1D vertical transport through SLs requires a more detailed modeling, basically due to the presence of large band-gap layers (barriers), difficult to overcome or to tunnel through by carriers depending on offset and thickness. For our studies we used 5 SRON/SiO_2 SLs deposited on B-doped Si substrates (p^{++}-type substrates). The nominal thickness of the SiO_2 barriers was fixed at 1 nm. This is the smallest value that we have tried successfully and it is the one that yields optimum transport properties, that is,

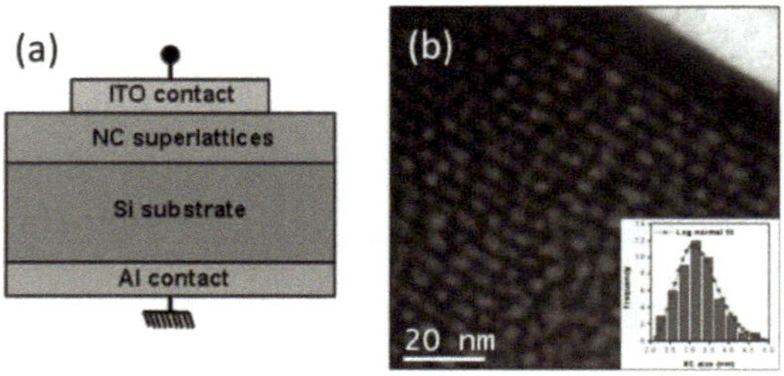

Figure 8.20 (a) Cross-sectional scheme of the devices employed for the vertical electrical characterization of SRON/SiO_2 superlattices. (b) Cross-sectional EFTEM image for a 125× SRON (4 nm)/SiO_2 (1 nm) superlattice structure. The inset presents size statistics on 50 NCs, revealing log-normal distribution with a mean diameter of $d = 3.1$ nm and a broadening of $\sigma = 0.1$. Reprinted with permission from Ref. [70]. Copyright [2013], AIP Publishing LLC.

higher conductivity (see Ref. [69]). The thickness corresponding to the SRON layers was varied, with values of 2.5, 3.5, and 4.5 nm, the total thickness of the SL structures being 17.5, 22.5, and 27.5 nm, respectively. A postdeposition annealing treatment at 1150°C was applied to the samples, in order to precipitate and crystallize the Si excess of the SRON layers. The stoichiometry of the SRON layers was held constant at $SiO_{0.93}N_{0.23}$, corresponding to a Si excess of 16.8 at.%. A top ITO and bottom Al electrodes were implemented to achieve the final device structure, whose scheme is shown in Fig. 8.20a. An energy-filtered transmission electron microscopy (EFTEM) image in cross section of the SL structure is presented in Fig. 8.20b. The bright spots correspond to the silicon precipitates and there is clearly an ordered SL system for the particular processing conditions performed [23]. The limited thickness of the SRON layers allows obtaining a narrow distribution Si NC size along the sample (see the inset of Fig. 8.20b).

The main results of the vertical $I(V)$ characterization are shown in Fig. 8.21 for dark conditions and different SRON layer thickness. We observe a strong rectification under substrate inversion ($V > 0$), due to the injection of minority carriers (holes in this particular case); therefore, we focus hereafter in the accumulation regime ($V < 0$). Both progressive and regressive voltage sweeps are presented and there is a shift between both curves. Additionally, current

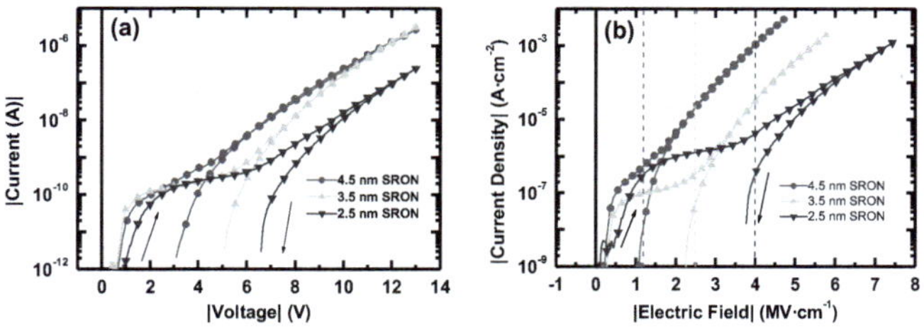

Figure 8.21 (a) Intensity vs. voltage and (b) current density vs. electric field characteristics of devices with SRON layer thickness ranging from 2.5 to 4.5 nm. A special nonpassivated device is also presented. Arrows indicate the progressive and regressive voltage sweeps. The vertical dashed lines in (b) indicate the threshold voltage/field for current conduction to occur. Reprinted with permission from Ref. [72]. Copyright [2013], AIP Publishing LLC.

goes to zero and changes sign for voltages between 3 and 7 V, which means important charge trapping. This charge trapping is, in turn, more noticeable for thinner SRON layers (2.5 and 3.5 nm), which may be associated to longer trapping times in smaller Si NCs [71].

From the $I(V)$ curves it is clearly seen a region at low voltages with an almost constant intensity, which is due to the displacement current (it changes depending on sweep velocity, that is, $I = C\mathrm{d}V/\mathrm{d}t$) or charging current of the capacitor. We define a threshold voltage, V_{th}, for which the conduction current starts to dominate, and this usually occurs at medium-high fields depending on the sample (see Fig. 8.21b). SLs with SRON thickness of 4.5 nm present V_{th} <1 MV/cm, while for those of 2.5 nm V_{th}>4 MV/cm, scaling with the inverse of the capacitance and dominating at low voltages. We also speculate on the fact that the density of states available for tunneling in the thin layers is significantly reduced due to QCEs. This, in turn, can be interpreted as a decrease in the effective barrier (ϕ_{eff}) that electrons must overcome when the energy of the confined states inside the quantum dots increases, which happens at smaller nanostructures, that is, thinner SRON layers (see Fig. 8.22).

From the fits of the $I(V)$ curves at high fields we obtain that the conduction proceeds via a kind of P-F-type mechanism (Eq. 8.14). This is a bulk limited conduction in contrast with that shown in the previous section where we had an electrode limited Fowler–

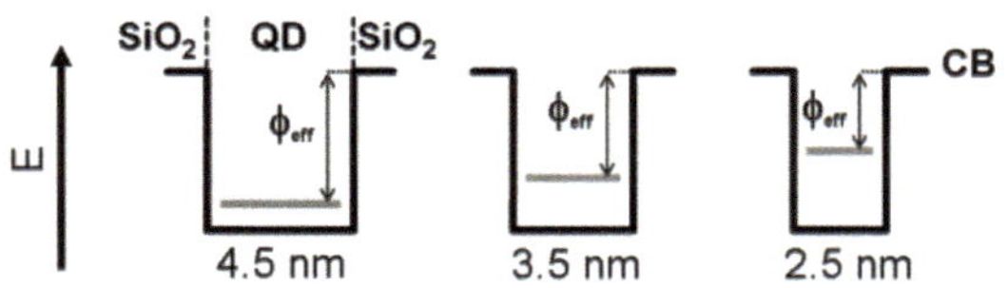

Figure 8.22 Energy band diagram corresponding to semiconductor QDs embedded in a high-band-gap matrix (SiO_2 in this case) for different QD sizes (corresponding to the different SRON layer thicknesses under study). Gray lines represent the quantum-confined electron states, whereas ϕ_{eff} is the effective energy difference between these states and the energy continuum of the conduction band.

Nordheim type of limiting current at very high voltages while still current was bulk limited at low voltage. In fact, the current here is much higher in concomitance with the barrier thickness much lower, and this is a must for a proper conduction at low voltage in solar cells. The thermal analysis of conductivity introduced below will confirm this hypothesis.

We make use of the $I{\cdot}V^{-1}$ versus $V^{1/2}$ (P-F) plot for the electrical data to validate the model. Figure 8.23 shows the obtained P-F representation for the device containing 3.5 nm SRON layers. A clear linear region appears at medium-high voltages, covering a wide range of more than four decades. As displayed in the inset of the figure, this linear part could be fitted by means of the abovementioned law. In addition, for a consistency check, we estimated the effective relative permittivity of the whole SL, resulting in 8.7, a physically reasonable value laying in between the ones for pure SiO_2 and Si (4 and 12, respectively).

The $I(V)$ characteristics have been measured for a range of temperatures between 50°C and 300°C. The conductivity was the compared at an applied voltage of 13 V, in the region where the bulk limited transport occurs see Fig. 8.21a. The conductivity data at different temperatures were interpreted using the Arrhenius law (see Eq. 8.26), whose plots are displayed in Fig. 8.24.

The Arrhenius fits displayed as solid lines in the figure are in very good accordance with the experimental data, releasing E_{A} values ranging from 47 to 165 meV. These are indeed low values when comparing with reported works on different matrices, such as Si_3N_4 [73], which we attribute to the presence of shallow traps

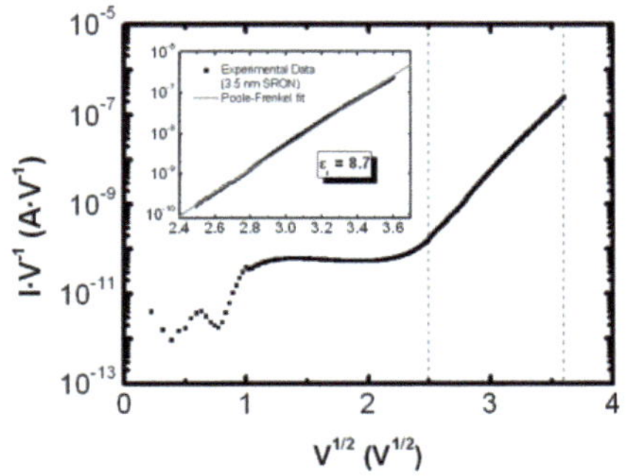

Figure 8.23 Poole–Frenkel representation of the $I(V)$ data corresponding to the device containing 3.5 nm passivated SRON layers. Vertical dashed lines show the plot's linear range. The inset displays the fit performed on the linear region, yielding a relative permittivity value of 8.7. Reprinted with permission from Ref. [72]. Copyright [2013], AIP Publishing LLC.

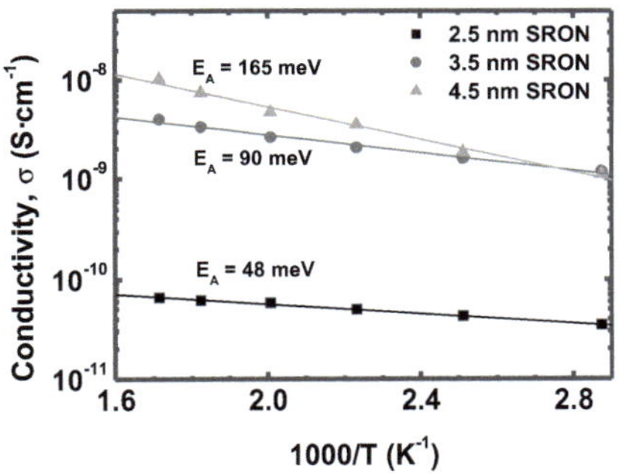

Figure 8.24 Conductivity versus the inverse of temperature for the devices under study. Solid lines represent the fits according to an Arrhenius law, whose estimated activation energies are indicated in the graph. Reprinted with permission from Ref. [72]. Copyright [2013], AIP Publishing LLC.

in our materials (low activation energies are required to allow for the carrier transport through traps). Actually, the activation energy scales with the threshold voltage (see $I(V)$ curves shown in Fig. 8.21a). Furthermore, the fact that the activation energy scales with the SRON layer thickness confirms the P-F mechanism as the most suitable one for our devices.

An electro-optical characterization was also carried out on the same devices. The electroluminescence (EL) spectra obtained for a

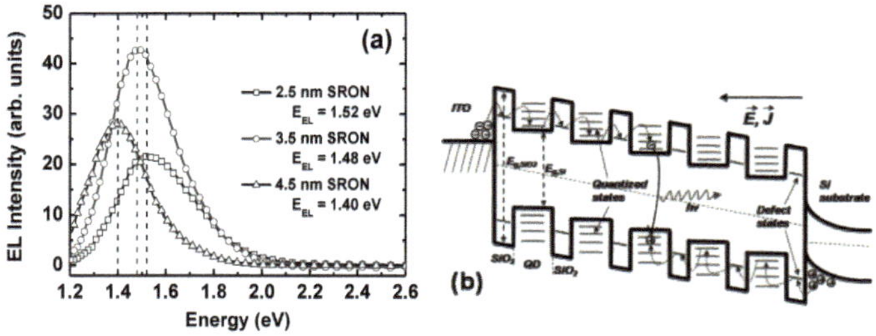

Figure 8.25 (a) Electroluminescence spectra of the devices under study. (b) Energy band diagram of the superlattice structure under applied voltage polarization, revealing electron–hole recombination for electroluminescence emission. Reprinted with permission from Ref. [72]. Copyright [2013], AIP Publishing LLC.

current intensity of 1 μA are displayed in Fig. 8.25a. The spectra show a clear peak-like emission feature at around 1.5 eV that is attributed to excitonic recombination inside the NCs. Moreover, no emission has been observed at higher energy (see high-energy region of Fig. 8.25a), demonstrating the optical inactivity of defects from the matrix. Another interesting issue is the behavior of the EL intensity as a function of the SRON layer thickness, reaching a maximum emission for 3.5 nm thickness of SRON, that is, at medium NC sizes. This result has also been observed in the PL obtained from equivalent SL samples, were NC sizes between 3 and 4 nm were shown to present a maximum emission [24]. This result is also in agreement with peak features of single layers as reported elsewhere [74].

In addition, the EL spectra present a peak red shift at increasing NC size, from 1.54 to 1.39 eV. This shift is a clear consequence of the variation, from sample to sample, in the electronic quantum confinement inside the NCs, and verifies that the origin of EL is NC related. For a better comprehension of the physical processes occurring in the SL system under study, a band diagram of the structure is presented in Fig. 8.25b. The low applied electric field allow for a resistive-like bulk conduction (electrons from the ITO gate, holes from the Si substrate) through the defect-related allowed energy levels in the SiO_2 matrix. Carriers consequently hop into the quantum confined levels inside the Si NCs (QDs), where the electron–hole (exciton) radiative recombination takes place. The fact that EL was not observed under inversion polarization confirms this mechanism, as the injected carrier concentration into the SL is strongly substrate limited, and thus the recombination probability dramatically decreases.

8.7 Conclusions

In this chapter, the band structure and electrical transport in amorphous SLs, with special emphasis in NC SLs, have been reviewed. These SLs are composed by alternative layers of two different amorphous materials, and at least in one of them, there appear nanometric crystalline inclusions. The other layers have a wider band-gap and act as barrier layers that provide carrier confinement into the QWs. Therefore, in studying their electrical transport different transport mechanisms were taken into account depending on whether the carrier transport is perpendicular or parallel to the growth direction of the SL. For perpendicular or vertical transport, the miniband models for SLs together with some particular coherent effects, like negative differential conductivity and/or resonant and/or sequential tunneling, were introduced. Whether transport is in the matrix and/or the NC conduction bands can be assessed from the field dependence of the currents (Fowler–Nordheim injection or P-F conduction). Regarding the horizontal transport in dense arrays of nanoclusters inclusions, percolation dependent models were introduced. Therein, below the percolation threshold density, conduction proceeds mainly via P-F or direct tunneling, whereas for high densities of silicon nanoclusters above percolation, conduction is mostly through resistance ohmic shunt pathways.

In the second part of the chapter, recent experimental results on NC SLs were presented and discussed for two different systems, Si NCs/SiC and Si NCs/SiO_2, considering both vertical and horizontal transport. All the review on structure and transport properties presented in the first part of the chapter is used to understand and model the real SL behavior. The structure of nc-SLs allows for layer engineering in which (i) NCs in the horizontal plane are closely packed, while (ii) their size distribution and vertical transport are limited by the nonstoichiometric barrier spacing between a Si NC ordered layer. Those SLs are meant for application in photovoltaics and light emission. In conclusion, the control of the NC location and distribution, its crystalline quality, the interfacial quality in terms of defects and geometrical data, like barrier thickness, are essential for tailoring their transport and luminescence properties, making it possible their use either in tandem solar cells or light emitting devices, both fully compatible with the mainstream Si technology.

Appendix A. Band Structure of Nanocrystal Superlattices

Let us try in the following to deduce some useful and insightful expressions for the electronic structure of amorphous and NC SLs. First of all, we write the Schrödinger equation in 3D for a QW within two barriers (similar to Eq. 8.1):

$$\left[-\frac{\hbar^2}{2m_A^*}\frac{d^2}{dx^2}+V_x(x)\right]\psi(x,y,z)+\left[-\frac{\hbar^2}{2m_A^*}\frac{d^2}{dy^2}+V_y(y)\right]\psi(x,y,z)$$
$$\left[-\frac{\hbar^2}{2m_A^*}\frac{d^2}{dz^2}+V_z(z)\right]\psi(x,y,z)=\left(E_x+E_y+E_z\right)\psi(x,y,z) \tag{8A.1}$$

The potential is additive in x, y, and z and thus the Schrödinger equation is separable. The electron (or hole) is free in x and y and is confined in z [$u(z)$ is the step function]:

$$\begin{aligned} V_x(x) &= 0 \\ V_y(y) &= 0 \\ V_z(z) &= V_0 u(z-a)+V_0 u(-z) \end{aligned} \tag{8A.2}$$

Consequently, the solution can be written as a product of functions of x, y, and z:

$$\psi(x,y,z)=\psi_x(x)\,\psi_y(y)\,\psi_z(z)=\sqrt{\frac{2}{a}}\sin\left(n\frac{\pi}{a}\left(z+\frac{a}{2}\right)\right)e^{ik_x x}e^{ik_y y} \tag{8A.3}$$

And finally the energy is given by the following equation in which we see that due to quantum confinement there appear subbands, each of them is labeled by n:

$$E = E_x + E_y + E_z = \frac{\hbar^2 k_x^2}{2m_A^*} + \frac{\hbar^2 k_y^2}{2m_A^*} + \frac{n^2\hbar^2\pi^2}{2m_A^* a^2} \tag{8A.4}$$

This is for a single QW. For a SL what we have is that each of the subbands develops into a miniband and hence we can write for the n-subband of the SL [27]:

$$E_n = \frac{\hbar^2 k_x^2}{2m_A^*} + \frac{\hbar^2 k_y^2}{2m_A^*} + \frac{n^2\hbar^2\pi^2}{2m_A^* a^2} - \beta_n - \Delta_n \cos(k_z d) \tag{8A.5}$$

The energy width of the miniband is essentially controlled by the thickness of the sublayers and the energy offset between them. In the tight binding model there is a direct relationship between the width of the miniband and the overlap integrals [45]:

$$\begin{aligned} \Delta_n &= -\int \psi_n^*(x)\, \Delta\psi_n(x-d) \\ \beta_n &= \int \Delta V\, |\psi_n(x)|^2 \end{aligned} \tag{8A.6}$$

where $V = V_0 + \Delta V$ is the potential of the SL, V_0 is the potential of the single well, d is the period of the SL and ψ_n is the unperturbed wavefunction of the single well. We have considered the interaction of a given well with only the two neighboring wells. Additionally, we have assumed that there is considerable energy difference between the levels of the single well so there is only hybridization of levels with the same n. In other words, there is not mixing of states with different n; this would be true if and only if the energy width of the miniband is small enough so that there is still energy gap between minibands.

When calculating transport properties we need to account for the number of carriers in the bands which for electrons can be calculated for a given band as:

$$n = \int N(E) \frac{1}{1 + \exp\left(\frac{E - E_F}{k_B T}\right)} dE \tag{8A.7}$$

where $N(E)$ is the density of states, that is, $N(E)dE = N(k)d^3k$, that refers to the number of k states with energy between E and $E+dE$. E_{Fn} is the quasi-Fermi energy of the electrons, which is used to

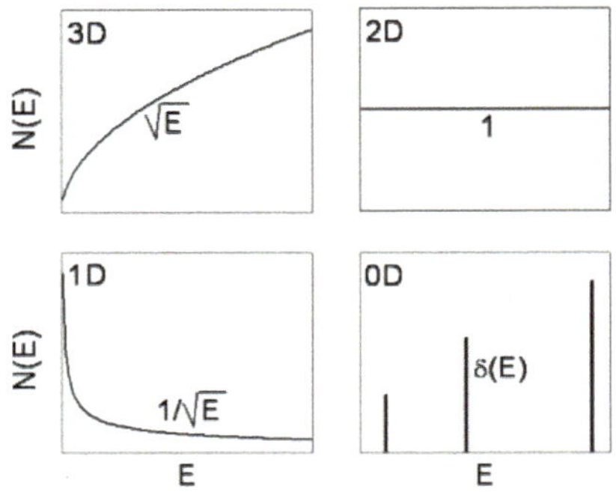

Figure 8A.1 Density of states for parabolic bands (3D) and confined subbands for quantum wells (2D), quantum wires (1D), and quantum dots (0D).

describe nonequilibrium transport by using the equilibrium Fermi–Dirac distribution function with a shifted Fermi energy for electrons. For QW layers we need to calculate the density of states for quasi-2D systems which is done by using the definition of $N(E)dE = N(k)d^2k$ and considering that there are subbands due to confinement in the z direction. If we further assume that the bands are parabolic and with a single effective mass, the density of states in 3D and in 2D per unit volume and per unit area, respectively, for a definite subband with minimum at $E = E_0$ is given by (Fig. 8A.1):

$$N_{3D} = \frac{1}{2\pi^2}\left(\frac{2m^*}{\hbar^2}\right)^{\frac{3}{2}}\sqrt{E - E_0}N_{2D} = \frac{m^*}{\pi\hbar^2}u\left(E - E_0\right) \qquad (8A.8)$$

Then, using the dispersion relation from Eq. 8.7 and the definition of the density of states for a SL miniband in the z direction (i.e., $N_{SL}(E)dE = N_{SL}(k_z)dk_z$) we can calculate easily the variation of the density of states of one SL miniband in that direction:

$$N_{SL}\left(E\right) = \frac{1}{\pi}\frac{1}{\left|\frac{dE}{dk_z}\right|} = \frac{1}{\pi\Delta_n d}\frac{1}{\left|\sin\left(k_z\left(E\right)d\right)\right|} \qquad (8A.9)$$

which will be similar to that of the quasi-2D QW, but having a sinusoidal dependence within the extend of the energy width of the miniband, instead of the abrupt step of the QW. The flat regions correspond to the gaps between minibands (see Fig. 8A.2).

We can get some insight into NC SLs by using a simple model (following Jiang and Green approach [31]) in which Si NCs are cubic

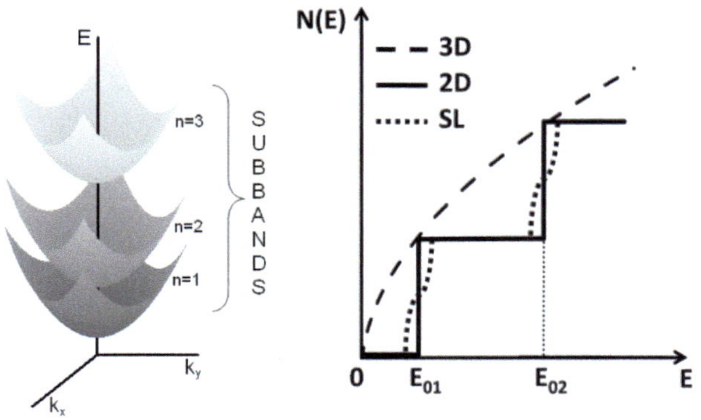

Figure 8A.2 $E(k)$ dispersion relations and density of states for parabolic bands and confined subbands for 3D, quantum wells (quasi-2D), and superlattices (SLs).

potential wells so the Schrödinger equation is separable into the three dimensions in a similar way as we did for the QW (see Fig. 8A.3 for details). The crystalline quantum dots have (100) orientated surfaces. Due to the similar averaged transversal-longitudinal electron effective mass in the cubic symmetry ($0.259m_0$) than in the spherical symmetry ($0.264m_0$) we expect the cubic dot wells to produce similar results that the spherical dots [31]. The effective mass of electrons in the dielectric matrix is taken as $0.4m_0$ (m_0 is the mass of free electrons).

Due to the separation of the Schrödinger equation and using the same procedure as for the unidimensional SL we can write the dispersion relation as:

$$E(\vec{k}) = \frac{2\hbar^2}{m^*}\left(\frac{v_{n_x}^2}{L_x^2} + \frac{v_{n_y}^2}{L_y^2} + \frac{v_{n_z}^2}{L_z^2}\right) - \left(\beta_{n_x} + \beta_{n_y} + \beta_{n_z}\right)$$
$$- \left(\Delta_{n_x}\cos(k_x d_x) + \Delta_{n_y}\cos\left(k_y d_y\right) + \Delta_{n_z}\cos(k_z d_z)\right) \tag{8A.10}$$

The energy levels are those for the finite wells and the resulting values will be depending of the band offset between the silicon NCs and the surrounding matrix (SiO_2, Si_3N_4, SiC, ...). This straightforward expression is a consequence of the simple cube Bravais lattice and the symmetry of the SL potential.

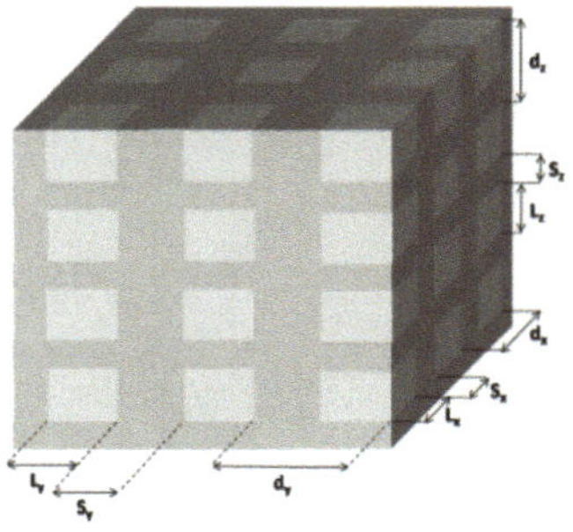

Figure 8A.3 Silicon nanocrystal superlattice consisting of arrays of regularly spaced, equally sized cubic silicon nanocrystals (in light gray) in a dielectric matrix (in dark gray) [31]. Reprinted with permission from [C.W. Jiang, M.A. Green, *J. Appl. Phys.* **99**, 114902 (2006).]. Copyright [2006], AIP Publishing LLC.

Following this model, it is possible to compute the band dispersion relations of nc-SLs as a function of the (i) matrix (through band offsets), (ii) NC size (through Ls), and (iii) distance between NCs (through Ss). As reported in Refs. [16, 31] this model is capturing the essential physics of nc-SLs and can be used for developing compact models for transport. The calculation of the density of states can be done using the general expression for a lattice with dispersion relation $E_n(k)$ [45]:

$$N(E) = \frac{1}{4\pi^3} \int \frac{dS}{\left|\nabla E_n(\vec{k})\right|}, \tag{8A.11}$$

where the integration is over a constant energy surface $E =$ constant and for having the whole $N(E)$ the integration has to be performed for each E throughout the Brillouin zone. Only if the dispersion relation is isotropic (spherically symmetric, that is, depending only on k magnitude) it can be inverted to yield $k(E)$ and then integrated over a volume $4\pi k^2 dk$. We show in Fig. 8A.4 the band structure and density of states of a Si NC SL with cubes of 2 nm of side which are 0.5 nm apart in a matrix of silicon nitride [16]. As can be seen, it is necessary to have a very high density of NCs in close proximity to have extended Bloch states. The band offset of conduction band is 1.9 eV in this case. For a SiO_2 matrix the NC density has to be larger as band offset is over 3 eV and

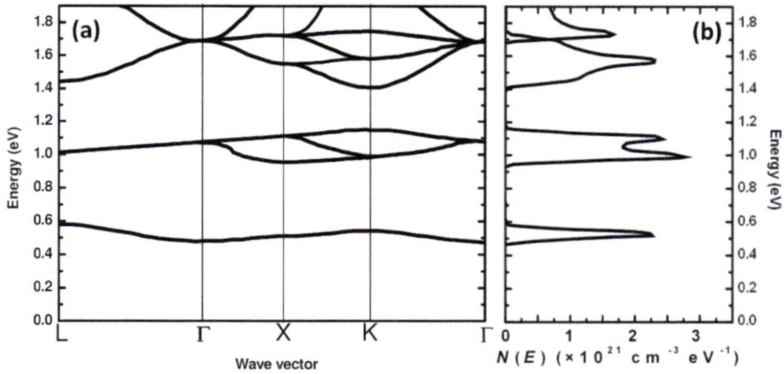

Figure 8A.4 (a) Band structure for 2 nm silicon nanocrystals with distance of 0.5 nm; the matrix is silicon nitride. The conduction-band edge of bulk silicon is taken as $E = 0$. (b) Density of states of the previous superlattice (adapted from Ref. [16]).

thus overlap integrals are small. Figure 8A.5 is showing variation of miniband position and width as a function of barrier height, dot size and interdot distance, as taken from Ref. [31].

An analytical approximation of the density of states can be deduced for small ks close to the Brillouin zone center (Γ) if we develop the cosine according to a truncated Taylor expansion:

$$\cos(kd) \approx 1 - \frac{1}{2}(kd)^2 \tag{8A.12}$$

This way the constant energy surfaces are the ellipsoids [E_c in this case is the minimum of the miniband which takes place at $k = 0$ for, for example, the lowest band (111)]:

$$E - E_c = \frac{\Delta_{n_x}}{2}(k_x d_x)^2 + \frac{\Delta_{n_y}}{2}(k_y d_y)^2 + \frac{\Delta_{n_z}}{2}(k_z d_z)^2 \tag{8A.13}$$

From this expression it is easy to see that the effective masses are given by:

$$m_x^* = \left(\frac{1}{\hbar^2}\frac{\partial^2 E}{\partial k_x^2}\right)^{-1} = \frac{\hbar^2}{\Delta_{n_x} d_x^2} \tag{8A.14}$$

being equivalent for y and z directions. As expected, only if the interdot distance is the same for x, y, and z, the effective mass will be a scalar. The effective mass is inversely proportional to the period

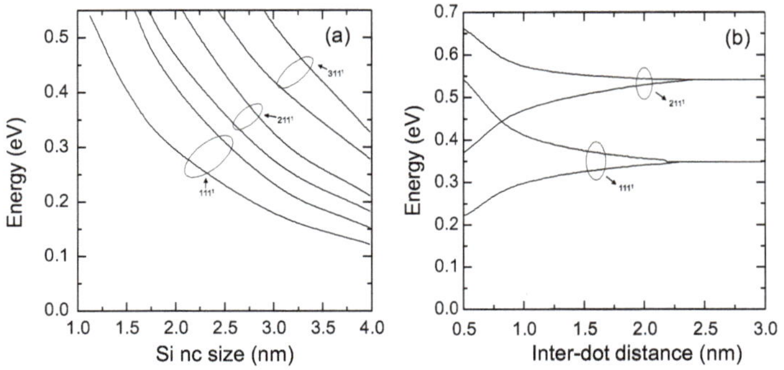

Figure 8A.5 Variation of position and width of minibands of a silicon nanocrystal superlattice with silicon nitride as a matrix. When fixed, the dot size is that of a 2 nm cube, the interdot distance is 1 nm, and the insulator barrier height is 1.9 eV. The bottom of the Si bulk conduction band is taken as $E = 0$. The miniband labeling is $n_x n_y n_z$ and the superscript is degeneracy [31]. Reprinted with permission from [C.W. Jiang, M.A. Green, *J. Appl. Phys.* **99**, 114902 (2006)]. Copyright [2006], AIP Publishing LLC.

of the SL and to the overlap integral, that is, the width of the band. Finally, for the density of states:

$$N(E) = \frac{1}{2\pi^2}\left(\frac{2m^*_{eff}}{\hbar^2}\right)^{\frac{3}{2}}(E - E_c)^{\frac{1}{2}} \tag{8A.15}$$

With the effective mass for the density of states defined by:

$$m^*_{\text{eff}} = \left(m^*_x m^*_y m^*_z\right)^{\frac{1}{3}} \tag{8A.16}$$

The procedure is similar to that followed for calculating the density of states and the effective masses of common bulk semiconductors (see Ref. [6], for example).

Appendix B. Semiclassical Conduction in the Extended States of a Superlattice

The Tsu–Esaki negative differential conductance (NDC) model can be applied to both crystalline and amorphous SLs and starts with the semiclassical 1D equations of motion without scattering:

$$\hbar\frac{dk_z}{dt} = qF,\ v_z = \frac{1}{\hbar}\frac{\partial E}{\partial k_z} \tag{8B.1}$$

If scattering is added as a mean collision time $\tau$and introduced through a classical exponential temporal decay, the mean drift velocity can be expressed as:

$$v_{\text{drift}} = \int_0^{\infty} \exp\left(\frac{-t}{\tau}\right) dv_z = \frac{qF}{\hbar^2}\int_0^{\infty} \frac{\partial^2 E}{\partial k_z^2}\exp\left(\frac{-t}{\tau}\right) dt$$

$$z(t) = \int v(k(t))\, dt = \frac{\Delta}{qF}\cos\left(\frac{qFd}{\hbar}t\right)$$

$$v_{\text{drift}} = \frac{\mu F}{1+\left(\frac{F}{F_c}\right)^2},\ \mu = \frac{q\tau\Delta d^2}{2\hbar^2} = \frac{q\tau}{m^*}F_c = \frac{\hbar}{q\tau d} \quad (8B.2)$$

where m^* is the electron effective mass at the bottom of the miniband and the other parameters have been defined in the previous sections. This simple model predicts NDC, that is, the decrease of the electron velocity (i.e., decrease of current) beyond a critical electric field F_c (Fig. 8B.1a). In spite of its simplicity, this model retains the basic physics. Treatments using the Boltzmann equation in the relaxation time approximation and the full Boltzmann equation by Montecarlo simulations show only small corrections to the formulae shown above. Equations above are quantitatively correct provided that $k_B T << \Delta/2$ (see, for example. Sibille in Ref. [41]).

The velocity decreases and finally its sign changes for $k_z d = \pi$, which is the limit of the first Brillouin zone of the SL, that is, the electron is backscattered. Then, after another semiperiod the electron will be reflected back again and the whole picture is that the electron is oscillating back and forth with energies between the limits of the miniband describing Bloch oscillations. In fact this oscillation is the source of the NDC. Neglecting scattering of the electrons within at least a few oscillations:

$$z(t) = \int v(k(t))\, dt = \frac{\Delta}{qF}\cos\left(\frac{qFd}{\hbar}t\right) \quad (8B.3)$$

Then it shows that the position is oscillating at a period:

$$T_{\text{osc}} = \frac{2\pi\hbar}{qFd} \quad (8B.4)$$

where the period T_{osc} is supposed to be large compared to the scattering time. In principle, Bloch oscillations could be thought off for atomic lattice periodic potentials but due to the smallness of the

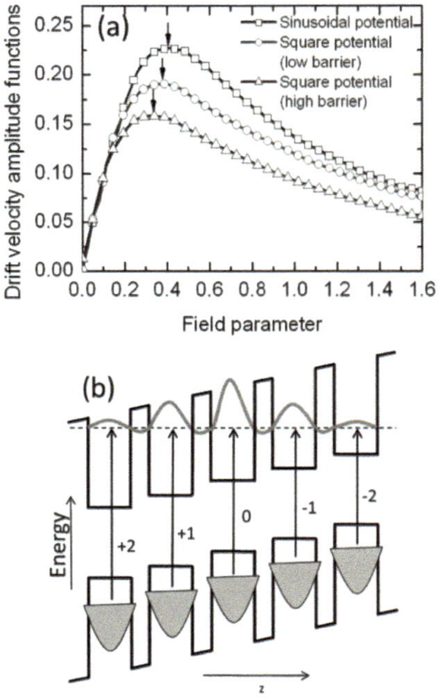

Figure 8B.1 (a) Drift velocity versus reduced field parameter $\xi = \frac{eF\tau}{\hbar k_d} k_d = \frac{\pi}{d}$ for different periodic potential (sinusoidal (black squares) and square with low (red circles) and high (blue triangles) barrier height relative to miniband width [1]. (b) Transitions between the delocalized electron wavefunctions and the localized wavefunctions of the holes (Wannier–Stark ladder [75]).

crystal lattice parameter, the period of the oscillations is so large that the condition of neglecting scattering never holds. In fact, these oscillations are a concept that Bloch and Zener already introduced in the context of transport in bulk crystals [45]. However, as for SLs d is much larger, such condition can be met. For example, if we consider $d = 10$ nm and $F = 10$ kV/cm, the oscillation frequency is 2.5 THz, so in principle SLs can be used as THz generators. The practical realization of a Bloch oscillator requires relatively wide minibands (i.e., very thin barriers) and high current densities (i.e., large electric fields). Large electric fields induce strong localization and NDC that in turn lead to electric field instabilities. Although several reports on

Bloch oscillation have been published, no efficient SL THz radiation generator has been yet achieved [76, 77].

We cannot finish this treatment of miniband transport without resorting to the Wannier–Stark ladders. This is a phenomenon which is intimately related to the application of a strong electric field, the localization that it induces and the appearance of Bloch oscillations. Let us consider electron and holes in a SL and both of them giving tight binding minibands:

$$E_n(k_z) = E_n - \Delta_n \cos(k_z d) \tag{8B.5}$$

The typical miniband widths of the electrons are of few tens of meV ($\Delta/2$). Nevertheless, for heavy holes with much larger effective mass, the $\Delta/2$ widths are only of few meV. At a given electric field F the carriers extend over the localization length given by the amplitude of the Bloch oscillations (see Fig. 8B.1b) and the equation for $z(t)$:

$$\lambda \approx \frac{\Delta}{qF} \tag{8B.6}$$

This is called the *Wannier–Stark localization*. It is essentially coming from the fact that the electric field tilts the miniband and the states cannot extend over the whole SL anymore. If tilting is larger than the intrinsic broadening of the levels, the miniband splits in a series of levels (the Wannier–Stark ladder, see Fig. 8B.1b). The potential energy in a SL period is reduced by eFd and this will be the energy separation of the Wannier–Stark ladder. This energy separation is nothing more than $\Delta E = \hbar\omega_{\mathrm{B}}$, and it is linked to the frequency of the Bloch oscillation. In the high-field limit, when the tilting eFd in one period becomes comparable to the width $\Delta/2$ of the miniband, the state is localized to a single well. The Wannier–Stark ladders have been detected nicely in many experiments in which the electric field is varied so that it induces complete localization of the hole wavefunction, while still leaving the electron delocalized through several wells. This makes a Wannier–Stark ladder for holes while not for electrons and turn into a series of optical electron–hole transitions separated by eFd [75, 78].

Appendix C. Generalized Trap-Assisted Tunneling Model

TAT is well known to be the best model to explain the conduction mechanism at medium fields (5–8 MV/cm). Figure 8C.1 schematically shows the energy band diagram of polysilicon-SiO_2-Si under two different tunneling processes depending on the electric field across the oxide. Particularly, the tunneling electrons meet either triangular barrier height (denoted as TAT triangular) or trapezoidal barrier (denoted as TAT trapezoidal). In that case, the electrons coming from the polysilicon electrode are injected into the traps existing in the oxide with tunneling probability P_1. Subsequently, these captured electrons tunnel again through the oxide up to conduction band with tunneling probability P_2.

Therefore, we can begin the generalized trap-assisted tunneling (GTAT) calculation by means of the Wentzel–Kramers–Brilloun (WKB) approximation for the tunneling probability, P_1 and P_2, which is given as follows:

$$P_i = \exp\left(-2\int |k(z)|\, dz\right), i = 1 \text{ or } 2, \quad (8C.1)$$

where $k(z)$ is given by:

$$k(z) = \left[\frac{2m_{\text{ox}}}{\hbar^2}(\phi_B - Fqz - E_{\text{e}})\right]^{1/2}, i = 1 \text{ or } 2, \quad (8C.2)$$

where m_{ox} is the effective mass in the oxide and E_{e} is the total electron energy in metal (taken as 0.2 eV). Integrating $k(z)$ and substituting suitable boundary conditions,

$$P_i = \exp\left(-\frac{4\sqrt{2m_{\text{ox}}}}{3\hbar qF}\left(\Psi_i^{3/2} - \psi_i^{3/2}\right)\right), i = 1 \text{ or } 2, \quad (8C.3)$$

Ψ and ψ_i depends on which barrier is considered:

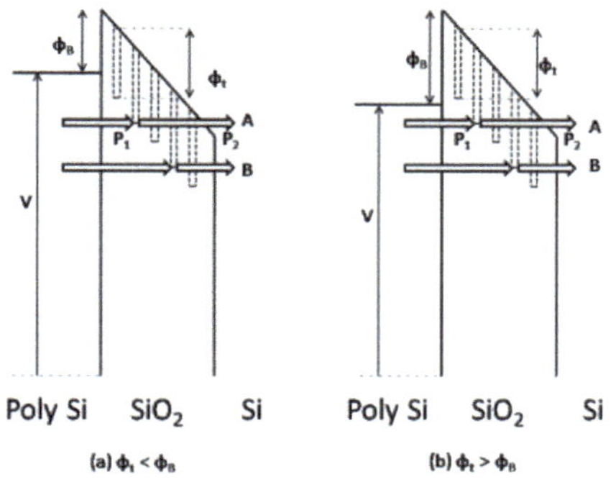

Figure 8C.1 Schematic energy band diagram of a $Si\text{-}SiO_2\text{-}Si$ structure for the case (a) $\phi_t < \phi_B$ and (b) $\phi_t > \phi_B$.

(i) For a triangular barrier (process A in Fig. 8C.1)

$$\Psi_1 = \phi\,(z)\,;\, \psi_1 = \phi_t$$
$$\Psi_2 = \phi_t;\, \psi_2 = 0 \tag{8C.4}$$

(ii) For a trapezoidal barrier (process B in Fig. 8C.1)

$$\Psi_1 = \phi\,(z)\,;\, \psi_1 = \phi_t$$
$$\Psi_2 = \phi_t;\, \psi_2 = \phi\,(z) - V \tag{8C.5}$$

With $\phi\,(z) = \phi_t + Fqz - E_e$

The tunneling current is calculated from:

$$J_{tat} = \int_0^{Z_1} \frac{qC_t N_t P_1 P_2}{P_1 + P_2} dz \tag{8C.6}$$

where N_t is the trap concentration and N_t is a function of ϕ_t and E_e (see Ref. [79]). Also, $Z_1 = \frac{V\text{-}\phi_t}{qF}$. Thus, C_t is given by:

$$C_t = \left(\frac{m^*_{\text{poly}}}{m^*_{\text{no}}}\right)^{5/2} \left(\frac{8E_e^{3/2}}{3\hbar\sqrt{\phi_t - E_e}}\right) \tag{8C.7}$$

If we assume that $\phi_t + Fqz \gg E_e$, we can get analytical expressions for the current:

$$\begin{aligned}
J &= J_{\text{triangular}} + J_{\text{trapezoidal}} \\
J_{\text{triangular}} &= C_1 \exp\left(-\frac{C_2}{qF}\right)\left\{\left(C_3 - \frac{3C_2}{2qF}\right)\right. \\
&\qquad \left. - \ln\left[\frac{1+\exp\left(C_3 - {}^{5C_2/2qF}\right)}{1+\exp\left(-{}^{C_2/qF}\right)}\right]\right\} \\
J_{\text{trapezoidal}} &= -C_1 R_1 \left\{\tan^{-1}\left(\frac{R_2}{R_1}\right)\tan^{-1}\right. \\
&\qquad \left. \times \left[\frac{\exp\left(-C_3 + {}^{3A\phi_t^{3/2}/2qF}\right)}{R_1}\right]\right\},
\end{aligned} \tag{8C.8}$$

where:

$$\begin{aligned}
A &= \frac{4\sqrt{2m^*_{\text{ox}}}}{3\hbar} \\
C_1 &= \frac{2C_t N_t}{3A\sqrt{\phi_t}} \\
C_2 &= \phi_t^{3/2} A,\ \text{for} \phi_t > \phi_B \\
C_2 &= \frac{1}{2} A\sqrt{\phi_t}\,(5\phi_t - 3\phi_B),\ \text{for}\ \phi_t < \phi_B \\
C_3 &= \frac{3}{2} A T_{\text{ox}} \sqrt{\phi_t} \\
R_1 &= \exp\left(-{}^{C_3}/_2\right) \\
R_2 &= \exp(C_3)
\end{aligned} \tag{8C.9}$$

However, for practical reason the relationship of J and F_{ox} is often approximated as:

$$J_t \approx \exp\left(-\frac{4\sqrt{2m^*_{\text{ox}}}}{3\hbar qF}\phi_t^{3/2}\right), \tag{8C.10}$$

where ϕ_t can be directly derived from the slope of the linear region in the $\ln(J)$ versus $1/F_{\text{ox}}$ plot.

References

1. L. Esaki, R. Tsu, *IBM J. Res. Dev.* **14**, 61 (1970).
2. D.L. Smith, C. Mailhiot, *Rev. Mod. Phys.* **62**, 173 (1990).

3. M.C. Burt, *Appl. Phys. Lett.* **65**, 717 (1994).
4. K. Seeger, *Semiconductor Physics*, 9th edition, p. 12, Springer (2004).
5. L. Esaki, A perspective in superlattice development, in *Symposium on Recent Topics in Semiconductor Physics*, World Scientific, Singapore, pp. 1–71 (1983).
6. T.P. Pearsall (Ed.), *Strain-Layer Superlattices: Physics, Semiconductors and Semimetals*, Vol. 32, Academic Press, New York (1990).
7. B. Abeles, T. Tiedje, *Phys. Rev. Lett.* **51**, 2003 (1983).
8. G. Allan, C. Delerue, M. Lannoo, *Appl. Phys. Lett.* **71**, 1189 (1997).
9. M.J. Estes, G. Moddel, *Phys. Rev. B* **54**, 14633 (1996).
10. Y. Li, M. Kondo, M. Yamaguchi, K. Morigaki, *J. Non-Cryst. Solids* **137**, 1115 (1991).
11. C.B. Roxlo, B. Abeles, *Phys. Rev. B* **34**, 2522 (1986).
12. K. Hattori, T. Mori, H. Okamoto, Y. Hamakawa, *Appl. Phys. Lett.* **51**, 1259 (1987).
13. G.V.M. Willians, *J. Appl. Phys.* **70**, 901 (1991).
14. S.R.P. Silva, Rusli, K.P. Homewood, G.A.J. Amaratunga, *J. Non-Cryst. Solids* **227–230**, 1137 (1998).
15. M. Zacharias, J. Heitmann, R. Scholz, U. Kahler, M. Schmidt, J. Blasing, *Appl. Phys. Lett.* **80**, 661 (2002).
16. P. Löper, R. Muller, D. Hiller, T. Barthel, E. Malguth, S. Janz, J.C. Goldschmidt, M. Hermle, M. Zacharias, *Phys. Rev. B* **84**, 195317 (2011).
17. S. Gutsch, A.M. Hartel, D. Hiller, N. Zakharov, P. Werner, M. Zacharias, *Appl. Phys. Lett.* **100**, 233115 (2012).
18. C.W. Jiang, M.A. Green, *J. Appl. Phys.* **99**, 114902 (2006).
19. G. Conibeer, M.A. Green, R. Corkish, Y. Cho, E.C. Cho, C.W. Jiang, T. Fangsuwannarak, E. Pink, Y. Huang, T. Puzzer, T. Trupke, B. Richards, A. Shalav, K.L. Lin, *Thin Solid Films* **511–512**, 654 (2006).
20. D. Song, E.C. Cho, G. Conibeer, Y. Huang, C. Flynn, M.A. Green, *J. Appl. Phys.* **103**, 083544 (2008).
21. E.C. Cho, S. Park, X. Hao, D. Song, G. Conibeer, S.C. Park, M.A. Green, *Nanotechnology* **19**, 245201 (2008).
22. I. Perez-Wurfl, X. Hao, A. Gentle, D.H. Kim, G. Conibeer, M.A. Green, *Appl. Phys. Lett.* **95**, 153506 (2009).
23. A.M. Hartel, D. Hiller, S. Gutsch, P. Löper, S. Estradé, F. Peiró, B. Garrido, M. Zacharias, *Thin Solid Films* **520**, 121(2011).

24. J. López-Vidrier, S. Hernández, A.M. Hartel, D. Hiller, S. Gutsch, P. Löper, L. López-Conesa, S. Estradé, F. Peiró, M. Zacharias, B. Garrido, *Energy Proc.* **10**, 43 (2011).
25. J. López-Vidrier, S. Hernández, J. Samà, M. Canino, M. Allegrezza, M. Bellettato, R. Shukla, M. Schnabel, P. Löper,L. López-Conesa, S. Estradé, F. Peiró, S. Janz, B. Garrido, *Mater. Sci. Eng. B* **178**, 639 (2013).
26. A. Anopchenko, A. Marconi, E. Moser, S. Prezioso, M. Wang, L. Pavesi, G. Pucker, P. Bellutti, *J. Appl. Phys.* **106**, 033104 (2009).
27. F. Gourbilleau, C. Dufour, B. Rezgui, G. Brémond, *Mater. Sci. Eng. B* **159–160**, 70 (2009).
28. C. Summonte, R. Rizzoli, R. Galloni, F. Giorgis, F. Giuliani, C.F. Pirri, E. Tresso, A. Desalvo, F. Zignani, P. Rava, *J. Non-Cryst. Solids* **227–230**, 1127 (1998).
29. R. Shukla, C. Summonte, M. Canino, M. Allegrezza, M. Bellettato, A. Desalvo, D. Nobili, S. Mirabella, N. Sharma, M. Jangir and I. P. Jain, *Adv. Mater. Lett.* **3**, 297–304 (2012).
30. A. Marconi, A. Anopchenko, M. Wang, G. Pucker, P. Bellutti, L. Pavesi, *Appl. Phys. Lett.* **94**, 221110 (2009).
31. C.W. Jiang, M.A. Green, *J. Appl. Phys.* **99**, 114902 (2006).
32. N.F. Mott, E.A. Davies, *Electronic Processes in Non-Crystalline Materials*, 2nd edition, Clarendon Press, Oxford (1971).
33. R.A. Street, *Hydrogenated Amorphous Silicon*, Cambridge University Press (1991).
34. J. Singh, K. Shimakawa (Eds.), *Advances in Amorphous Semiconductors*, Taylor and Francis (2003).
35. S. Baranovski (Ed.), *Charge Transport in Disordered Solids with Applications in Electronics*, John Wiley (2006).
36. P.G. LeComber, W.E. Spear, in *Semiconductors and Semimetals*, Vol. 21, p. 89 (Eds. R. K. Willardson and A. C. Beer), Academic Press, New York (1984).
37. S.M. Sze, Section 7.3.4, in *Physics of Semiconductor Devices*, 2nd edition, John Wiley (1981).
38. R. Tsu, *Microelectron. J.* **38**, 959–1012 (2007).
39. A.C. Gossard, *IEEE J. Quantum Electron.* **22**, 1649 (1986).
40. S. Miyazaki, Y. Ihara, M. Hirose, *Phys. Rev. Lett.* **59**, 125 (1987).
41. A. Sibille, J. F. Palmier, H. Wang, F. Mollot. *Phys. Rev. Lett.* **64**, 52 (1990).
42. A. Sibille, in *Semiconductor Superlattices* (Ed. H.T. Grahn), World Scientific (1995).

43. I. Pereyra, M.N.P. Carreño, R.K. Onmory, C.A. Sassaki, A.M. Andrade, F. Alvarez, *J. Non-Cryst. Solids* **97–98**, 871 (1987).
44. E. Vateva, I. Gegorkieva, *J. Non-Cryst. Solids* **164–166**, 865 (1993).
45. N.W. Ashcroft, N. David Mermin, *Solid State Phys.*, HRW International (1976).
46. E.S. Daniel, B.K. Gilbert, J.S. Scott, S.J. Allen, *IEEE Trans. Electron Devices*, **50**, 2434 (2003).
47. E. Vateva, I. Georkieva, *J. Non-Cryst. Solids* **114**, 124 (1989).
48. E. Cartier, F.R. McFeely, *Phys. Rev. B* **44**, 10689–10705 (1991).
49. F. Capasso, K. Mohammed, A.Y. Cho, *Appl. Phys. Lett.* **48**, 478 (1986).
50. F. Jerome, F. Capasso, D.L. Sivco, C. Sirtori, A.L. Hutchinson, A.Y. Cho, *Science* **264** (5158), 553–556 (1994).
51. Rainer Waser (Ed.), *Nanoelectronics and Information Technology*, Wiley, 2005.
52. J.M. Ramírez, F. FerrareseLupi, Y. Berencén, A. Anopchenko, J.P. Colonna, O. Jambois, J.M. Fedeli, L. Pavesi, N. Prtljaga, P. Rivallin, A. Tengattini, D. Navarro-Urrios, B. Garrido, *Nanotechnology* **24**, 115202 (2013).
53. J.M. Ramírez, F. FerrareseLupi, O. Jambois, Y. Berencén, D. Navarro-Urrios, A. Anopchenko, A. Marconi, N. Prtljaga, A. Tengattini, L. Pavesi, J.P. Colonna, J.M. Fedeli, B. Garrido, *Nanotechnology* **23**, 125203 (2012).
54. Mohsen Razavy, *Quantum Theory of Tunneling*, World Scientific (2003).
55. A. Mehonic, S. Cueff, M. Wojdak, S. Hudziak, O. Jambois, c. Labbé, B. Garrido, R. Rizk, A.J. Kenyon, *J. Appl. Phys.* **111**, 074507 (2012).
56. R. Dingle, H.L. Stoermer, A.C. Gossard, W. Wiegmann, *Appl. Phys. Lett.* **33**, 665 (1978).
57. H. Oheda, *J. Non-Cryst. Solids* **137–138**, 1147 (1991).
58. D. Nesheva, Chapter 10, in *Handbook of Semiconductor Nanostructures and Nanodevices*, American Scientific (2006).
59. N. Ibaraki, H. Fritzche, *Phys. Rev. B* **30**, 5791 (1984).
60. D. Nesheva, C. Raptis, Z. Levi, *Phys. Rev. B* **58**, 7913 (1998).
61. M. López, PhD thesis, University of Barcelona (2003).
62. I. Balberg, *J. Appl. Phys.* **110**, 061301 (2011).
63. H. Scher, R. Zallen, *J. Chem. Phys.* **53**, 3759 (1970).
64. A. Ron, D.J. DiMaria, *Phys. Rev. B*, **30**, 807 (1983).
65. J.M. Ramírez, O. Jambois, Y. Berencén, D. Navarro-Urrios, A. Anopchenko, A. Marconi, N. Prtljaga, N. Daldosso, L. Pavesi, J.-P. Colonna, J.-M. Fedeli, B. Garrido, *Mater. Sci. Eng. B* **177**, 734–738 (2012).

66. J.M. Ramírez, Y. Berencén, L. López-Conesa, J.M. Rebled, F. Peiró, B. Garrido, *Appl. Phys. Lett.* **103**, 081102 (2013).
67. Y. Berencén, O. Jambois, J.M. Ramírez, J. M. Rebled, S. Estradé, F. Peiró, C. Domínguez, J.A. Rodríguez, B. Garrido, *Opt. Lett.* **36**, 2617 (2011).
68. Y. Berencén, J.M. Ramírez, B. Garrido. *IEEE Xplore Electron Devices (CDE) Spanish Conference on Electron Devices*, 245–248 (2013).
69. S. Gutsch, J. Laube, A.M. Hartel, D. Hiller, N. Zakharov, P. Werner, M. Zacharias, *J. Appl. Phys.* **113**, 133703 (2013).
70. J. López-Vidrier, S. Hernández, D. Hiller, A.M. Hartel, S. Gutsch, L. López-Conesa, S. Estradé, F. Peiró, M. Zacharias, B. Garrido, *J. Appl. Phys.* **116**, 133505 (2014).
71. V.I. Turchanikov, A.N. Nazarov, V.S. Lysenko, J. Carreras, B. Garrido, *J. Phys.: Conf. Ser.* **10**, 409 (2005).
72. J. López-Vidrier, Y. Berencén, O. Blázquez, S. Hernández, S. Gutsch, J. Laube, D. Hiller, P. Löper, M. Schnabel, S. Janz, M. Zacharias, B. Garrido, *J. Appl. Phys.* **114**, 163701 (2013).
73. Y. Berencén, J.M. Ramírez, O. Jambois, C. Domínguez, J.A. Rodríguez, B. Garrido, *J. Appl. Phys.* **112**, 033114 (2012).
74. M. Perálvarez, J. Barreto, J. Carreras, A. Morales, D. Navarro-Urrios., Y. Lebour, C. Domínguez, B. Garrido, *Nanotechnology* **20**, 405201 (2009).
75. E.E. Méndez, G. Bastard, *Phys Today* **June 1993** 34.
76. Y. Shimada, K. Hirakawa, M. Odnoblioudov, K.A. Chao, *Phys. Rev. Lett.* **90**, 046806 (2003).
77. V. Sankin, A. Andrianov, A. Petrov, A. Zakharin, A. Lepneva, P. Shkrebiy, *Nanoscale Res. Lett.* **7**, 560 (2012).
78. E.E. Méndez, F. Agulló-Rueda, J.M. Hong, *Phys. Rev. Lett.* **60**, 2426 (1988).
79. M.P. Houng, Y.H. Wang and W.J. Chang, *J. Appl. Phys.* **86**, 1488 (1999).

Chapter 9

Ge Nanostructures for Harvesting and Detection of Light

Antonio Terrasi,[a,b] Salvatore Cosentino,[a,b] Isodiana Crupi,[b] and Salvo Mirabella[b]

[a] *Department of Physics and Astronomy, University of Catania, via S. Sofia 64, Catania 95123, Italy*
[b] *CNR-IMM UOS Catania (Università), via S. Sofia 64, Catania 95123, Italy*
antonio.terrasi@ct.infn.it

9.1 Introduction

Germanium (Ge) played a role of primary importance since the very beginning of the solid-state electronics age. The first transistor, invented at Bell Laboratories in 1947 by William Shockley, John Bardeen, and Walter Brattain, was made with a Ge crystal [1]. On the other hand, the subsequent fast and huge development of microelectronics and integrated circuits was based on silicon (Si). The reasons why Si dominates the microelectronics industry are mostly related to its abundance, low cost and excellent properties of its oxide, the SiO_2. Ge has been then mostly applied in the field of infrared photodetectors, due to the lower band gap with respect to Si. In recent years, the use of Ge in microelectronics strongly

Nanotechnology and Photovoltaic Devices: Light Energy Harvesting with Group IV Nanostructures
Edited by Jan Valenta and Salvo Mirabella

ISBN 978-981-4463-63-8 (Hardcover), 978-981-4463-64-5 (eBook)
www.panstanford.com

Table 9.1 Properties of intrinsic bulk Ge at room temperature

Crystal structure	Density (atoms/cm^3)	Lattice parameter (nm)	Energy gap (eV)	Melting point (°C)	Bohr exciton radius (nm)
Diamond	4.42×10^{22}	0.5658	0.67	938	24

increased with the development of new devices with high carrier mobility and low commutation time [2].

Similarly to Si, bulk Ge is produced as polycrystalline blocks and slabs, or monocrystalline wafers cut from ingots grown by Czochralski and floating-zone methods. Due to the high cost of this element with respect to Si, the use of bulk Ge as substrate has been quite limited. Ge wafers, for example, are employed for the fabrication of GaInP/GaAs/Ge multijunction high-efficiency solar cells [3], while many microelectronics devices are based on the use of Ge thin layers [4, 5]. In thin-film technology Ge is deposited by physical processes such as thermal evaporation (e.g., molecular beam epitaxy) and sputtering (for amorphous and polycrystalline films) or by chemical vapor deposition. Compared to Si, Ge is characterized by lower band-gap energy, melting temperature, thermal conductivity and atomic density but higher carrier mobility, lattice parameter and Bohr exciton radius (i.e., the critical dimension at which quantum confinement effects [QCEs] take place) [6, 7].

Table 9.1 reports some fundamental properties of Ge.

Nowadays, with the advent of nanotechnology, all materials are experiencing a new life. Nanostructures (NSs), in fact, show different and appealing behaviors with respect to bulk materials. Due to strong changes of structural, electrical, optical, and chemical properties of materials at the nanoscale length, nanotechnology opened the way to a new era in many fields, most of which have a huge impact on the daily life, from nanoelectronics to nanomedicine. The study and use of materials and devices based on NSs has shown huge growth, giving origin to what is now known as *nanoscience*.

Among the elements of group IV of the periodic table, carbon was probably the first to be used for NSs. In 1985 the discovery of fullerenes opened the way to carbon-based NSs, as carbon nanotubes and, more recently, graphene [8]. The study and development of Si-based NSs were strongly supported by the advent of optoelectronics. The tailoring of the electron energy bands with the dimension of the NSs and the challenge to fabricate Si-based light-emitting devices are behind of the enormous interest in this field since its beginning [9]. Soon after Si, also Ge NSs became very attractive and the study of Si-based materials containing Ge nanoclusters (NCs) (quantum dots [QDs] and nanoislands), nanometric thin films (quantum wells [QWs]), and nanowires increased very fast, being today a very promising topic for new devices in the field of energy conversion, micro-, and optoelectronics [4, 5, 10].

Ge is a narrow-band-gap semiconductor with a quasi-direct band gap of 0.67 eV, very high carrier mobility, a large light absorption coefficient ($\sim 2 \times 10^5$ cm^{-1} at 2 eV) and a wide compatibility with existing Si technology [6, 7]. The absorption coefficient of crystalline (c-) and amorphous (a-) Ge, together with the absorption coefficient of c-Si, are reported in Fig. 9.1 as a function of the photon energy. The two broad shoulders at about 2.2 eV and 4 eV in the spectra of c-Ge are associated to the direct E_1 and E_2 transitions occurring in bulk

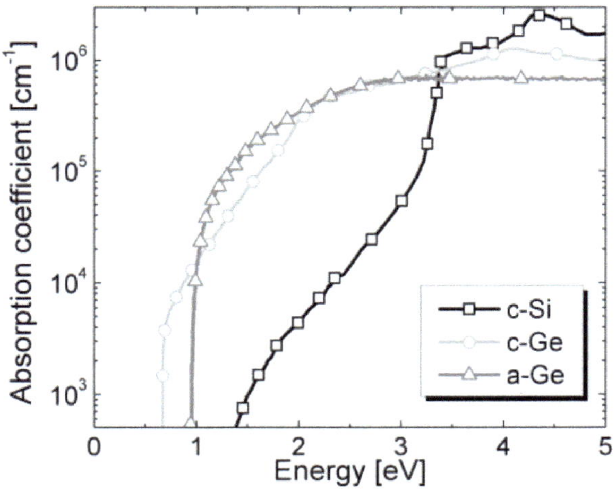

Figure 9.1 Absorption coefficient of crystalline (c-) and amorphous (a-) germanium and silicon.

Ge [12, 13], while the amorphous phase has a featureless shape due to the less defined electronic band structure.

Many properties relevant to light harvesting applications, such as the band gap, the efficiency of luminescence and the oscillator strength of the optical transitions, can be easily tuned by exploiting the QCE. Since Ge shows a Bohr exciton radius (~24 nm) much larger than Si (~5 nm), the quantum confinement regime in Ge NSs is obtained more easily with respect to Si. In addition, the absorption coefficient of Ge is more than one order of magnitude higher than Si up to photon energies of about 3 eV (see Fig. 9.1). Because of this, Ge NSs became very attractive as active absorbers for the fabrication of efficient light harvester, solar cells, and novel optoelectronic devices [3, 4, 5, 11].

This chapter aims to discuss the effects of the quantum confinement on the optical behavior of two kinds of Ge NSs, Ge QDs embedded in SiO_2 and Si_3N_4 matrices and Ge QWs confined between SiO_2 barriers. Fabrication, optical properties, and possible applications for light harvesting and detection of these materials will be described.

9.2 Light Absorption, Confinement Effects, and Experimental Methods

The optical properties of a semiconductor are defined by the interband transitions in the 1–10 eV range and can be studied within the semiclassical theory of light–matter interaction (see Chapter 2). We consider a radiation of frequency ω, wave vector $\mathbf{q}$, and amplitude A_0:

$$A(\mathbf{r}, t) = A_0 e^{i(\mathbf{q}\cdot\mathbf{r}-\omega t)} \tag{9.1}$$

One of the most important figures of merit to study the optical behavior of a material is the absorption coefficient. When a radiation of intensity $I_0 = |A|^2$ passes through a material of thickness x and absorption coefficient α, it comes out with a lower intensity I given by the Beer–Lambert law:

$$I = I_0 e^{-\alpha x} \tag{9.2}$$

The absorption coefficient α is defined as the ratio between the energy absorbed per unit volume and time and the incident flux of electromagnetic energy $u\,(c/n)$:

$$\alpha\,(\omega) = \frac{\hbar\omega W\,(\omega)}{u\,(c/n)} \tag{9.3}$$

where n is the refractive index of the medium and $W(\omega)$ is the rate of interband transitions per unit volume.

The transition probability P of an electron from the initial state $\boldsymbol{k}_{\mathrm{i}}$ in the valence band to the final state $\boldsymbol{k}_{\mathrm{f}}$ in the conduction band, induced by light of frequency ω, is calculated from the Fermi golden rule [6, 15]

$$P_{vk_{\mathrm{i}}\to ck_{\mathrm{f}}} = \frac{2\pi}{\hbar}\left(\frac{eA_0}{mc}\right)^2 \left|\left\langle \varphi_{ck_{\mathrm{f}}}\right| e^{iqr}\boldsymbol{e}_0\cdot\boldsymbol{p}\left|\varphi_{vk_{\mathrm{i}}}\right\rangle\right|^2 \delta\,(E_{\mathrm{f}}-E_{\mathrm{i}}-\hbar\omega) \tag{9.4}$$

The total rate of interband transitions per unit volume $W(\omega)$ is obtained by summing for all the allowed $\boldsymbol{k}$ in the Brillouin zone (BZ) between the valence and conduction bands:

$$W\,(\omega) = \sum_{\mathrm{v,c}} \int_{BZ}^{\forall k} \frac{2dk}{(2\pi)^3} P_{vk_{\mathrm{i}}\to ck_{\mathrm{f}}} \tag{9.5}$$

While the density of electromagnetic energy $u(c/n)$ is given by

$$u = \frac{n^2 A_0^2 \omega^2}{2\pi c^2} \tag{9.6}$$

From Eqs. 9.2–9.5 the absorption coefficient of a semiconductor material is

$$\alpha\,(\omega) = \frac{4\pi^2 e^2}{ncm^2\omega} \sum_{\mathrm{v,c}} \int_{\mathrm{BZ}}^{\forall k} \frac{2dk}{(2\pi)^3} \left|e_0\cdot M_{\mathrm{cv}}(k)\right|^2 \delta\,(E_{\mathrm{f}}-E_{\mathrm{i}}-\hbar\omega) \tag{9.7}$$

where $M_{\mathrm{cv}}\,(k) = \left\langle \varphi_{ck_{\mathrm{f}}}\right| e^{iqr} e\cdot p \left|\varphi_{vk_{\mathrm{i}}}\right\rangle$ is the optical matrix element and describes the effective probability of the electronic transition. Since $\left|\left\langle \varphi_{ck_{\mathrm{f}}}\right| e^{iqr} e\cdot p\left|\varphi_{vk_{\mathrm{i}}}\right\rangle\right|^2$ is slowly varying with $\boldsymbol{k}$, it is convenient to neglect the $\boldsymbol{k}$ dependence of $M_{\mathrm{cv}}\,(k)$. Then, Eq. 9.7 can be rewritten in a simplified form as

$$\alpha\,(\omega) = \frac{4\pi^2 e^2}{ncm^2\omega}\cdot J_{\mathrm{cv}}(k)\cdot\left|e_0\cdot M_{\mathrm{cv}}(k)\right|^2 \tag{9.8}$$

where $J_{\mathrm{cv}}\,(k) = \int_{\mathrm{BZ}}^{\forall k}\frac{2dk}{(2\pi)^3}\delta\,(E_{\mathrm{f}}-E_{\mathrm{i}}-\hbar\omega)$ is the joint density of states (JDOS) in valence and conduction bands involved in the absorption of a photon with energy $\hbar\omega$. Under the assumptions of parabolic band edges for valence and conduction bands and optical transitions between extended states from valence band toward the conduction band (valid for α values larger than 10^4 cm^{-1}), one gets $J_{c,v}(h\nu) \propto (h\nu - E_g)^2$ [6, 14, 15]. Thus, α can be written as:

$$\alpha = \frac{B}{\hbar\omega}(\hbar\omega - E_g)^2 \tag{9.9}$$

where E_g is the optical band gap of the material. Equation 9.9 is known as Tauc's law and is successfully used to describe the higher part of α ($>10^4$ cm^{-1}) in amorphous semiconductors [14, 16]. The Tauc coefficient B is proportional, through the optical matrix element M^2, to the oscillator strength (O_s) of the optical transition and it measures the magnitude of the coupling between states in valence and conduction bands involved in the light absorption process. Finally, Eq. 9.9 can be easily linearized by plotting $\sqrt{\alpha \cdot h\nu}$ vs. $h\nu$ to extract the values of E_g and B.

It is worth noting that the comparison among samples having different amount of absorbing centers can be misleading as far as the absorption coefficient is concerned, since a different α can be related to a different amount of absorbing centers. This problem, assuming that all the Ge atoms are involved in the formation of NSs, can be overcome using the absorption cross section σ. Such a quantity is defined as the absorption coefficient α normalized to the density of the atoms involved in the photon absorption process [17, 18]. The absorption cross section can be then used within the Tauc formalism as

$$\sigma = \frac{B^*}{\hbar\omega}(\hbar\omega - E_g)^2 \tag{9.10}$$

where B^* is a modified Tauc coefficient, having the same physical meaning of B, only scaled to the Ge atomic density of the specific sample. It should be emphasized that the use of σ allows a rigorous comparison between samples with different density of absorbing centres, while the B^* coefficient suitably compares samples with different optical band gap.

As described above, the measurement of α allows to determine two of the most critical and important optical parameters of a semiconductor: the energy gap E_g and the magnitude of the light absorption transition (through the value of B constant). The same approach holds for NSs, where a further degree of freedom must be considered: the increase of the forbidden energy gap due to QCEs. This is taken into account using the simplest effective mass approximation by the formula [6]

$$E_g\,(\mathrm{NS}) = E_g\,(\mathrm{bulk}) + \frac{\hbar^2 \pi^2}{2m_{\mathrm{eh}} L^2} \tag{9.11}$$

showing as the gap of an NS is increased with respect to the bulk material by a quantity which is inversely proportional to the square of its size L and to the effective mass of the electron–hole pair m_{eh}. Depending on the type of NS, the L value can be either the diameter of a QD or the thickness of a QW. Quantum confinement effects become important when the size L of the NS is smaller than the Bohr exciton radius and the electron–hole pair is "confined" within the NS. The gap tailoring with size is one of the most important and potentially useful property of a semiconducting NS, in particular for optical devices, since the absorption properties of a semiconductor primarily depend on the forbidden energy gap. In addition, also the oscillator strength of the optical transition increases by shrinking the NS, due to the reduced exciton dimension [19, 20]. This, in turn, leads to an enhancement of the efficiency of the interband transitions induced by the light absorption. Therefore, by using the Tauc approach it is possible to experimentally evaluate the dependence of α on E_g and B (or B^*), that is, on the energy separation between bands or on the coupling of the electromagnetic field with electrons in the solid and also estimate the confinement effects occurring at the nanoscale.

Many experiments are then devoted to measure the variation of E_g and oscillator strength with the size of NSs, as well as any other physical parameter showing QCE. From the optical point of view, the absorption coefficient is the key parameter to be determined. One of the most frequent and simple experimental configuration used to measure α is to deposit a thin film of thickness d_f onto a transparent (glass or quartz) substrate, whose transmittance T_S (the percentage of transmitted light, I/I_0) has been previously measured (see Fig. 9.2).

The transmittance (T) and the reflectance (R) of the sample are then measured and the α_f coefficient can be extracted by using the

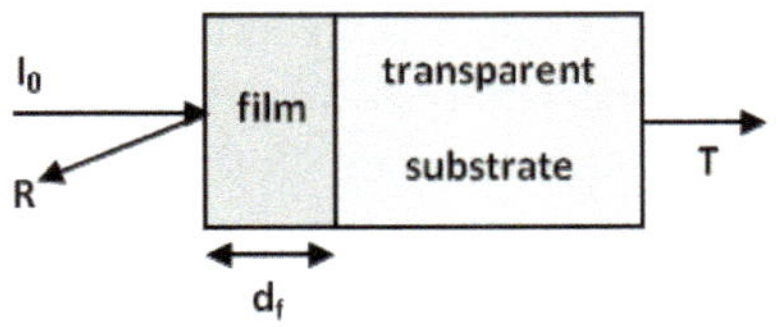

Figure 9.2 Schematic representation used to determine the absorption coefficient α of a thin film from reflectance and transmittance measurements.

formula [21]:

$$\alpha_f = \frac{1}{d_f}\ln\frac{T_s(1-R)}{T} \tag{9.12}$$

once known the thickness of the film d_f. In Eq. 9.12 multiple reflections at the film/substrate interface are not considered, since they are fairly irrelevant due to the typical low values of R [21–23]. Despite the above-mentioned assumption and including the errors on d, T, and R (see Chapter 2 for more details on optical absorption measurements), the overall error on α using Eq. 9.12 is typically lower than 10%.

9.3 Synthesis of Ge Nanostructures

A material is defined as a "nanostructure" when at least one of the three spatial dimensions is smaller than the Bohr exciton radius. It is a convention to name 2D (QWs), 1D (quantum wires), and 0D (QDs) those NSs having, respectively, 2 or 1 or 0 macroscopic spatial dimensions. Nowadays there are many techniques to grow Ge NSs, both by chemical and physical approaches. Sol–gel, chemical vapor deposition, physical vapor deposition (sputtering or evaporation), and ion implantation are some of the possible ways to form 0D, 1D, and 2D NSs. Depending on the preparation technique, Ge NSs can be synthesized on a free surface or embedded in a suitable matrix (e.g., oxides, nitrides, carbides), in both amorphous or crystalline phases. Each method presents advantages and disadvantages in terms of costs, capability of large area production, quality and structural control of the material.

The results presented in this chapter deal with films of Ge QWs and QDs confined in SiO_2 and Si_3N_4 matrices, synthesized by cosputtering deposition or ion implantation [18, 24, 25]. In most cases the formation of Ge QD requires a heating process during or after the deposition or the implantation process. The thermal energy has the aim to promote Ge diffusion for nucleation, size increasing and, eventually, amorphous to crystal phase change of QD.

The synthesis of Ge QDs (sometimes also referred to NCs) by ion implantation can be realized by implanting Ge^+ ions in Si_3N_4 or SiO_2 matrices. In Ref. [25], an implantation energy of 100 keV and fluences between 2.9 and 9.6×10^{16} Ge ions/cm^2 were chosen to induce the NC precipitation within 90 nm from the surface. Even though the energy loss for the incoming ions is slightly larger in Si_3N_4 than in SiO_2, the projected ranges came out to be $\sim$45 nm in both cases. After implantation, the matrices were subjected to furnace annealing processes (1 hour, N_2 ambient, 600°C–900°C). The matrix plays a dominant role in the formation of NSs, affecting both Ge diffusion and nucleation rates. Large differences arise in fact in the formation of Ge NCs in Si_3N_4 or in SiO_2, as evidenced by the scanning transmission electron microscopy (STEM) images in Fig. 9.3 (samples implanted with a dose of 7.3×10^{16} Ge/cm^2 and annealed at 850°C). Ge NCs appear as bright spots, due to the higher Z-contrast of the NCs with respect to both the matrices. The NC diameter ($2r$) is much larger in SiO_2 ($2r \approx$ 3–24 nm) than in Si_3N_4 ($2r \approx <$2 nm). Moreover, as evidenced by electron diffraction analysis (shown in the insets of Fig. 9.3), Ge NCs in SiO_2 are crystalline, contrary to the Si_3N_4 case, where no diffraction spots are observed. The same holds also after annealing at 900°C. Raman scattering was also employed to analyze the crystalline state of the aggregates, finding good agreement with STEM results.

Despite the high annealing temperature (Ge melting point is 938°C), Ge NCs into Si_3N_4 matrix are amorphous and small in size. Studying Ge NCs in Si_3N_4, obtained by magnetron sputtering technique, Lee et al. observed that large and crystalline NCs are only formed for very high Ge concentration and for temperatures as high as 900°C [26]. On the other hand, it was shown that Ge clusters in SiO_2 during annealing undergo Ostwald ripening and crystallization for temperatures higher than 700°C [23]. These evidences clearly prove that the embedding matrix significantly affects the formation of Ge NCs. Actually, the Ge diffusivity in Si_3N_4 can be much smaller

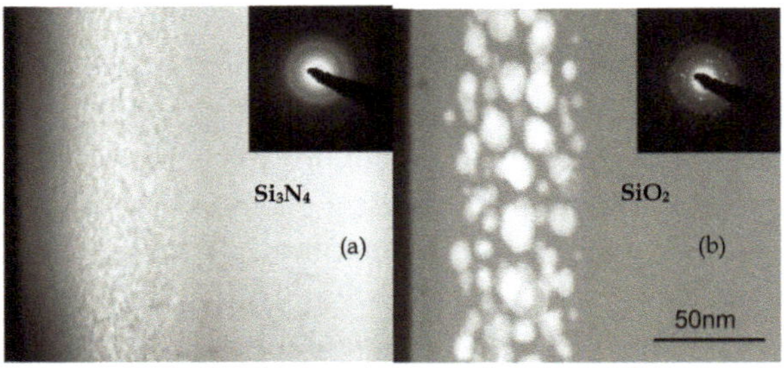

Figure 9.3 Cross-sectional high-angle annular dark-field STEM images of the Ge nanoclusters embedded in (a) Si_3N_4 or in (b) SiO_2 matrices, obtained after Ge ion implantation and annealing at 850°C, 1 h. Larger Ge nanoclusters are observed in the SiO_2 matrix. Diffraction images of the two samples are shown as insets. Reprinted with permission from [S. Mirabella, S. Cosentino, A. Gentile, G. Nicotra, N. Piluso, L. V. Mercaldo, F. Simone, C. Spinella and A. Terrasi (2012). *Matrix role in Ge nanoclusters embedded in Si_3N_4 or SiO_2*, Appl. Phys. Lett., **101**, 011911]. Copyright [2012], AIP Publishing LLC.

than in SiO_2, as it occurs for Si diffusivity ($\sim 3 \times 10^{-13}$ cm^2/s at 800°C in SiO_2 [27], $\sim 10^{-24}$ cm^2/s at 840°C in Si_3N_4 [28]). After annealing neither the implanted Ge fluence nor its profile has changed in Si_3N_4. In SiO_2 matrix a small loss (~6%) of Ge was measured. By using high-resolution Rutherford backscattering spectrometry (RBS) it was verified that Ge diffusivity in Si_3N_4 at 850°C cannot be larger than 7×10^{-17} cm^2/s [25]. These results point out that at 850°C Ge easily migrates in SiO_2, leading to NC ripening in the inner part of the film and Ge out-diffusion in the surface near region, while at the same temperature Ge diffusion in Si_3N_4 is very low, limiting the NC ripening in Si_3N_4.

The lack of crystalline phase in Ge NCs in Si_3N_4 can be related to their small size. According to the classical nucleation theory, a critical radius exists above which the amorphous to crystalline transition lowers the free energy, since for large nuclei the extra interfacial energy (γ) is compensated by the gain in the internal free energy (ΔG_{phase}) due to crystallization:

$$r^* = \frac{-2\gamma}{\Delta G_{\text{phase}}} \tag{9.13}$$

Recently, the Ge/Si_3N_4 interface was shown to have a larger γ in comparison to the Ge/SiO_2 one [29]. This supports a larger critical radius for Ge NC in Si_3N_4 (since ΔG_{phase} is not affected by the matrix, at a first approximation), justifying the lack of c-Ge NCs in the Si_3N_4 samples. Thus, the larger interfacial energy and the reduced diffusivity of Ge in Si_3N_4 limit the NC ripening and crystallization. This means that the kinetics of NC formation and crystallization is much slower in Si_3N_4 than in SiO_2, evidencing as the hosting matrix is another critical factor in NS fabrication.

The formation of Ge QDs in a SiO_2 matrix, both as single layers or multilayered systems of QDs separated by thin SiO_2 barriers, can be successfully done also via sputtering and chemical vapor deposition [30–34]. In the case of cosputtering from pure SiO_2 and Ge targets (using an Ar atmosphere and a nominal deposition temperature of 400°C) [18, 23] a mixed layer of SiGeO is formed. The thermal budget supplied during the deposition is high enough to promote a partial phase separation of the SiGeO film and the nucleation of small a-Ge QDs (due to the precipitation of Ge in excess). A post-thermal annealing in the range of 600°C to 800°C (for 1 hour, in a N_2 ambient) promotes a further phase separation of SiGeO film into SiO_2, GeO_2, and the growth of larger Ge NCs.

Multilayered samples of Ge QDs can be realized by repeating the barrier/film/barrier (SiO_2/SiGeO/SiO_2) structure. This approach, first used by Zacharias et al. [32, 33] in SiO_x/SiO_2 systems is extremely efficient to control the size of the QD in the growth direction and to produce a well-ordered array of NSs (see Chapter 4 for details on multilayer structures). In this kind of structures, the size of the QD is limited by the distance between two SiO_2 barriers (i.e., the thickness of the SiGeO layer), while their density can be varied by changing either the excess of Ge in the SiGeO layer or the thickness of the SiO_2 barrier. In particular, by varying the thickness of the SiO_2 barriers it is possible to produce multilayers with tightly packed or fairly isolated Ge QDs films, at least in the growth direction, perpendicular to the surface [18].

Bright-field TEM images, reported on Fig. 9.4, show the multi-layered structure of the films with SiO_2 barriers (brighter layers) embedding very thin SiO_2 films (~4 nm in thickness) containing ~3 nm Ge QDs (darker spots). The thickness ($d_\perp$) of the SiO_2

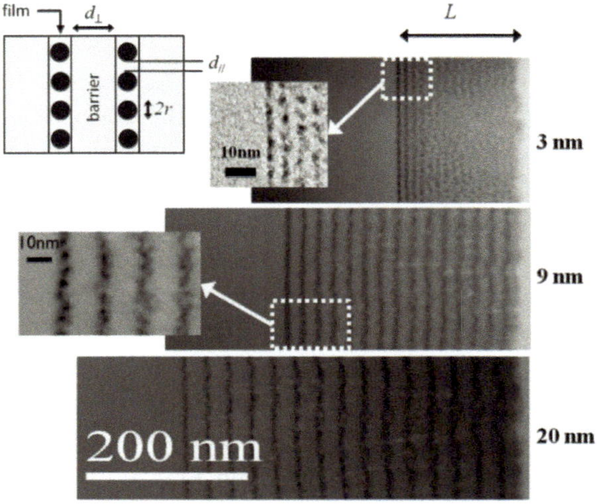

Figure 9.4 (a) Schematic and cross-sectional BF-TEM images of Ge QD multilayered samples with different thicknesses of the SiO_2 barrier. The BF-TEM images marked by the white arrows show the multilayer of Ge QDs with the two thinnest SiO_2 barriers. Reprinted with permission from [S. Mirabella, S. Cosentino, M. Failla, M. Miritello, G. Nicotra, F. Simone, C. Spinella, G. Franzò and A. Terrasi (2013). *Light absorption enhancement in closely packed Ge quantum dots,* Appl. Phys. Lett., **102**, 193105]. Copyright [2013], AIP Publishing LLC.

barrier was 3 nm for the tightest QDs configuration (106 nm total sample thickness), 9 nm for the intermediate packaging (245 nm total thickness), and 20.4 nm for the most spaced one (439 nm total thickness). The multilayered configuration also allows a better control of the size and vertical order distribution of Ge QDs.

Finally, in the case of Ge QWs confined in SiO_2, RF magnetron sputtering was used to deposit a sequence of SiO_2/Ge/SiO_2 layers. All the depositions were done at room temperature to ensure the formation of an a-Ge film whose thickness was changed from 3 nm to 125 nm to study the effect of an 1D quantum confinement [24]. Top and bottom SiO_2 films (approximately 10 nm thick each) were used as barriers for the QW structure, as schematized in Fig. 9.5.

It is worth noting that the ability to grow 2D NSs such as a QW is a non trivial aspect in nanotechnology. Films in the nanometric range of thickness show a real possibility of porous or discontinuous

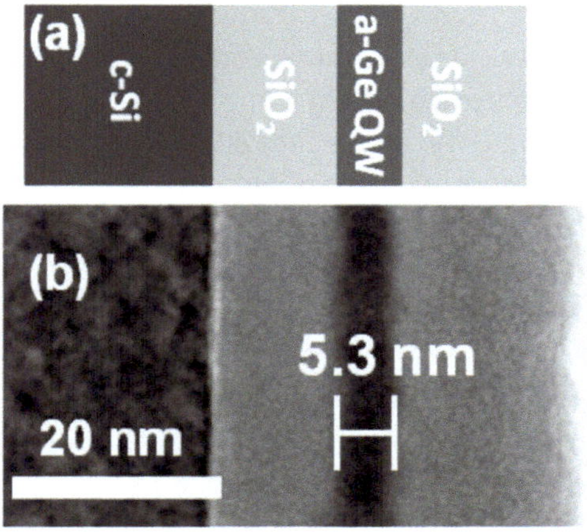

Figure 9.5 (a) Schematic of sample structure and (b) cross-sectional bright-field Z-contrast TEM images of a 5 nm thick a-Ge QW sample. Figure reprinted with permission from Ref. [24].

structure due to mechanical strain, roughness or intermixing at the interfaces. Any deviation from an ideal NS can destroy or hide QCE with detrimental consequences on technological applications. The deposition of a continuous film, with the density of the correspondent bulk material, is then a mandatory requirement to study QCE in this kind of 2D NSs. In our case we were able to confirm this aspect by using RBS and TEM analyses to measure Ge dose and film thickness of the QW.

9.4 Light Absorption in Germanium QWs

Although the most extreme level of quantum confinement is that of QD (i.e., 0D systems), the role of QWs has always been of strong relevance in many applications as, for example, 2D electron gases in microelectronics devices, optical modulators and lasers [4, 35–38]. In this sense, Ge QWs show an easier achievement of the quantum confinement properties. For example, quantum confined Stark effect has been recently demonstrated in optical modulators based on strained c-Ge multiquantum walls (MQWs) operating at 1550 nm [4, 39]. Similar systems have also shown a photoluminescence

(PL) emission at room temperature that has been attributed to the thermal excitation of carriers from the confined states of c-Ge MQWs [40]. Despite the number of studies on the optoelectronic properties of c-Ge QWs, only a limited literature is present for this material in the amorphous phase. Moreover, with respect to c-Ge, a-Ge QWs could allow the reduction of fabrication costs due to their lower temperature of fabrication. In the past years, optical studies on a-Ge thin films demonstrated an optical energy gap of ~0.8–0.9 eV and also a larger absorption coefficient than the one of c-Ge thin films [41]. However, few studies have been performed on single a-Ge films at the nanoscale regarding the possibility of having QCE at room temperature. It is clear that the possibility to exploit QCE also in a-Ge, if any, could make this material very promising for the fabrication of low-cost optoelectronic devices operable at specific tailored wavelengths, or for the development of efficient light harvesters and solar cells able to absorb a larger portion of solar spectrum via size-dependent tuning of a-Ge QW band gap. In this regard, it is interesting to study the optical behavior of a single a-Ge QW deposited at room temperature onto fused silica substrates. Accurate T and R measurements (some of which are reported in the inset of Fig. 9.6a have been performed at room temperature in the wavelength range from 200 to 2000 nm to extract the absorption coefficient α of such thin Ge films (as described above in Eq. 9.2). Figure 9.6a shows the α spectra of the a-Ge QWs and of an a-Ge film (125 nm thickness) used as a reference for bulk, unconfined film.

The absorption coefficient of the 30 nm a-Ge QW is similar to that of the 125 nm a-Ge sample, both evidencing an absorption edge at about 0.8 eV, typical of a bulk a-Ge [41]. On the contrary, by decreasing the thickness of the a-Ge QW from 12 to 2 nm, an evident blue shift occurs in the onset of the absorption spectrum. Moreover, in the 12 nm a-Ge QW, the α spectrum is higher than in the 30 nm a-Ge QW sample, despite the similar onset. Therefore, for layers thinner than 30 nm, the thickness of the a-Ge QW clearly affects the photon absorption mechanism, as an effect of spatial confinement on the electronic energy bands. Actually, the Bohr radius for excitons in Ge is about 24 nm [42, 44] and the observed variation in the absorption spectra can be thought as a QCE on the energy band

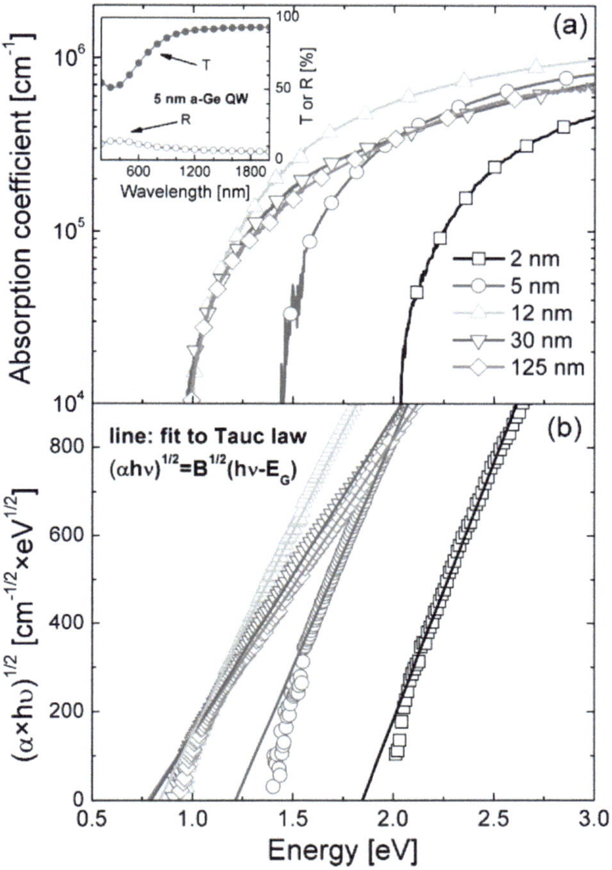

Figure 9.6 (a) Transmittance and reflectance spectra of a 5 nm a-Ge QW (inset). Absorption coefficient of an a-Ge QW of different thicknesses along with the one of a bulk-like 125 nm a-Ge film. (b) Tauc plots (symbols) and relative linear fits according to the reported Tauc law (lines). Figure reprinted with permission from Ref. [24].

in a-Ge QWs. To further analyze this point, a proper description of the light absorption mechanism is needed in the framework of the Tauc's model described above in Eq. 9.9. If the Tauc's law properly describes the light absorption also in amorphous NSs, $(\alpha h\nu)^{1/2}$ versus $h\nu$ (Tauc's plot) should give a linear trend in the energy range for which $\alpha > 10^4$ cm^{-1}, as it clearly occurs for all the a-Ge

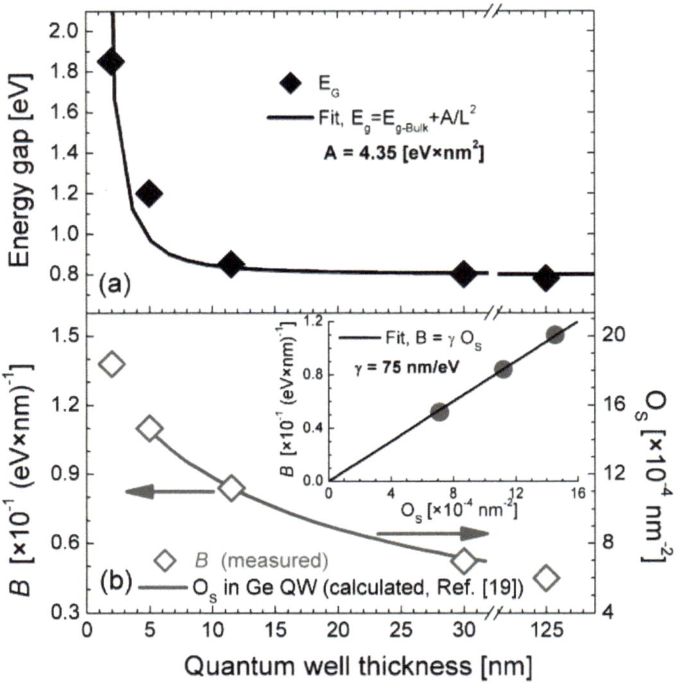

Figure 9.7 (a) Experimental values (diamonds) of the energy gap in a-Ge QW versus thickness, fitted through effective mass theory (solid line). (b) Experimental values of B (diamonds, left axis) compared with the calculated trend [19] for the oscillator strength (O_S) in Ge QWs (line, right axis). Inset shows the linear correlation between B and O_S. Figure reprinted with permission from Ref. [24].

QWs (Fig. 9.6b). The application of Tauc's law to a-Ge QWs allows to determine B and E_g through linear fitting procedures (lines in Fig. 9.6b).

To claim that the differences reported in Fig. 9.6 arise from QCE in the QW, a direct relationship between QW size and light absorption must be shown. Figure 9.7a demonstrates the dependence of the optical band gap (diamonds) on the QW thickness, evidencing a blue shift up to 1 eV for the 2 nm sample. Our E_g data have been fitted (solid line) within the effective mass theory assuming an infinite barrier, according to Eq. 9.11 (with $A = \hbar^2\pi^2/2m_{eh}$ being the only fit parameter). The band-gap energy of bulk a-Ge, $E_g(bulk)$, was fixed at 0.8 eV [41], which is also in good agreement with our value for 30 nm QW, for which a weak QC is expected. The good fitting of the experimental data confirms that the shift in the band-gap energy arises from QCE and that SiO_2 layers act as an infinite potential barrier, ensuring a strong confinement of electrons within the Ge QWs. Moreover, the experimental confinement parameter in a-Ge QWs results to be 4.35 eV·nm^2, which is not so far from the theoretical value of 1.97 eV·nm^2 reported by Barbagiovanni et al. for a strong quantum confinement in c-Ge QW [45]. Our value of A for a-Ge QWs is also much larger than that measured in a-Si QWs (0.72 eV·nm^2 [46]), evidencing the larger effect of quantum confinement in Ge NSs.

Figure 9.7b reports the increase in the light absorption efficiency as a function of the QW thickness. In fact, apart from the energy blue shift, another interesting effect of the spatial confinement is the enhanced interaction of light in confined systems. On the left axis of Fig. 9.7b, the variation of B with QW thickness is plotted, as extracted from fits in Fig. 9.6b. This quantity significantly increases up to three times going from bulk to the thinnest QW, evidencing the noteworthy increase of the light absorption efficiency. In fact, the thinner the QW thickness, the smaller the Bohr exciton radius is, thus giving rise to a larger oscillator strength (O_s) [6]. Such an effect was predicted and observed for c-Ge QWs [19], but now, for the first time, it is experimentally assessed also in a-Ge QWs. Since the B parameter in Eq. 9.9 includes the matrix element of optical transition M (which is related to O_s), the increase in B can be thought as the evidence of the enhanced oscillator strength in the confined system. Indeed, on the right axis of Fig. 9.7b the variation of O_s with thickness in the c-Ge QW is reported, as calculated in the 5 to 35 nm thickness range by Kuo and Li, using a 2D exciton model and infinite barrier [19]. The good agreement between the experimental values of B and calculated values of O_s is the confirmation that the enhanced absorption efficiency observed at room temperature in a-Ge QWs is actually due to the excitonic effect. The inset of Fig. 9.7b evidences the linear correlation between B (measured at 5, 12, and 30 nm) and the expected O_s (for those thicknesses), allowing for the estimation of the factor of proportionality ($\gamma = B/O_s$, which accounts for the absorption efficiency normalized to the oscillator strength). Thus, a proper modeling applied to light absorption measurements at room temperature allowed to quantify the extent of size effect in a-Ge QWs and to disentangle the oscillator strength increase and the band-gap widening in these structures.

As soon as one moves from an almost ideal 2D confined system, as single QW is, to a 0D confined system (i.e., QD), one expects a stronger quantum confinement of the excitons, as theoretically and experimentally observed. Indeed, the light absorption process in more complex structures, as QDs are, cannot be modeled only by their size, since other relevant effects can strongly contribute to the photoconversion process. As described in the following paragraphs, the type of insulating matrix where QDs are embedded or the QD–QD distance can effectively modify both the optical band gap and the absorption efficiency.

9.5 Confining Effects in Germanium QDs

9.5.1 *Matrix Effects: SiO_2 vs. Si_3N_4*

Besides the NS size, one of the main parameter contributing to QCE is represented by the potential barrier of the matrix where NSs are embedded. According to theory, by reducing the height V_0 of the potential barrier a lower confinement of the electron–hole pair should occur. In fact, Eq. 9.11 is valid in the case of an infinite confining barrier potential. In a real dielectric matrix, with a finite barrier height V_0, the value of E_g given in Eq. 9.11 is reduced by the factor $\left[1+\frac{\hbar}{r\sqrt{2m^* V_0}}\right]^2$. Hence, lower potential barriers reduce the effectiveness of quantum confinement. In addition, the optical behavior of NC can also be largely affected by other matrix effects, for example, defects, unpassivated bonds, NC/matrix interface states [23, 31, 43]. For this reason, understanding the influence of the hosting matrix on the QCE is a crucial step toward light-harvesting applications. To this aim, once the Ge NC formation in SiO_2 and Si_3N_4 matrices (as described in the paragraph 1.2.2) has been studied, the optical properties of these materials are compared to evidence the matrix role, if any, on the photon absorption mechanism. As the NC formation is affected by the matrix, the comparison of light

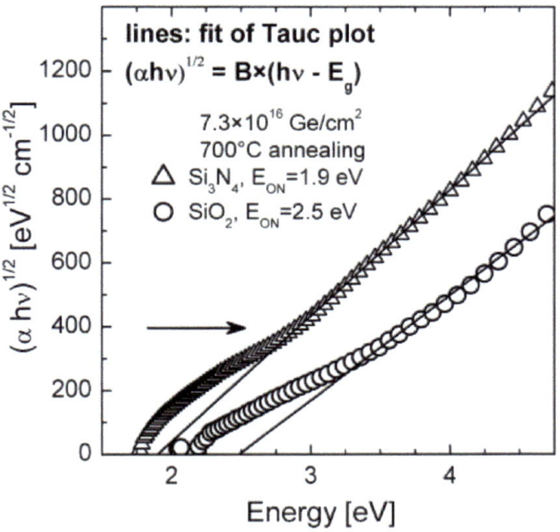

Figure 9.8 Tauc plot for Ge NCs in Si_3N_4 (triangles) or in SiO_2 (circles) and corresponding linear fit (defects induced absorption, below the arrow, causes deviation from the fit). Adapted and reprinted with permission from [S. Mirabella, S. Cosentino, A. Gentile, G. Nicotra, N. Piluso, L. V. Mercaldo, F. Simone, C. Spinella and A. Terrasi (2012). *Matrix role in Ge nanoclusters embedded in Si_3N_4 or SiO_2*, Appl. Phys. Lett., **101**, 011911]. Copyright [2012], AIP Publishing LLC.

absorption was done for samples annealed at 700°C, which are expected to give comparable Ge NCs (amorphous and 2–4 nm in size) in both kind of matrices. Tauc plots are reported in Fig. 9.8 for samples implanted at the medium fluence of 7.3×10^{16} Ge/cm^2 in Si_3N_4 (triangles) or SiO_2 (circles) matrices. The plots show a kink (indicated by an arrow for the Si_3N_4 sample) above which the trend is linear, according to the Tauc model. Below the kink, transitions can occur involving electronic states in the band tails or in the midgap, for which the Tauc law is no longer valid. Implantation damage can create midgap or tail states which account for the not linear absorption trend observed in the Tauc plots below the kink [25]. Because of this, a univocal determination of E_g^{opt} cannot be performed but still Tauc plots can be fitted above the kink (lines in Fig. 9.8), giving an onset energy for light absorption (E_{ON}). Ge NCs in Si_3N_4 (triangles) show a smaller E_{ON} (~1.9 eV) than in SiO_2

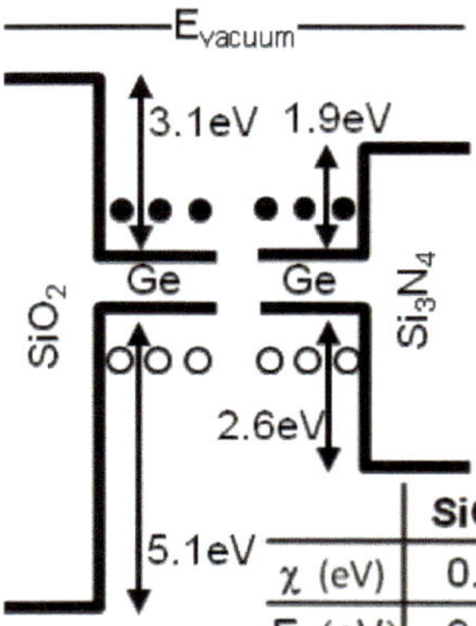

	SiO_2	Ge	Si_3N_4
χ (eV)	0.9	4.0	2.1
E_g(eV)	9.0	0.8	5.3

Figure 9.9 Band alignment scheme for Ge NCs embedded in SiO_2 or in Si_3N_4, with relative electron affinity (χ) and band gap (E_g).

($\sim$2.5 eV). In both matrices NCs exhibit E_{ON} larger than that of not-confined a-Ge ($\sim$0.8 eV), which is due to QCEs.

To explain the different E_{ON} of Ge NCs, we should consider the barrier heights seen by electrons and holes in the Ge NCs embedded in the high gap matrix, as drawn in Fig. 9.9. The offsets between conduction and valence band edges can be computed relating their position to the vacuum level using, at a first order approximation, the electron affinities (χ) and band gaps (E_G) of bulk materials (reported in the table in Fig. 9.9 [25]).

Assuming an infinite barrier (in the framework of the effective mass approximation), the variation of E_g with the size L of QDs is given by Eq. 9.11: $E_G = E_G^{bulk} + A/(L)^2$, where E_G^{bulk} is the band gap of the bulk semiconductor (0.8 eV for a-Ge), and A is the confinement parameter ($\sim$ 7.8 eV $\times$ nm^2 in strongly confined Ge QD [45]). Assuming 2 nm sized Ge NCs, E_G is 2.7 eV, in good agreement with E_{ON} of Ge QDs in SiO_2 (whose barrier can be assimilated to an infinite one). Instead, Si_3N_4 offers a lower barrier to carriers (sum of offsets $V_0 \approx 4.5$ eV) so that a finite barrier calculation is needed, where band-gap widening is reduced by the factor $\left[1 + \frac{\hbar}{r\sqrt{2m^* V_0}}\right]^2$ (effective mass of exciton, m^*, is about 0.1 for Ge [44]). This factor lowers the expected E_g to 2.0 eV, compatible with the observed E_{ON} of Ge QDs in Si_3N_4. A crucial role of the matrix is then demonstrated, pointing out that Si_3N_4 matrix allows Ge NCs to absorb light much more efficiently than in the case of SiO_2 matrix.

9.5.2 *QD–QD Interaction Effects*

As described in the previous paragraphs, the interaction of light with QD may depend on many factors such as hosting matrix, QD size, structural phase and, probably, also on the interaction among QD. A multilayer structure (as shown in Section 9.3) gives the possibility to produce QD whose size is very well controlled by the thickness of the SiGeO mixed layer (where Ge QD are formed upon thermal annealing during or after the deposition), but also the distance between two layers of QD can be varied by changing the thickness of the SiO_2 barriers. This makes possible to study any eventual role of the distance among QD on the light absorption in NSs. Samples used to this aim consist of 15 layers of Ge QD whose average diameter is 3 nm, separated by SiO_2 barriers, 3, 9 or 20 nm thick [18].

TEM analysis has been used to estimate the QD size, while RBS analysis has been used for the Ge content in the films. From the RBS spectra we found that the Ge content (D) is fairly the same, being around 6.5×10^{16} Ge/cm^2 and giving out an areal density of $\sim 4.3 \times 10^{15}$ Ge/cm^2 within each QDs layer. On the basis of TEM evidences, we can assume spherical QDs with an average diameter ($2r$) of 3 nm. Thus, each QD layer has a mean QD areal density of $\sim 7 \times 10^{12}$/cm^2, corresponding to a surface-to-surface distance ($d_{||}$) of about 1 nm between adjacent Ge QDs in the same layer. As $d_{||}$ is fixed and well lower than the vertical spacing $d_{\perp}$ due to SiO_2 barriers, the multilayer approach allows us to play only with the distance between Ge QDs films along the growth direction. In other words, in multilayer samples the QD–QD distance can be varied only in the vertical direction, while it is fixed in the plane of the QDs film.

To compare the light absorption of Ge QDs arranged in the three multilayer samples, we used the atomic absorption cross section [17], extracted as follows:

$$\sigma = \frac{\alpha d}{D} \tag{9.14}$$

where α is the absorption coefficient spectrum, d is the sample thickness, and D the Ge atomic content. As d is different in the three samples, a comparison between α spectra can be misleading, while σ represents the photon absorption probability normalized to the Ge atomic content present in the sample. Thus, if the QD configuration and distance do not play any role, σ should be the same in our samples. Instead, this is not strictly the case, as shown in Fig. 9.10a, where σ is reported for multilayered samples with Ge QDs films spaced by barriers of 3, 9, and 20 nm (squares, circles, and triangles, respectively). The measured σ spectrum for not-confined a-Ge is added for comparison (diamonds). In all cases, the absorption cross section of Ge QDs is clearly lower than for bulk Ge, in agreement with Ref. [17]. This is related to the different onset energy, much lower for bulk Ge, as expected for QCE. All the σ spectra of Ge QDs show a similar onset at about 2 eV, while the multilayer with the largest barrier clearly reports the worst performance, in terms of absorption. By using this procedure we compare the experimentally measured absorption efficiency of Ge QDs ordered in a different configuration, more and less spaced, in order to put in evidence the effects related to the QD–QD spacing, if any. It should be also noted that the rate of increase in the σ spectra is higher for the smaller barrier multilayered sample. This is an experimental evidence that a closely packed array of Ge QDs produces an enhancement of the light absorption.

To the aim of better explaining this aspect, the modified version of Tauc model can be used to describe the photon absorption process. If the Eq. 9.10 properly describes the light absorption in these confined systems, we should get a linear trend of the experimental quantity $(\sigma h\nu)^{1/2}$ plotted versus $h\nu$ (a sort of modified Tauc plot). This is what occurs for all our samples, in a wide range of energy, as reported in Fig. 9.10b with symbols, confirming that the photon absorption process can be suitably depicted by Eq. 9.10. A simple linear fitting procedure (lines in Fig. 9.10b) allows us to determine B^* and E_g, which are the only two parameters describing the spectrum according to Eq. 9.10. For all the samples, the fitting line satisfactorily agrees with the experimental data. Figure 9.11 summarizes the extracted optical parameters (E_g and B^*) of the multilayer samples as a function of the barrier thickness. Figure 9.11a shows that all the multilayer samples (square data) exhibit the same optical energy gap (1.9 ± 0.1 eV), mostly independent of

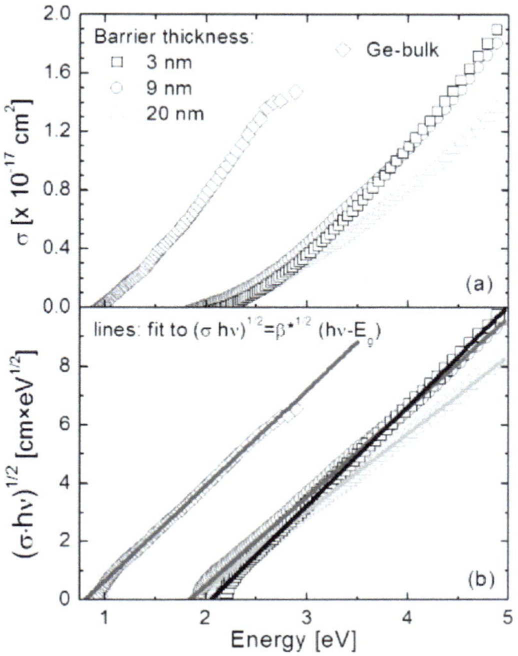

Figure 9.10 (a) Atomic absorption cross section (σ) spectra for Ge QD multilayers with different SiO_2 barriers, together with the σ spectra of amorphous bulk Ge. (b) Tauc plot (symbols) and corresponding linear fits (lines) for Ge QDs arranged in multilayers and an amorphous bulk Ge film. Reprinted with permission from [S. Mirabella, S. Cosentino, M. Failla, M. Miritello, G. Nicotra, F. Simone, C. Spinella, G. Franzò and A. Terrasi (2013). *Light absorption enhancement in closely packed Ge quantum dots,* Appl. Phys. Lett., **102**, 193105]. Copyright [2013], AIP Publishing LLC.

the barrier thickness. This evidence is in agreement with the QCE of size tuning of E_g, as all the Ge QDs are similar in size ($2r \approx 3$ nm), and the E_g value is well larger than that of not-confined a-Ge ($\sim$0.8 eV, reported as black rectangle in Fig. 9.11a). According to Eq. 9.11, for Ge QDs of $\sim$3 nm in diameter, E_g should be around $\sim$1.7 eV, in reasonable agreement with the experimental data [45]. Thus, the enhanced light absorption cross section in closely packed Ge QDs cannot be ascribed to different optical band gaps.

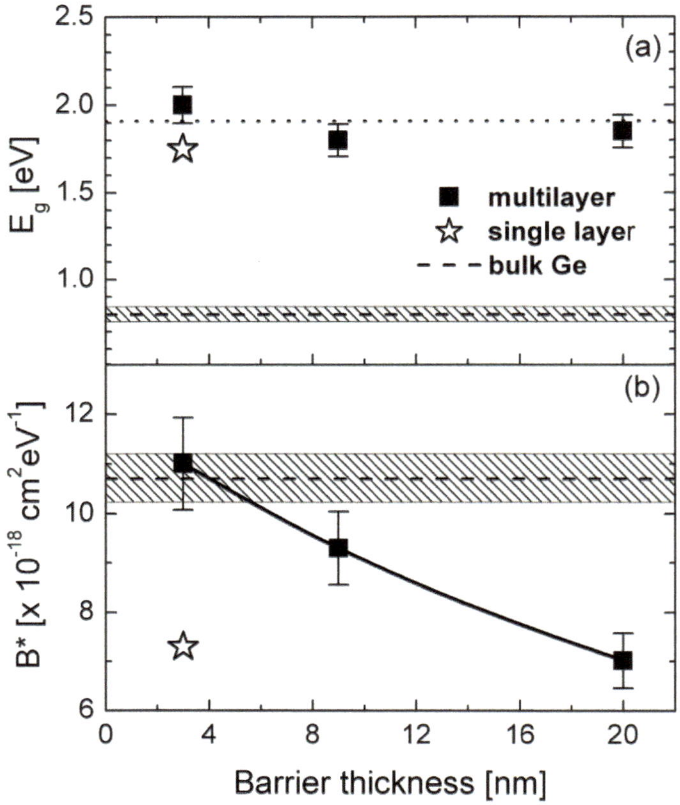

Figure 9.11 (a) Experimental values of energy gap (E_g, squares) and (b) absorption efficiency (B^*, squares) of Ge QDs arranged in multilayers with different SiO_2 barrier thicknesses. For comparison the E_g and B^* values are reported for amorphous bulk Ge (shadowed regions) and Ge QDs in a single-layer (star) configuration. Adapted and reprinted with permission from [S. Mirabella, S. Cosentino, M. Failla, M. Miritello, G. Nicotra, F. Simone, C. Spinella, G. Franzò and A. Terrasi (2013). *Light absorption enhancement in closely packed Ge quantum dots,* Appl. Phys. Lett., **102**, 193105]. Copyright [2013], AIP Publishing LLC.

In Fig. 9.11b, we report the barrier thickness dependence of the modified Tauc coefficient B^*, clearly showing a strong decrease in the sample with the more spaced Ge QDs array. This evidence is directly linked to the observed enhanced light absorption cross section in closely packed Ge QDs, as B^* represent a sort of absorption efficiency, independent of the optical band gap. In particular, in the closest packed configuration such absorption efficiency is almost twice, with respect to the case of the most spaced Ge QDs. To get a basis for comparison, we experimentally extracted the modified Tauc coefficient B^* for not-confined a-Ge film, and reported it as shadowed region in Fig. 9.11b. Such a value is in agreement with that for the closest packed Ge QDs, while other samples show a lower absorption efficiency in comparison to bulk. Even if this experimental approach allows us to modify the QD–QD distance only in the vertical direction, some general consideration can be drawn. Thus, as far as the light absorption mechanism is concerned, the quantum confinement in Ge QDs clearly increases the optical band gap with respect to the bulk, but it does not give a clear advantage on the light absorption efficiency. Instead, by using largely spaced Ge QDs an evident loss occurs in the efficiency of the absorption process. Such an effect has been observed in our samples up to 20 nm of QD–QD vertical spacing.

To account for the effect of a 3D QD–QD spacing on the absorption efficiency, a sample with a single layer (200 nm thick) of SiO_2:Ge was fabricated and characterized in the same way [23]. Ge QDs of similar diameter have been found, with a surface-to-surface distance (now in all directions) of 3 nm. This single-layer sample can be compared with the multilayer sample with a barrier thickness of 3 nm, to account for the modulation of $d_{||}$, the in-plane QD–QD distance. The single-layer sample shows an optical band gap of 1.7 eV (star in Fig. 9.11a), as expected for the QD diameter, still an absorption efficiency (star in Fig. 9.11b), lower than the multilayer sample with the smaller barrier thickness, and comparable with the largest barrier thickness. In other words, when Ge QDs are spaced by 3 nm in three dimensions, they absorb as much as in a multilayer configuration with 20 nm of vertical spacing and 1 nm of in plane spacing. These data evidence that the QD–QD spacing plays a key role in the photon absorption process. Therefore, some long range interaction between QDs has to be assumed to account for the observed effect. Actually, the presence of electronic coupling between semiconductor nanoparticles has been theoretically described [47, 48], for which energy transfer occurs between semiconductor nanocrystals up to 10–20 nm apart, mainly by means of dipole–dipole interactions. This effect is typically observed in light emission spectroscopy [49]. If some electron coupling occurs between closely packed Ge QDs, it should affect the electron transition probability and then the light absorption mechanism. On the other hand, as the QD–QD distance within the film is quite small ($d_{||} \approx 1$ nm), a kind of collective behavior cannot be excluded at all, as if the interaction responsible for the light absorption enhancement occurs between the ensembles of Ge QDs contained in each film. Anyway, this effect can be further exploited for enhancing the absorption of NS materials for photovoltaics devices.

So far we have shown how photoconversion in confined Ge NSs can offer very promising chances for the fabrication of more efficient solar cells and light detectors. For example, the large QCE on the optical band gap and on the oscillator strength occurring at room temperature in single a-Ge QW make these systems very attractive toward the fabrication of energy tunable solar cells. As soon as one moves from almost ideal system, as single QWs are, to more complex structures, the light absorption process cannot be modeled only by their size, since other relevant effects strongly contribute to the photoconversion process. For example, the presence of oxide defects and surface states at the QD/SiO_2 interface can dominate over the QCE of Ge QDs embedded in SiO_2. In addition, the different barrier height offered by SiO_2 or Si_3N_4 effectively modifies the optical band gap of Ge QDs. Finally, also the QD–QD spacing can significantly change the effectiveness of the light absorption in these systems.

9.6 Light Detection with Germanium Nanostructures

Despite several photovoltaics potentialities of NSs have been clearly proven, the fabrication of solar cells based on NSs is still a subject of research. The main challenges on the way to a large-scale production are costs of NSs, the precise control of their structural parameters that define the above mentioned QCE, and, more importantly, the efficient transport of the photogenerated carriers through a dielectric-like matrix. In this scenario, Ge NSs embedded in a dielectric-like matrix seem to be even more promising as active material for the fabrication of light detecting devices, due to the more efficient transport of carriers under applied bias [50–53].

In recent years, Si QD-based photodetectors have been shown to achieve relatively high responsivity (defined as the photocurrent generated in a photodetector divided by the incident optical power), with peak values in the range of 0.4–2.8 A/W and optoelectronic conversion efficiencies as high as 200% [53, 54]. Recently it has been demonstrated that Ge QDs embedded in SiO_2 can be used as active material for the fabrication of broadband photodetectors with even higher efficiency. These Ge QDs were synthesized by magnetron cosputtering of SiO_2 and Ge targets on (001) *n*-doped Si substrate. The low substrate temperature during the deposition (400°C) allows the nucleation of small a-Ge QDs with a size of about 2–3 nm. A metal-insulator-semiconductor (MIS) configuration (schematically drawn in the inset of Fig. 9.12) was pursued after sputter deposition at room temperature of a transparent gate electrode (In-doped ZnO, 3 mm in diameter) onto the SiO_2 film

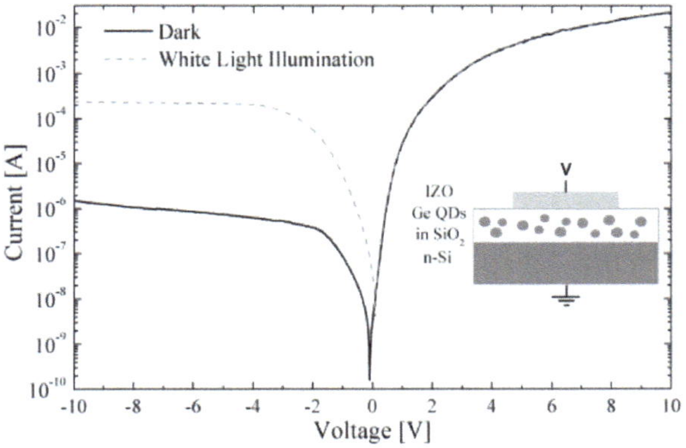

Figure 9.12 Current–voltage *I*(V) characteristics in dark and under white light illumination of a metal-insulator-semiconductor (MIS) photodetector (PD) with Ge QDs embedded in a silicon dioxide layer. Please note that the current is in absolute value. The inset shows a schematic cross section of the device. Adapted and reprinted with permission from [S. Cosentino, Pei Liu, Son T. Le, S. Lee, D. Paine, A. Zaslavsky, S. Mirabella, M. Miritello, I. Crupi, A. Terrasi and D. Pacifici (2011), *High-efficiency silicon-compatible photodetectors based on Ge quantum dots*, Appl. Phys. Lett. **98**, 221107]. Copyright [2011], AIP Publishing LLC.

containing Ge QD. Finally, silver paste was used to ensure the electrical back contact [50].

Figure 9.12 compares the $I(V)$ curves in the dark and under white light illumination of the MIS device with Ge QDs embedded in the insulating (SIO_2) matrix. The dark $I(V)$ shows a rectifying behavior, with a small current in reverse bias and an exponential increase of current in forward bias, typical for MIS devices on a n-type semiconductor substrate [7]. Upon white light illumination, the forward bias $I(V)$ remains largely unaffected, but there is a strong increase of the reverse current by a factor larger than 10^2 due to the contribution of photogenerated carriers.

To clarify the mechanism of photoinduced conduction, $I(V)$ measurements have been performed by illuminating the device at various incident λ in the 400–1100 nm range under continuous-wave (CW) radiation. As shown in Fig. 9.13a, a clear dependence on λ in the 500–1000 nm range is observed, indicating a clear wavelength dependence of the carrier photogeneration. A reference device, with the same oxide thickness on the same substrate but without Ge QDs, exhibits no response for any λ (see Fig. 9.13b), indicating the key role of the Ge QDs in the photoconduction.

Figure 9.14a shows the responsivity of a device containing a-Ge QDs as a function of λ obtained by measuring the photogenerated current from Fig. 9.13a (defined as the difference between the total current under illumination minus the dark current at a given reverse bias) and normalizing it to the incident optical power calibrated using a Si reference cell. The peak responsivity shows a broad spectrum peaked at $\lambda \approx 900$ nm reaching a value of $\sim$4 A/W and $\sim$1.75 A/W at -10 V and -2 V bias voltages, respectively. Such values of responsivity are much higher than those of commercially available Si-based detectors, as shown in Fig. 9.14a. The internal quantum efficiency (IQE) of a-Ge QDs MIS photodiode can be calculated by measuring the reflectance R at normal incidence, shown in Fig. 9.14b, and then normalizing the number of photogenerated carriers by the number of absorbed photons (i.e., by [1–R] times the number of incident photons) for any given λ, according to the formula:

$$\mathrm{IQE} = \frac{hc}{\lambda} \frac{\left(I_{\mathrm{light}} - I_{\mathrm{dark}}\right)}{(1-R) \cdot \mathrm{Power}} \tag{9.15}$$

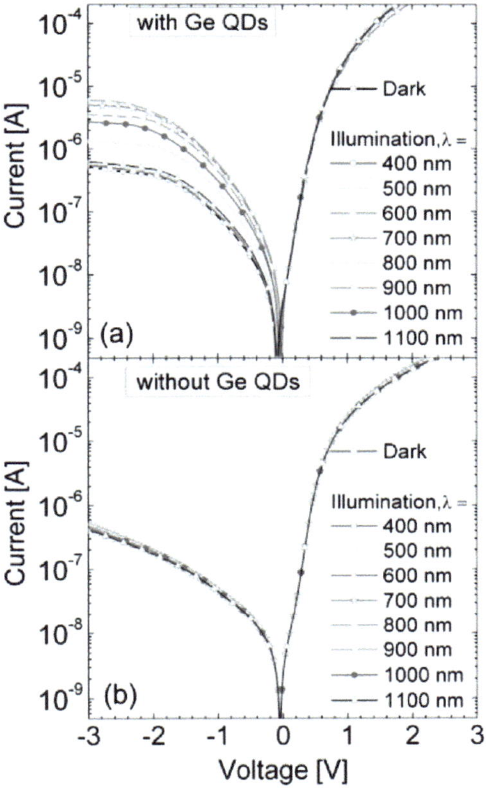

Figure 9.13 MIS PD $I(V)$ characteristics as a function of excitation wavelength in the 400–1100 nm range for a device with (a) and without (b) a-Ge QDs. Please note that the current is in absolute value. Reprinted with permission from [S. Cosentino, Pei Liu, Son T. Le, S. Lee, D. Paine, A. Zaslavsky, S. Mirabella, M. Miritello, I. Crupi, A. Terrasi and D. Pacifici (2011), *High-efficiency silicon-compatible photodetectors based on Ge quantum dots*, Appl. Phys. Lett. **98**, 221107]. Copyright [2011], AIP Publishing LLC.

The results are summarized in Fig. 9.14c, which shows IQE as high as 700% at −10 V and 300% at a lower bias of −2 V. These results evidence the existence of a large photoconductive gain due to the presence of Ge QDs. Since V drops almost only over the thick insulating layer and since gain is present at V as low as $-2V$, impact

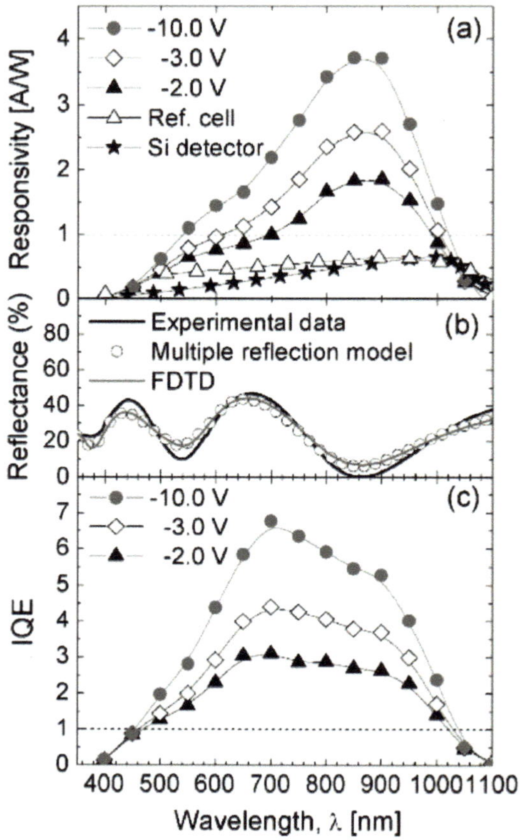

Figure 9.14 (a) Spectral responsivity of our MIS PD versus reverse bias; stars and open triangles indicate, respectively, the responsivity of a commercial Si PD and of an NREL-calibrated silicon reference cell. (b) Measured reflectance spectra and simulations using a multiple-reflection model and FDTD analysis. (c) IQE. Reprinted with permission from [S. Cosentino, Pei Liu, Son T. Le, S. Lee, D. Paine, A. Zaslavsky, S. Mirabella, M. Miritello, I. Crupi, A. Terrasi and D. Pacifici (2011), *High-efficiency silicon-compatible photodetectors based on Ge quantum dots*, Appl. Phys. Lett. **98**, 221107]. Copyright [2011], AIP Publishing LLC.

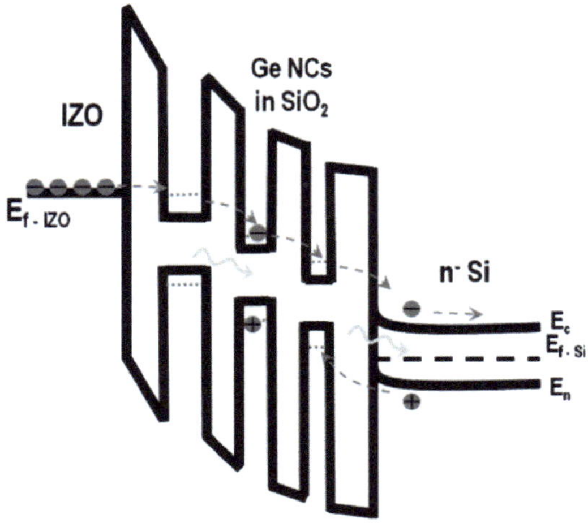

Figure 9.15 Schematic representation of the energy band diagram of an MIS PD with Ge QDs and mechanism of conduction under illumination and reverse bias. Reprinted from *Thin Solid Films*, **548**, 551–555, (2013), S. Cosentino, S. Mirabella, Pei Liu, Son T. Le, M. Miritello, S. Lee, I. Crupi, G. Nicotra, C. Spinella, D. Paine, A. Terrasi, A. Zaslavsky and D. Pacifici, Role of Ge nanoclusters in the performance of Ge-based photodetectors, Copyright (2013), with permission from Elsevier.

ionization in the Ge QDs or in the substrate (typically observed at higher bias) is ruled out as the dominant gain mechanism.

Actually, the large photoresponse is ascribed to a mechanism of hole trapping mediated by the Ge NCs. According to this model, schematically shown in of Fig. 9.15, (1) electron–hole pairs are photogenerated both in the Ge QDs layer and in the Si substrate; (2) due to the large difference in the tunneling mass, the holes are exponentially slower than electrons to tunnel between QDs in the SiO_2; therefore (3) a net positive (hole) charge accumulates in the Ge QD layer; and to maintain charge neutrality, (4) additional electrons need to be supplied from the IZO reservoir, which tunnel through the SiO_2 and contribute to the observed photoconductive gain.

Therefore, the main role of Ge NCs is to trap photogenerated holes injected from the Si substrate, thus acting as a hopping conduction channel for the electrons injected from the IZO gate contact. The contribution of the Si substrate in the photogeneration of holes is crucial to explain the observed gain in our PDs, especially for wavelengths where Ge NCs do not play a significant role in the optical absorption of incident photons ($\lambda > 800$ nm). These results suggest that Ge QDs could be very promising for the fabrication of high-performance integrated optoelectronic devices, fully compatible with silicon technology in terms of fabrication and thermal budget.

9.7 Conclusions

For most of the people the great advantages and hopes offered by the advent of nanotechnology are related to the very small dimensions of objects and devices that can be manipulated and fabricated. This is certainly of extreme importance in many applications, but one of the most important consequences of nanoscience is that, below certain critical dimensions, physical systems enter into the quantum physics world, exhibiting new properties that we call QCE. In particular, QCE can be really effective in modifying the photon absorption process in NSs, opening new chances to increase the light harvesting up to unprecedented levels. In this chapter we described some of the main aspects regarding Ge NSs, from preparation methods to optical properties and integration into novel light detectors with much better performances than standard devices. The relationship between optical behavior and QCE has been reported for two different kinds of Ge NSs, namely, QDs and QWs, showing the role of size, distance, and embedding matrix. In fact, the energy threshold for light absorption (optical band-gap energy) can be increased, in a quasicontinuous way, from the Ge bulk value (0.8 eV) up to more than 2.6 eV (for Ge QDs in fused silica matrix). On the other hand, a clear excitonic effect is evidenced in very thin Ge QWs, which can enhance up to three times the light absorption efficiency. Moreover, a significant QD–QD interaction is also demonstrated to affect the light absorption process through some long-range electronic coupling. Beyond these evidences, many further aspects still need to be clarified before the technological transfer of these materials to commercial devices but the results obtained so far are extremely encouraging and promise that quantum effects will be part of our daily life in a near future.

Acknowledgments

The knowledge and results reported in this chapter are based on the efforts of many other people than simply the authors. We are indebted to Dr. Maria Miritello of the CNR-IMM Matis for sputtering depositions, to Dr. Corrado Spinella and Dr. Giuseppe Nicotra of the CNR-IMM for TEM analysis, to Prof. Domenico Pacifici and Dr. Pei Liu of Brown University (Providence, USA) for the collaboration on the fabrication of photodetectors, to Prof. Francesca Simone of the University of Catania for her support in the optical characterizations, and to Mr. Carmelo Percolla, Mr. Salvo Tatì, and Mr. Giuseppe Pantè of the CNR-IMM Matis for technical assistance. Some of the results shown in this chapter have been obtained in the framework of the project ENERGETIC, PON00355_3391233.

References

1. J. Bardeen and W. Brattain (1949). *Physical principles involved in transistor action*, Phys. Rev., **75**, 1208–1225.
2. C. Claeys and E. Simoen (2007): *Germanium-Based Technologies: From Materials to Devices* (Elsevier, Amsterdam).
3. W. Guter, Jan Schone, S. P. Philipps, Marc Steiner, et al. (2009). *Current-matched triple-junction solar cell reaching 41.1% Conversion efficiency under concentrated sunlight*, Appl. Phys. Lett., **94**, 223504.
4. Y. Kuo, Y. K. Lee, Y. Ge, S. Ren, J. E. Roth, T. I. Kamins, D. A. B. Miller and J. S. Harris (2005). *Strong quantum-confined Stark effect in germanium quantum-well structures on silicon*, Nature **437**, 1134–1136.
5. J. Liu, M. Beals, A. Pomerene, S. Bernardis, R. Sun, J. Cheng, L. C. Kimerling and J. Michel (2008). *Waveguide-integrated, ultralow-energy GeSi electro-absorption modulators*, Nat. Photon., **2**, 433–437.
6. F. Bassani and G. P. Parravicini (1975). *Electronic States and Optical Transitions in Solids*, Ed. R. A. Ballinger (Pergamon Press, Oxford).

7. Simon M. Sze (1981). *Physics of Semiconductor Devices*, 2nd Ed. (John Wiley & Sons, New York, NY).
8. H. W. Kroto, J. R. Heath, S. C. O'Brien, R. F. Curl and R. E. Smalley (1985). *C_{60}: buckminsterfullerene*, Nature (London), **318**, 162–163.
9. L. Pavesi, L. Dal Negro, C. Mazzoleni, G. Franzò and F. Priolo (2000). *Optical gain in silicon nanocrystals*, Nature, **408**, 440–444.
10. J. Liu, X. Sun, R. Camacho-Aguilera, L. C. Kimerling and J. Michel (2010). *Ge-on-Si laser operating at room temperature*, Opt. Lett., **35**, 679–681.
11. D. Ahn, C. Hong, J. Liu, W. Giziewicz, M. Beals, L. C. Kimerling, J. Michel, J. Chen and F. X. Kärtner (2007). *High performance, waveguide integrated Ge photodetectors*, Opt. Express, **15**, 3916–3921.
12. P. Tognini, L. C. Andreani, M. Geddo, A. Stella, P. Cheyssac, R. Kofman and A. Migliori (1996). *Different quantum behavior of the E1 and E2 spectral structures in Ge nanocrystals*, Phys. Rev. B, **53**, 6992–6995.
13. H. P. Philipp and E. A. Taft (1959). *Optical constants of germanium in the region 1 to 10 eV*, Phys. Rev., **113**, 1002–1005.
14. S. Knief and W. von Niessen (1999). *Disorder, defects and optical absorption in a-Si and a-Si:H*, Phys. Rev. B, **59**, 12940–12946.
15. M. S. Dresselaus (1999). *Solid State Physics: Part II; Optical Properties of Solids*, Lecture notes part 2. Available at http://web.mit.edu/course/6/6.732/www/6.732-pt2.pdf.
16. J. Tauc (1974). *Amorphous and Liquid Semiconductors*, Ed. J. Tauc (Plenum Press, London and New York), 175.
17. C. Uhrenfeldt, J. Chevallier, A. N. Larsen and B. B. Nielsen (2011). *Near-infreared-ultraviolet absorption cross sections for Ge nanocrystals in SiO2 thin films: effect of shape and layer structure*, J. Appl. Phys., **109**, 094314.
18. S. Mirabella, S. Cosentino, M. Failla, M. Miritello, G. Nicotra, F. Simone, C. Spinella, G. Franzò and A. Terrasi (2013). *Light absorption enhancement in closely packed Ge quantum dots*, Appl. Phys. Lett., **102**, 193105.
19. Y. H. Kuo and Y. S. Li (2009). *Variational calculation for the direct-gap exciton in the Ge quantum well systems*, Phys. Rev. B, **79**, 245328.
20. Y. H. Kuo and Y. S. Li (2009). *Direct-gap exciton and optical absorption in the Ge/SiGe quantum well system*, Appl. Phys. Lett., **94**, 121101.
21. S. Mirabella, R. Agosta, G. Franzò, I. Crupi, M. Miritello, R. Lo Savio, M. A. Di Stefano, S. Di Marco, F. Simone and A. Terrasi (2009). *Light absorption in silicon quantum dots embedded in silica*, J. Appl. Phys., **106**, 103505.
22. S. Mirabella, G. Di Martino, I. Crupi, S. Gibilisco, M. Miritello, R. Lo Savio, M. A. Di Stefano, S. Di Marco, F. Simone and F. Priolo (2010). *Light*

absorption and electrical transport in Si:O alloys for photovoltaics, J. Appl. Phys., **108**, 093507.

23. S. Cosentino, S. Mirabella, M. Miritello, G. Nicotra, R. Lo Savio, F. Simone, C. Spinella and A. Terrasi (2011). *The role of the surfaces in the photon absorption in Ge nanoclusters embedded in silica*, Nanoscale Res. Lett., **6**, 135.
24. S. Cosentino, M. Miritello, I. Crupi, G. Nicotra, F. Simone, C. Spinella, A. Terrasi and S. Mirabella (2013). *Room-temperature efficient light detection by amorphous Ge quantum wells*, Nanoscale Res. Lett., **8**, 128.
25. S. Mirabella, S. Cosentino, A. Gentile, G. Nicotra, N. Piluso, L. V. Mercaldo, F. Simone, C. Spinella and A. Terrasi (2012). *Matrix role in Ge nanoclusters embedded in Si_3N_4 or SiO_2*, Appl. Phys. Lett., **101**, 011911.
26. S. Lee, S. Huang, G. Conibeer and M. A. Green (2011). *Structural and optical study of Ge nanocrystals embedded in Si3N4 matrix*, Energy Proc., **10**, 20.
27. H. G. Chew, W. K. Choi, Y. L. Foo, F. Zheng, W. K. Chim, Z. J. Voon, K. C. Seow, E. A. Fitzgerald and D. M. Y. Lai (2006). *Effect of germanium concentration and oxide diffusion barrier on the formation and distribution of germanium nanocrystals in silicon oxide matrix*, Nanotechnology, **17**, 1964–1968.
28. H. Schmidt, U. Geckle and M. Burns (2006). *Simultaneous diffusion of Si and N in silicon nitride*, Phys. Rev. B, **74**, 045203.
29. J.E. Chang, P. H. Liao, C. Y. Chien, J. C. Hsu, M. T. Hung, H. T. Chang, S. W. Lee, W. Y. Chen, T. M. Hsu, T. George and P. W. Li (2012). *Matrix and quantum confinement effects on optical and thermal properties of Ge quantum dots*, J Phys. D: Appl. Phys., **45**, 105303.
30. M. Zacharias, F. Stolze, T. Drusedau and W. Bock (1995). *The Preparation of amorphous $Si_xGe_yO_z$ alloy films*, Phys. Status Solidi B, **189**, 409.
31. M. Zacharias and P. M. Fauchet (1997). *Blue luminescence in films containing Ge and GeO_2 nanocrystals: the role of defects*, Appl. Phys. Lett., **71**, 3.
32. M. Zacharias, R. Weigand, B. Dietrich, F. Stolze, J. Bläsing et al. (1997). *A comparative study of Ge nanocrystals in SixGeyOz alloys and SiOx/GeOy multilayers*, J. Appl. Phys., **81**, 2384.
33. M. Zacharias and P. Streitenberger (2000). *Crystallization of amorphous superlattices in the limit of ultrathin films with oxide interfaces*, Phys. Rev. B, **62**, 8391–8396.
34. T. Baron, B. Pelissier, L. Perniola, F. Mazen, J. M. Hartmann and G. Rolland (2003). *Chemical vapor deposition of Ge nanocrystals on SiO2*, Appl. Phys. Lett., **83**, 1444.

35. G. Abstreiter, H. Brugger, T. Wolf, H. Jorke and H. J. Herzog (1985). *Strain-induced two-dimensional electron gas in selectively doped Si/SixGe1-x superlattices*, Phys. Rev. Lett., **54**, 2441–2444.
36. S. Thiel, G. Hammerl, A. Schmehl, C. W. Schneider and J. Mannhart (2006). *Tunable quasi-two-dimensional electron gases in oxide heterostructures*, Science, **313**, 1942–1945.
37. O. Ambacher, J. Smart, J. R. Shealy, N. G. Weimann, K. Chu, M. Murphy, W. J. Schaff, L. F. Eastman, R. Dimitrov, L. Wittmer, M. Stutzmann, W. Rieger and J. Hilsenbeck (1999). *Two-dimensional electron gases induced by spontaneous and piezoelectric polarization charges in N-and Ga-face AlGaN/GaN heterostructures*, J. Appl. Phys., **85**, 3222.
38. E. A. Fitzgerald, Y.-H. Xie, D. Monroe, P. J. Silverman, J. M. Kuo, A. R. Kortan, F. A. Thiel and B. E. Weir (1992). *Relaxed Gex-Si1-x structures for III–V integration with Si and high mobility two-dimensional electron gases in Si*, J. Vac. Sci. Technol. B, **10**, 1807.
39. Yu-hsuan Kuo, Yong Kyu Lee, Yangsi Ge, Shen Ren, Jonathan E. Roth, Theodore I. Kamins, David A. B. Miller and James S. Harris (2006). *Quantum-confined Stark effect in Ge/SiGe quantum wells on Si for optical modulators*, IEEE J. Sel. Top. Quantum Electron., **12**, 1503.
40. E. Gatti, E. Grilli, M. Guzzi, D. Chrastina, G. Isella and H. von Känel (2011). *Room temperature photoluminescence of Ge multiple quantum wells with Ge-rich barriers*, Appl. Phys. Lett., **98**, 031106.
41. L. J. Pilione, K. Vedam, J. E. Yehoda, R. Messier and P. J. McMarr (1987). *Thickness dependence of optical gap and void fraction for sputtered amorphous germanium*, Phys. Rev. B, **35**, 9368.
42. Y. Maeda, N. Tsukamoto, Y. Yazawa, Y. Kanemitsu and Y. Masumoto (1991). *Visible photoluminescence of Ge microcrystals embedded in SiO_2 glassy matrices*, Appl. Phys. Lett., **59**, 3168–3170.
43. M. Zacharias, D. Hiller, A. Hartel and S. Gutsch (2012). *Defect engineering of Si nanocrystals interfaces and the temperature dependence of the band gap of size selected Si nanocrystals*, Phys. Status Solidi A, **209**, 2449.
44. A. G. Cullis, L. T. Canham and P. D. J. Calcott (1997). *The structural and luminescence properties of porous silicon*, J. Appl. Phys., **82** 909.
45. E. G. Barbagiovanni, D. J. Lockwood, P. J. Simpson and L. V. Goncharova (2012). *Quantum confinement in Si and Ge nanostructures*, J. Appl. Phys., **111**, 034307.
46. D. J. Lockwood, Z. H. Lu and J.-M. Baribeau (1996). *Quantum confined luminescence in Si/SiO2 superlattices*, Phys. Rev. Lett., **76**, 539–541.

47. G. Allan and C. Delerue (2007). *Energy transfer between semiconductor nanocrystals: validity of Forster's theory*, Phys. Rev. B, **75**, 195311.

48. Roi Baer and Eran Rabani (2008). *Theory of resonance energy transfer involving nanocrystals: the role of high multipoles*, J. Chem. Phys., **128**, 184710.

49. F. Priolo, G. Franzo', D. Pacifici, V. Vinciguerra, F. Iacona and A. Irrera (2001). *Role of the energy transfer in the optical properties of undoped and Er-doped interacting Si nanocrystals*, J. Appl. Phys., **89**, 264.

50. S. Cosentino, Pei Liu, Son T. Le, S. Lee, D. Paine, A. Zaslavsky, S. Mirabella, M. Miritello, I. Crupi, A. Terrasi and D. Pacifici (2011). *High-efficiency silicon-compatible photodetectors based on Ge quantum dots*, Appl. Phys. Lett., **98**, 221107.

51. Pei Liu, S. Cosentino, Son T. Le, S. Lee, D. Paine, A. Zaslavsky, D. Pacifici, S. Mirabella, M. Miritello, I. Crupi and A. Terrasi (2012). *Transient photoresponse and incident power dependance of high-efficiency germanium quantum dot photodetectors*, J. Appl. Phys., **112**, 083103.

52. S. Cosentino, S. Mirabella, Pei Liu, Son T. Le, M. Miritello, S. Lee, I. Crupi, G. Nicotra, C. Spinella, D. Paine, A. Terrasi, A. Zaslavsky and D. Pacifici (2013). *Role of Ge nanoclusters in the performance of Ge-based photodetectors*, Thin Solid Films, **548**, 551–555.

53. S. Cosentino, E. G. Barbagiovanni, I. Crupi, M. Miritello, G. Nicotra, C. Spinella, D. Pacifici, S. Mirabella, A. Terrasi, *Size tuning on the optical absorption and carrier transport in Germanium quantum dots for efficient light harvesting devices*, in press on Solar Energy Materials and Solar Cells DOI: 10.1016/j.solmat.2014.09.012.

54. J. M. Shieh, W. C. Yu, J. Y. Huang, C. K. Wang, B. T. Dai, H. Y. Jhan, C. W. Hsu, H. C. Kuo, F. L. Yang and C. L. Pan (2009). *Near-infrared silicon quantum dots metal oxide-semiconductor field-effect transistor photodetector*, Appl. Phys. Lett., **94**, 241108.

Chapter 10

Application of Surface-Engineered Silicon Nanocrystals with Quantum Confinement and Nanocarbon Materials in Solar Cells

Vladimir Svrcek[a] and Davide Mariotti[b]

[a]*Research Center for Photovoltaic Technologies, AIST, Tsukuba 305-8568, Japan*
[b]*Nanotechnology and Integrated Bio-Engineering Centre (NIBEC), University of Ulster, Newtownabbey, BT37 0QB, UK*
vladimir.svrcek@aist.go.jp

Since surface characteristics and functionalization determine the overall properties of silicon nanocrystal (Si NC), this chapter will highlight aspects that relate to the role of Si NC surfaces with respect to quantum confinement. Specifically, this chapter will focus on surface engineering approaches that rely on plasma-based and surfactant-free processing of doped Si NCs in liquid media by either pulsed ns-laser or direct current (DC)/radio frequency (RF) microplasmas. These techniques share a common characteristic whereby atmospheric pressure plasmas that are generated and confined within or in contact with liquids are capable of inducing nonequilibrium liquid chemistry to tune and stabilize the optoelectronic properties of Si NCs. The modified surface

Nanotechnology and Photovoltaic Devices: Light Energy Harvesting with Group IV Nanostructures
Edited by Jan Valenta and Salvo Mirabella

ISBN 978-981-4463-63-8 (Hardcover), 978-981-4463-64-5 (eBook)
www.panstanford.com

characteristics have allowed exploring the performance of Si NCs in photovoltaics. In particular Si NCs were electronically coupled with carbon and nanocarbon materials. Si NCs with engineered surface characteristics enhancing the electronic interactions with carbon nanomaterials, and at the same time serve as metal-free catalysts for the growth of multiwalled carbon nanotubes (CNTs). Entirely consisting of Si NC/nanocarbon solar cells represent an environmentally friendly potential solution for the large-scale manufacturing of energy-harvesting devices; here we therefore discuss the feasibility of prototype solar cells that consist of Si NCs combined with fullerenes or with semiconducting single-walled CNTs.

10.1 Introduction

Processing technologies that can change the characteristics of material surfaces and improve or enhance material properties or add new functions are becoming key fabrication steps in the photovoltaic (PV) industry. At the same time, quantum confinement effects in Si NCs have captured great attention within the PV community [1–6]. Particularly, new physical phenomena such as carrier multiplication that could lead to conversion efficiencies higher than 100% are of great importance and impact [7]. It is also well known that Si NC surface chemistry is critical and can strongly affect the observation of carrier multiplication [8–11]. Therefore research activities addressing the role of surface states of Si NCs with sizes exhibiting quantum confinement have dramatically increased in recent years [12–14]. Wet chemistry that generally uses organic-based steric stabilization has demonstrated the possibility of tuning the optoelectronic properties of Si NCs [15]. However organic terminating molecules can introduce additional challenges with respect to achieving efficient exciton dissociation and effective carrier transport as required within solar cell devices. It follows that improving the control over Si NC characteristics by alternative surface chemistries or without using large ligand molecules is highly desirable.

In recent years we have therefore focused on alternative approaches that could take advantage of nonequilibrium chemistry generated by confined plasmas interacting with Si NC colloids [14, 16–19]. This resulted in the investigation of nanosecond (ns)-laser-generated plasmas in liquid [18] and direct current (DC)/radio frequency (RF) atmospheric pressure microplasma in contact with liquid media [14, 16]; in both cases the processing techniques have shown efficient Si NC surface tunability leading to stable and enhanced Si NC optoelectric properties.

It is clear that beyond improved stability of Si NC properties, the surface chemistry has impact on the interactions between Si NCs with other application device components and in general with adjacent structured nanomaterials. A primary example is represented by junction interfaces formed between nanomaterials in advanced and novel solar cell architectures. In this context, the focus here is on hybrid silicon-based PV devices that include carbon structures as the complementary junction material. Carbon plays a vital role in many of successful current technologies; carbon can provide a large variety of nanoscale structures (e.g., nanotubes, fullerenes, graphene, etc.) with peculiar properties and characteristics [20–23]. Due to the significant modifications in the electron and hole wavefunctions, the combination of quantum-confined Si NCs with nanocarbons can reveal synergistic phenomena; for instance, these may help overcoming the negative impact of silicon indirect band-gap nature and enhancing phonon-less transitions [24, 25].

It is already well documented in the literature that carbon-based nanomaterials, for example, single-walled carbon nanotubes (SWCNTs) [21, 26] and C_{60}, exhibit peculiar properties very beneficial for solar cells [27]. C_{60} molecules possess superior electron-accepting properties already exploited in polymer-based solar cells, while semiconducting SWCNTs might offer an enlargement of the absorption down to the near infrared region ($\sim$0.9 eV), beyond the bulk Si limit of 1.1 eV [28]. It is believed that the combination of Si NCs and nanocarbon materials could bring synergic effects for the development of new types of multifunctional structures of interest for PVs [2, 7, 21, 28, 29]. Up to date the potential of C nanomaterials that are based on semiconducting SWCNTs and/or fullerenes (C_{60}) with Si NCs is not fully explored yet.

Here we firstly in Sections 10.2 and 10.3 show that plasma-based Si NC surface engineering approaches are capable of stabilizing doped Si NC properties without using surfactants. Secondly, in Section 10.4 we discuss electronic interactions and structural properties of quantum-confined Si NCs with carbon terminations and the role of surface engineering. Thirdly, in Section 10.5 we demonstrate that the Si NC surface engineering facilitates filling of cavity and allows metal-free catalyst growth of multiwalled CNTs. Finally, in Section 10.6 we report on the performance of prototype solar cells that consist of Si NCs and semiconducting SWCNTs and fullerenes. In all cases the combination of fullerenes or SWCNTs with Si NCs leads to energy conversion with promising features that include a simplified approach to the fabrication of PV devices.

10.2 Si NC Surface Engineering in Liquids

Silicon, as a covalent semiconductor, has very strong sensitivity to surface modification [30], and depending on the synthesis path, different surface conditions can be produced. When the size Si NCs is sufficiently small and quantum confinement effects take place, the wavefunctions of carriers are delocalized over the Si NC volume and include the surface states. The small volume also allows for carriers to easily diffuse into/from the nanocrystal core/surface that could result in surface-localized states and recombination processes [31] that can influence the Si NC luminescence [32]. Therefore, the optical properties are sensitive not only to the size of the nanocrystals but also to the Si NC surface and surface terminations. The availability of Si NCs in colloids (vs. embedded in solid matrices) is very advantageous for both decreasing the cost of solar cells fabrication (e.g., via inkjet printing and spray coating) and for facilitating the control of the Si NC surface chemistry. A well-known Si NC synthesis technique that allows subsequent Si NC dispersion in colloids is represented by electrochemical etching, widely used to produce porous silicon. The electrochemical etching process is a top-down approach that leads to the fabrication of high-purity Si NCs with the possibility of producing both *p*- and *n*-doped Si NCs. Colloidal suspensions of Si NCs are obtained by mechanical pulverization of electrochemically etched layers [36]. These colloidal solutions have multiple advantages for both fundamental investigations and for

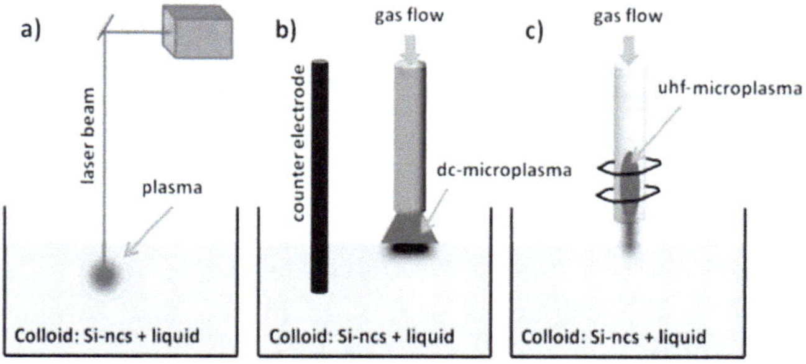

Figure 10.1 Schematic diagram representing three different plasma–liquid systems for surface engineering of silicon nanocrystals (Si NCs). (a) A laser beam is used to generate a plasma in the liquid medium; (b) an atmospheric pressure direct current (DC) microplasma is generated outside and coupled to the colloid via a counter–carbon electrode; (c) an ultrahigh-frequency (UHF) microplasma is generated in a quartz capillary and "jetted" out onto the colloid. Reproduced from Ref. [14] by permission of the Royal Society of Chemistry.

nanotechnology. In one extreme, surface engineering techniques can be applied to Si NCs dispersed in different colloidal solutions.

In this section we focus on three techniques for Si NC surface engineering directly in colloids which have shown the possibility of surface chemistries not accessible via traditional wet chemistry [33]. In all cases the colloids are formed by surfactant-free Si NCs dispersed in liquid media (e.g., ethanol or water). Figure 10.1 shows a schematic diagram representing three different plasma–liquid systems for surface engineering of Si NCs. Namely, Fig. 10.1a represents a pulsed laser beam to generate a confined plasma in a liquid medium. The chemistry induced on the surface of the Si NCs depends on both the laser-induced heat as well as the production of a range of radicals such as hydroxyl groups. Photothermal heating and Coulomb explosion are the major processes that induce peculiar environmental conditions enabling Si NC surface engineering [34]. In the second approach (Fig. 10.1b) an atmospheric pressure DC microplasma is generated outside and coupled to the colloid via a counter carbon electrode. Compared to laser-induced plasmas, the electrons are directly injected from the microplasma source to the liquid colloid resulting in nonequilibrium liquid chemistry processes [35]. The interactions of electrons with the Si NC colloid provide unique conditions for cascaded liquid chemistry and allow effective surface engineering of the Si NCs [16, 19]. However, compared to laser-based surface engineering, the DC microplasma process requires the colloid to be somewhat conductive and therefore limits the selection of the dispersion. To overcome this drawback a third setup has been developed where an RF microplasma, which does not require the counter electrode to be immersed in the colloid, is generated in a quartz capillary and impacted onto the colloid as depicted in Fig. 10.1c.

The three different approaches have been studied to modify the surface characteristics of Si NCs in colloids. The initial dry Si NC powder with stable H-terminations exhibits strong room-temperature photoluminescence (PL) that peaks in the range of 590–620 nm (2.0–2.1 eV) [19, 36]. Under the quantum confinement model, reducing the crystal core causes a widening of the band-gap. Quantum confinement in Si NCs smaller than 5 nm also results bright PL at room temperature. The high PL intensity can be justified due to confinement effects in real space that would, most likely under Heisenberg's uncertainty principle, cause sufficient spreading of the wavefunction in momentum space for direct-like band-to-band recombination. As mentioned above, the effects of surface states influence the quantum confinement and the PL, and therefore surface characteristics should be taken account to fully describe the behavior of Si NCs [14, 19].

Figure 10.2a presents a typical example of how the PL characteristics vary following surface modification due to either ns laser processing or DC microplasma. It is clearly seen that the surface engineering conducted by DC microplasma considerably enhance the PL compared to the ns laser treatment; mainly, it is important to observe a red shift of the PL maxima. The inset of Fig. 10.2a shows a typical transmission electron microscopy (TEM) image which confirms the presence of Si NCs after surface engineering. Si NCs with diameters that are expected to exhibit quantum confinement are clearly identified. Structural transformations of the Si NC core cannot be observed in the TEM images. Figure 10.2b plots the PL intensity maxima as a function of the processing time for a range of different processing conditions. For instance, the PL intensity enhancement is clear for Si NCs subjected to DC microplasma

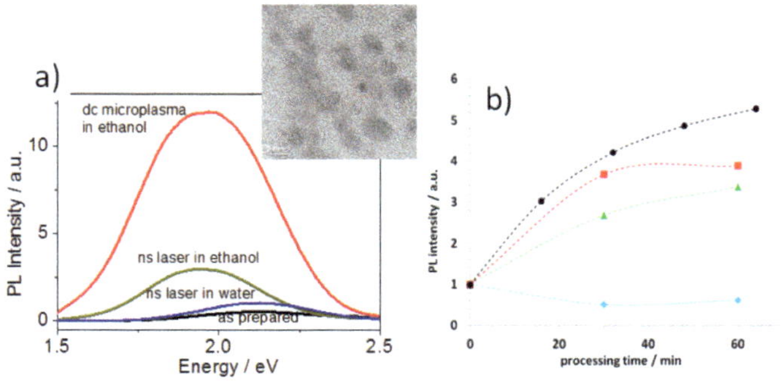

Figure 10.2 (a) Typical photoluminescence (PL) spectra from Si NCs after nanosecond laser processing in water/ethanol (blue/green) and after treatment by a DC microplasma in ethanol (red). PL spectra of as-prepared Si NCs dispersed in ethanol are shown for comparison (black). Inset shows typical transmission electrom image of Si NCs after surface engineering induced by DC microplasmas in water/polymer solutions. (b) Summary of the PL intensity for Si NCs processed for different times utilizing a DC microplasma in ethanol (black circles), in water (green triangles), or a UHF microplasma in water (red squares). PL properties of Si NCs stored in water for the same time are also reported (blue diamonds). The excitation wavelength was 400 nm for all samples.

in ethanol (black circles) and to UHF-microplasma in water (red squares). The degrading PL properties of Si NCs stored in water for the same time are also reported (blue diamonds).

These observations can be rationalized, for water-based colloids, as follows. Once the surface engineering process is initiated by the plasma (pulsed laser or RF/DC microplasmas)the replacement of Si-H bonds with Si-OH bonds is accelerated (Figs. 10.2 and 10.3) [14, 18, 19]. This is mainly due to electrons generated in liquid, which appear to be in higher numbers in the case of the RF microplasma approach [14]. In addition, in case of ns-laser-generated plasmas, where the generation of liquid soluble electrons is less effective, the Si NC aggregates are also fragmented through shock waves [34, 37]. As a result, Si NCs previously unexposed to the liquid environment are now subjected to the induced surface chemistry as well as to water-induced degradation. Therefore, laser-based surface engineering that is less efficient in surface reconstruction,

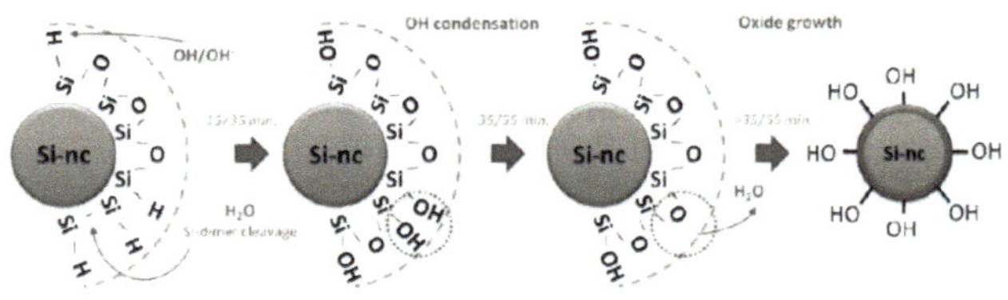

Figure 10.3 Schematic diagram depicting changes at the surface of Si NCs during surface engineering. Reprinted with permission from Ref. [18]. Copyright (2011) American Chemical Society.

suffers from the competitive degradation of the Si NC surface due to water cleavage resulting in weaker PL emission.

The PL intensity increases under DC microplasma treatment also because Si dimers are efficiently removed from the Si NC surface and better OH-terminations provide a higher degree of passivation by removing possible surface defects (Fig. 10.2b) [14, 18, 19]. As the surface engineering continues and OH surface coverage is complete, the wavelength of the PL maxima reaches a plateau (Fig. 10.2b). Correspondingly, a red shift is also observed (Fig. 10.2b) that can be justified by the replacement of a partial H coverage with a fully OH-terminated Si NC surface (Fig. 10.3); the smaller band gap that determines the red shift for OH-terminated Si NCs compared to H-terminated ones has been reported following theoretical calculations [10, 38]. Contrary to DC/RF microplasma, in the case of ns laser, oxidation induced by fragmentation leaves behind strained bonds and defects that can provide nonradiative paths to exciton recombination; this further decreases the PL emission as a function of treatment duration [17].

10.3 Surface Engineering of Doped Si NCs

Similarly to bulk materials, the availability of doped Si NCs offer a wider range of opportunities to design PV structures [39]. Device *p–n* junctions based on doped nanocrystals are expected to be far more efficient compared to Schottky-type or bulk heterojunction devices [40]. Bottom-up synthesis of doped Si NCs has so far encountered considerable challenges [41]; the inclusion of dopants has often resulted in their nonuniform segregation either at the surface or in the core where they act as recombination centers. For these reasons, top-down electrochemical etching is still the preferential route for producing doped Si NCs.

The question is whether the surface engineering induced by microplasma affect B or P dopant of Si NCs. To monitor doping of Si NCs similar to bulk Si low-temperature PL and electron paramagnetic resonance (EPR) analysis is commonly used [42, 43]. Therefore in our case PL under weak laser intensity and EPR at 4 K were applied to confirm the doping [42–44]. Structured PL emission spectra in the case of *p*-type doped Si NCs were observed at low temperatures (<20 K). A boron bound exciton peak at 1134.5 nm was identified, accompanied by two PL bands from a boron-bound multiple-exciton complex at 1128.9 and 1084 nm, respectively (Fig. 10.4a).

On the other hand, the PL intensities from phosphorous bound excitons were generally weaker (by more than 10 times) and it did not allow phosphorous-related exciton detection; therefore the low-temperature EPR analysis at 4 K was applied. For both *n*- and *p*-type Si NCs a narrow symmetric line shape with g = 2.0050 with a peak line width of 11 G was observed Fig. 10.4b. This signal is characteristic of nonbonding electrons from silicon dangling bonds on three-coordinated silicon atoms [45–48]. Both spectra showed a broader and intense asymmetric line shape with

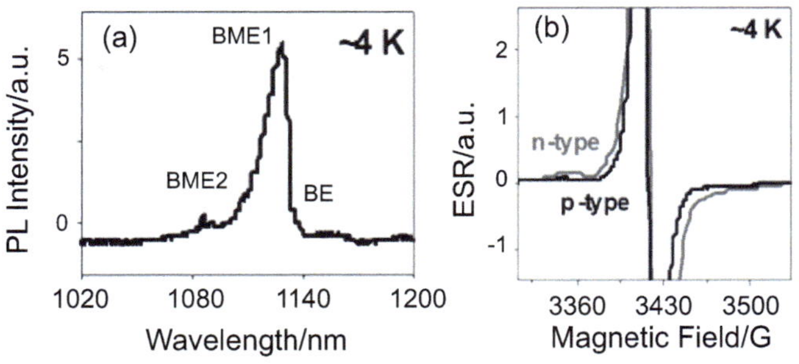

Figure 10.4 (a) PL spectra of boron (*p*-type) doped Si NCs at 4 K with laser excitation at 733 nm. (b) EPR spectra of boron (*p*-type) and phosphorus (*n*-type) doped free-standing Si NCs taken at 4 K. Adapted from Ref. [28]. Copyright (2011) American Chemical Society.

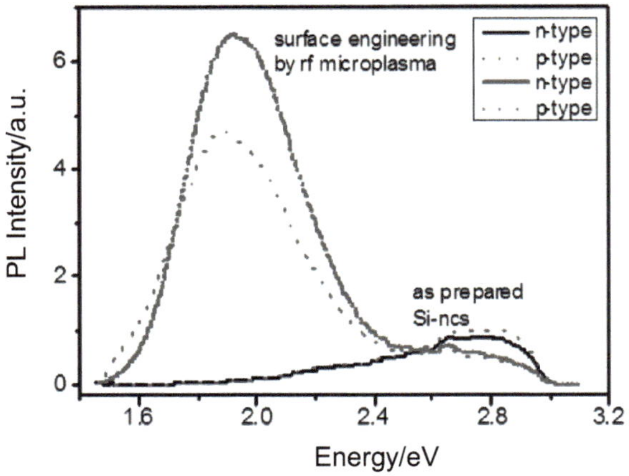

Figure 10.5 Typical PL spectra evolution from Si NCs after RF microplasma processing in ethanol for *n*- (solid lines) and *p*-type (dotted lines) Si NCs. PL spectra of as-prepared doped Si NCs are shown for comparison.

an effective $g = 2.0026$ for *p*-type and $g = 2.0037$ for *n*-type Si NCs. The higher signal intensity and spectra broadening are consistent with material doping [43]. The line shape and inhomogeneous broadening originated from g-anisotropy and from "spin–spin" interactions of direct bonds between the doping atoms (P and B).

These observations also showed that for surface engineering, the dopants did not play a significant role and the same PL properties for both doped Si NCs were recorded. Figure 10.5 reports typical PL spectra evolution from Si NCs after RF microplasma induced surface engineering in ethanol for *n*- (solid line) and *p*-type (dotted line) Si NCs. PL intensity of as-prepared Si NCs is for both types very weak. After surface engineering, a clear increase in PL intensity at 1.9 eV is recorded with more pronounced increase for *n*-type doped Si NCs.

10.4 Tuning Optoelectronic Properties of Si NCs by Carbon Terminations

Surface engineering without surfactants allows Si NCs to interact directly with other elements. As-prepared Si NCs produced by electrochemical etching are mostly terminated by H, and a degree

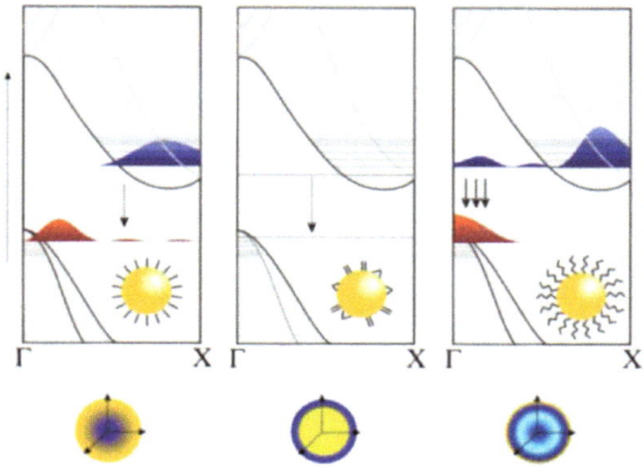

Figure 10.6 Schematic illustration of radiative channels in (a) H-, (b) O-, and (c) C-terminated Si NCs. Hole (red) and electron (blue) wavefunctions are onto the bulk Si indirect energy band-gap structure. Reprinted by permission from Macmillan Publishers Ltd: [*Light: Science & Applications*] (Ref. [25]), copyright (2013).

of oxygen-based terminations are formed upon exposure to air and water. C-based functionalization is achievable via a range of surface functionalization approaches. Figure 10.6 represents a sketch of the dominant radiative channels in H-, O-, and C-terminated Si NCs [24, 25]. Hole and electron density of states in **k**-space are depicted in red and blue, respectively, by projecting them onto the bulk Si band-gap diagram. In H-terminated Si NC slow dynamics and microsecond PL decay is observed. Slow radiative rate PL originates from phonon-assisted quasidirect excitonic recombination. In O-terminated Si NCs, slow PL can be related to oxygen surface defects [9] where electrons (and holes) are considered trapped on the defect state. In C-terminated Si NCs, radiative rate is dramatically enhanced as a result of direct phonon-less recombination [25] resulting from electron density that is more homogeneously distributed through the Si NC.

As it can be seen, direct termination with carbon can significantly modify Si NC optoelectronic properties. Thus it can be explained in terms of the distribution of the electron and the hole state densities in the lowest excited state in the real and the **k**-space. Since carbon atoms are smaller than those of silicon atoms, stronger surface confinement broadens the k-space distribution of the wavefunction and the electron density in the lowest excited state (Fig. 10.6c). Consequently, the probability of radiative nonphonon transitions grows as compared to both H-terminated Si NCs (Fig. 10.6a) and O-terminated Si NCs (Fig. 10.6b) where electron–hole recombination involves O-related defect centers. This is due to the higher degree of overlap between electron and hole densities along the C–X direction, when C-terminated Si NCs are compared with H-terminated ones [25]. The resulting radiative transitions between the lowest excited states of electron and hole can therefore exhibit phonon-free direct-like features with enhancement the in radiative rates, dynamics, and efficient fast light emission.

10.5 Functionalization of Surface-Engineered Si NCs with Carbon Nanotubes

Relations described by Einstein connecting emission and absorption show that a higher radiative rate always enhances the absorption cross section; therefore the enhancement of radiative rates in C-terminated Si NCs enlarges the band-edge absorption cross section, in comparison with H- and O-terminated Si NCs. Enhanced absorption of C-terminated Si NCs increases the attractiveness of hybrid PV systems that are based on Si/C nanostructures. Among the large variety of carbon-based nanostructure CNTs have received enormous attention due to great scientific and technological interest [49–51]. Control of CNT and Si NC interaction through surface engineering may allow producing novel Si NC/CNT nanocomposites with enhanced absorption in a large spectral region with unique synergic phenomena.

The tubular structure of CNTs has allowed the insertion of a wide range of materials [52–54]. The properties of the materials inserted within CNTs could be significantly modified as a result of the interactions with the surrounding carbon wall and due to the confinement effect. The combination of CNTs with Si NCs can thus offer a unique possibility for the fabrication of very solid, as well as flexible, Si NC–based nanodevices with C-tuned electronic properties. It is clear that the surface characteristics are key to both promoting electronic interactions as well as in determining

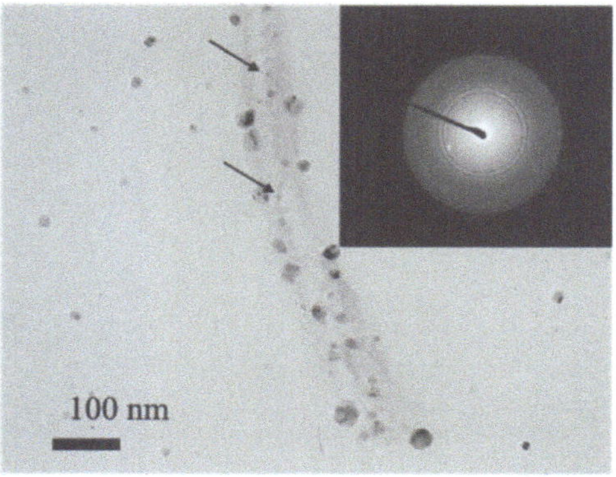

Figure 10.7 Transmission electron microscopy (TEM) image of Si NCs prepared by ablation of an immersed silicon wafer in a CNT/water suspension by laser fluence of 1.1 mJ/pulse. The inset shows the corresponding selected area electron diffraction pattern revealing the rings of a diamond lattice of silicon. Reprinted with permission from Ref. [60]. Copyright (2008) American Chemical Society.

the insertion in the CNT cavity [29, 55]. H-terminated silicon or Si NC surfaces are hydrophobic, but once Si surfaces see oxygen, they become hydrophilic. Producing Si NCs by laser ablation of crystalline silicon targets in de-ionized water results in hydrophilic surface, which at the same time help considerably the cavity filling process [18, 56, 57] via shock waves that propagate through the liquid solution [37, 58–60]. Detailed TEM analysis was performed to confirm the presence of Si NCs within the CNT cavities. Figure 10.7 shows a typical image of a filled multiwalled CNT with an inner diameter of 50 nm when the Si target is immersed in CNT/water colloid and irradiated by a laser [60]. It is observed that some spherical Si NC agglomerates are in the CNT cavity with a diameter around 25 nm on average (Fig. 10.7, indicated by arrows). The inset shows the corresponding electron diffraction pattern taken in the CNT cavity. Discrete spots on the circles indicate the presence of crystalline silicon in the cubic phase.

To enhance electronic interaction between Si NCs and CNTs it is preferable to directly covalently bond both nanomaterials. However up to date most of the growth CNTs is mediated by metalic catalyst. Although the catalyzing efficiency remains to be improved in case of

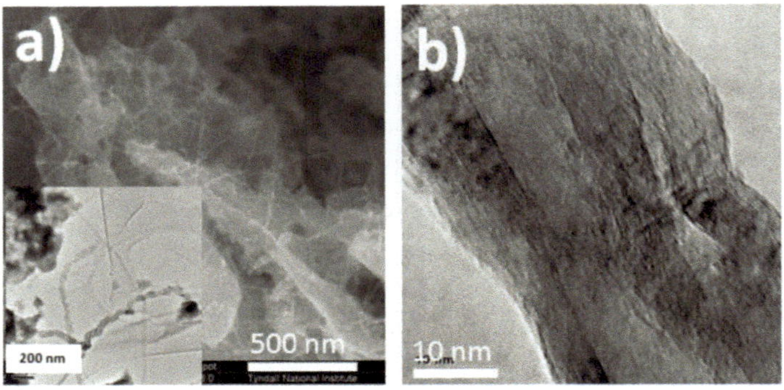

Figure 10.8 (a) Scanning electron microscopy images of Si NCs after plasma treatment, showing the presence of filamentary structures; the inset shows the corresponding TEM image. (b) Typical TEM images of multiwalled carbon nanotubes (CNTs) grown from samples of Si NCs that were produced by electrochemical etching and subsequently surface-engineered by ns laser treatment. CNTs were grown without any metal catalyst by exposing the Si NCs to a microwave plasma-enhanced chemical vapor deposition process. Adapted from Ref. [61].

nonmetallic particles to be suitable for CNT growth, some exciting attempts have been made [61–65]. Our results have confirmed that diverse Si NC surface features are a key factor to determine the growth of CNTs using Si NCs as catalyst particles [61]. Specific surface engineering of Si NCs is essential to activate the nucleation and growth of CNTs without using any metal catalyst [62]. In particular, only Si NCs that were surface engineered by an ns laser process in water have allowed the growth of multiwalled CNTs by a CH_4 low-pressure plasma treatment [61]. The formation of fiber-like structures with lengths exceeding 1 μm (inset of Fig. 10.8a) with diameter in the range of about 30 nm is observed. A more detailed TEM analysis has shown that the fibrous structures are for the most part represented by multiwalled CNTs (Fig. 10.8b). The spacing between the walls corresponds to 0.34 nm correlating with the (0 0 2) *d*-spacing of graphite.

Although the phenomenon that allows CNT growth on nonmetallic catalysts is not yet fully understood, two main differences can be indentified when compared with metal-activated nucleation: i) due to the reduced catalytic activity of nonmetals, carbon precursors need to reach the surface of the nanoparticles largely decomposed either through higher processing temperatures or through plasma-induced decomposition where the change of Gibbs free energy is more favorable to nucleate and form the graphene cap [63]; ii) because a molten layer would be difficult to produce on the nonmetallic catalyst nanoparticles, carbon atoms can only adsorb and diffuse on the solid surface [64].

CNT growth on Si NCs may present similarities with the growth on SiO_2 nanoparticles [65] as several surface engineering techniques can produce an oxide-based shell structure as it is the case for the Si NCs. In particular it was found that oxygen atoms can increase the capture of $-CH_x$ and consequently facilitate the growth of SWCNTs on oxygen-containing SiO_x nanoparticles [65]. Figure 10.8 whiteness multiwalled CNT growth without any metal catalyst by exposing of ns laser surface-engineered in water and partially O-terminated Si NCs to a microwave plasma-enhanced chemical vapor deposition process. These results indicate a promising research direction that could lead to indeed the fabrication of nanoscale junctions between two nanostructures with unique quantum confinement effects. The possibility of synthesizing Si NC/CNT nanodevices is an exciting scientific opportunity with a wide range of applications.

10.6 Solar Cells Based on Si NCs and Nanocarbon Materials

To demonstrate the electronic interactions with nanocarbons and the possibility to generate photocurrent, solar cells based on a bulk heterojunction architecture have been fabricated. Although the synthesis, surface engineering, and device fabrication processes are not optimized, Si NC/nanocarbon experimental devices allow us to assess initial and fundamental PV functionalities. One such example is represented by a solar cell devices formed with ns laser surface-engineered Si NCs and fullerenes (C_{60}). Firstly, a colloid of engineered Si NCs has been drop-casted on a PEDOT:PSS/ITO/glass substrate. The formation of the bulk heterojunction with the Si NC layer has been achieved by further depositing a layer of fullerenes [17, 18]. A second device architecture was made by spray-coating

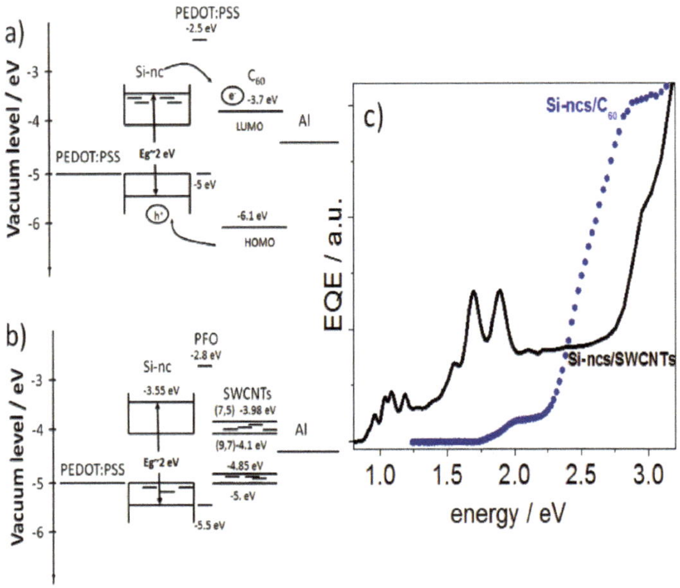

Figure 10.9 (a) Energy levels of fullerenes (C_{60}), Si NCs with quantum confinement, Al, and PEDOT:PSS; (b) energy levels of semiconducting SWCNTs with different chiral indexes shown together with Si NCs, Al. and PEDOT:PSS; (c) normalized external quantum efficiency (EQE) of a device based on Si NCs/C_{60} (blue circles) and of a device based on Si NCs/SWCNTs (black line). Si NCs were surface-engineered by ns laser in water. Adapted with permission from Refs. [18, 28]. Copyright (2011) American Chemical Society.

a mixture of Si NCs/SWCNTs on a PEDOT:PSS/ITO/glass substrate [28]. In both cases, aluminum was used as counter electrode. Figures 10.9a and 10.9b report the corresponding energy band diagrams of these two devices which suggest the possibility of forming type II bulk heterojunctions. Because the size distribution of the Si NCs is relatively broad, Fig. 10.9a depicts multiple energy levels above about −4.1 eV up to about −3.55 eV, which would result from the quantum confinement induced band-gap widening.

The conduction band of Si NCs is higher than any of the C_{60} LUMO values [17] and similarly for the Si NC/SWCNT devices (Fig. 10.9b). In the first case the excitons are created by photon absorption in both the Si NCs and fullerenes and the difference in electron affinity anodionization potential between the nanocrystals and fullerenes provide the energy driver for exciton dissociation. Nanotubes are members of the fullerene structural family while sheets are rolled at specific and discrete ("chiral") angles. The combination of the rolling angle and radius decides the nanotube properties. Indeed, in the second case, the different chirality of the SWCNTs and offsets are sufficient for excitons dissociation allowing consequent photoconductivity generation. In both solar cells the nanocarbon material serve as electrontransporting material when electronically coupled with Si NCs.

The results indicate that Si NCs can be electronically coupled with both C_{60} and SWCNTs [18, 28]. Both devices showed $I-V$ characteristic confirming the formation of a bulk-like heterojunction solar cell. The electrical characteristics confirm the presence of the heterojunction with a typical rectified diode characteristic with short-circuit current and open circuit voltage. Normalized external quantum efficiency (EQE) as a function of the photon energy of both devices is shown in Fig. 10.9c. Circles show the typical EQE of the device based on C_{60}.

As it can be seen, the EQE in the visible region starts to increase from about 1.75 eV where the absorption and electronic coupling of Si NCs with C_{60} occurred EQE of devices based on Si NCs/SWCNTs is plotted by the black line. Importantly, in the case of SWCNTs, the conversion active region is considerably enlarged (0.9–3.1 eV). EQE shows an increased efficiency going beyond the infrared region. Enhanced EQE around 0.9–1.2 eV and 1.5–2 eV correspond to the optical transitions of semiconducting SWCNTs with chiral indexes (9.7), (8,6), (8,7), (7,5), and (7,6) [28]. We also observe a considerable increase in EQE (>2 times) in the spectral region where the absorption of the Si NC overlaps (~1.2–2.7 eV) with that of SWCNTs if compared to the spectral range where the only SWCNTs contribution is expected (0.9–1.2 eV). These results clearly support the presence of electronic interactions between Si NCs and SWCNTs. An optical band gap as low as 1.15 eV will give a valence band edge as low as 4.55 eV, which might allow for type II heterojuction, for example, with for SWCNTs that have the chiral index (7,5). The optical absorption peak intensity of (7,5) and (7,6) are smaller than that of (9,7), (8,7), and (8,6), but the EQE peak intensity are just

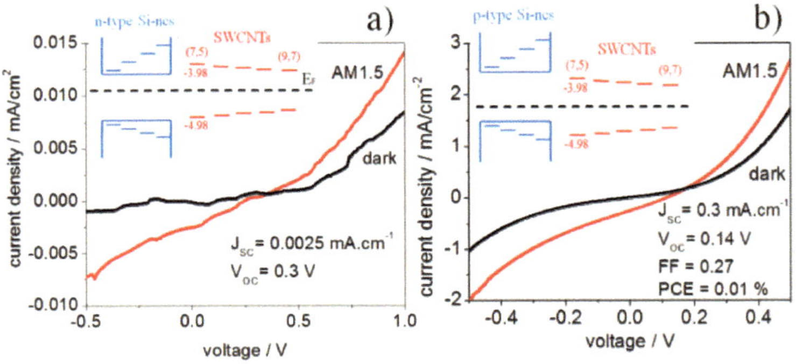

Figure 10.10 The I–V characteristics in the dark (black line) and under A.M. 1.5 illumination (red line) of prototype solar cells based on semiconducting SWCNTs and (a) n-type and (b) p-type doped Si NCs, respectively. Inset shows corresponding band alignment of n-type doped and p-type doped Si NCs with semiconducting SWCNTs, respectively. Adapted with permission from Ref. [28]. Copyright (2011) American Chemical Society.

the opposite. Therefore the EQE and exciton dissociation yield of (7,5) and (7,6) is larger than that of (8,7), (8,6), and (9,7) SWCNTs [28]. It is important to highlight that the enhancement of the charge generation with the composite p-type Si NCs/SWCNTs is larger by orders of magnitude in comparison with n-type Si NCs/SWCNTs. The I–V characteristics of the solar cells consisting of n-type and p-type doped Si NCs mixed with semiconducting SWCNTs are presented in Fig. 10.10 in the dark (black line) and under illumination (red line), respectively. I–V curves for both p-type Si NC/SWCNT and n-type Si NC/SWCNT devices show rectification and a diode like behavior. The n-type Si NC–based device has a lower I_{sc} (3 orders magnitude lower) than the p-type device.

From these results, it can be deduced that the presence of dopants does play the expected role when Si NCs are electrically coupled. Furthermore, these results suggest that the photocarrier generation intrinsically involve both SWCNTs and Si NCs, and it is not due to the possible formation of a Schottky-type junction at the electrodes interface [28]. The low short circuit current in the device with n-type Si NCs can be explained in terms of an excess of electrons, which might shift the conduction, valence, and Fermi levels, resulting in a larger energy barrier between the SWCNTs and the Si NCs (inset in Fig. 10.10a). However, at the same time, excess carriers represented by electrons in this case might partially recombine with the photogenerated holes and eventually lower the photocarrier generation efficiency. An opposite effect is observed for the devices with *p*-type Si NCs, whereby the excess of holes might shifts up the respective conduction and valence bands after Fermi levels equalization (inset in Fig. 10.10b). Consequently, the energy-level alignment of the Si NCs and SWCNTs enhances holes transfer into Si NCs, resulting in an overall better electronic configuration.

10.7 Conclusions and Outlooks

In this chapter we have discussed novel plasma-based approaches that allow controlling and stabilize Si NC surface characteristics without using large molecules and surfactants. In particular, ns-laser-generated plasma and DC/RF microplasmas that interact with Si NC colloids were presented. The possibilities arising from surface engineering of doped Si NCs were also discussed confirming that doped Si NC surface treatments can be efficiently achieved by the proposed approaches.

Since Si NC surface modification by C changes considerably nanocrystals energy structure and excitonic recombination dynamics we have discussed direct functionalization with carbon nanomaterials. The surface engineering of Si NCs induced in a liquid medium plays a key factor in filling and direct growth of CNTs. We showed that the surface of Si NCs produced by pulsed ns laser treatment in water is enough hydrophilic to allow not only insertion in the CNT cavities but also direct growth of multiwalled CNTs. Indeed, at this stage of the research we do not have sufficient evidence and we are unable to provide full details on the filling/growth mechanisms; therefore future work has to focus on an improvement and understanding of such mechanisms, including Si NC absorption tuning, and of the conformation of the Si NC/CNT junction.

The potential of surface-engineered doped Si NCs combined with carbon nanomaterials is demonstrated in view of prospective PV applications. The feasibility of PV solar cells made from surface engineered Si NCs with quantum confinement effects combined with fullerenes (C_{60}) and with purified semiconducting SWCNTs was also demonstrated. In the prototype solar cells, electronic interactions between Si NCs and both nanocarbon materials are evident and confirmed. In both cases we have shown that Si NCs can serve as electron-transporting material and where C_{60}/SWCNTs behave as hole-transporting material. Importantly, in the case of SWCNTs, a conversion efficiency region is considerably enlarged (0.9–3.1 eV) compared to devices made of Si NCs/C_{60}. We also argue that Si NC doping and SWCNT chirality plays an important role and greater opportunities for solar cell performance. The results suggest that the combination of *p*-type doped Si NCs and semiconducting SWCNTs such as (7,5) is electronically favorable for exciton dissociation and carrier (electrons/holes) generation. It is believed that the combination of Si NCs with nanocarbon materials such as C_{60} or semiconducting SWCNTs is a viable and promising approach for low-cost, environmentally friendly, and efficient solar cells. Since the investigation of Si NC/nanocarbon interactions is still largely unexplored, great improvements are expected with the possibility of achieving considerable improvements in the device performance.

Acknowledgments

This work was partially supported by a NEDO project (Japan), by DM's JSPS Invitation and Bridge Fellowship (Japan), by the Leverhulme International Network on "Materials Processing by Atmospheric Pressure Plasmas for Energy Applications" (Award n.IN-2012-136), and by the Royal Society International Exchanges scheme (Award n.IE120884).

References

1. Green, M.A. (2001). Third generation photovoltaics: ultra-high conversion efficiency at low cost, *Prog. Photovoltaics*, **9**, 123–135.
2. Nozik, A.J. (2008). Multiple exciton generation in semiconductor quantum dots, *Chem. Phys. Lett.*, **457**, 3–11.

3. Švrček, V., Slaoui, A., Muller, J.-C. (2004). Silicon nanocrystals as light converter for solar cells, *Thin Solid Films*, **451**, 384–391.
4. Liu, C.-Y., Holman, Z.C., Kortshagen, U.R. (2009). Hybrid solar cells from P3HT and silicon nanocrystals, *Nano Lett.*, **9**, 449–454.
5. Ley, M., Švrček, V., Kuznicki, Z., (2003). Photoluminescence of a superficial Si nanolayer and an example of its use, *Appl. Phys. Lett.*, **82**, 4056–4059.
6. Kortshagen, U.R., Gresback, R., Holman, Z., Ligman, R., Liu, C.-Y., Mangolini, L., Campbell, S.A. (2008). Plasma synthesis of group IV quantum dots for luminescence and photovoltaic applications, *Pure Appl. Chem.*, **80**, 1901–1907.
7. Beard, M.C., Knutsen, K.P., Yu, P.R., Luther, J.M., Song, Q., Metzger, W.K., Ellingson, R.J., Nozik, A.J. (2007). Multiple exciton generation in colloidal silicon nanocrystals, *Nano Lett.*, **7**, 2506–2510.
8. Veinot, J.G.C., (2006). Synthesis, surface functionalization, and properties of freestanding silicon nanocrystals, *Chem. Commun.*, **40**, 4160–4168.
9. Wolkin, M.V., Jorne, J., Fauchet, P.M., Allan, G., Delerue, C. (1999). Electronic states and luminescence in porous silicon quantum dots: the role of oxygen, *Phys. Rev. Lett.*, **82**, 197–200.
10. Konig, D., Rudd, J., Green, M.A., Conibeer, G. (2008). Role of the interface for the electronic structure of Si quantum dots, *Phys. Rev. B*, **78**, 035339–035346.
11. Reboredo, F.A., Galli, G. (2005). Theory of alkyl-terminated silicon quantum dots, *J. Phys. Chem. B*, **109**, 1072–1078.
12. Veinot, J.C. (2006). Synthesis, surface functionalization, and properties of freestanding silicon nanocrystals, *Chem. Commun.*, **40**, 4160–4165.
13. Hessel, C.M., Rasch, M.R., Hueso, J.L., Goodfellow, B.W., Akhavan, V.A. et al. (2012). Synthesis of ligand stabilized silicon nanocrystals with size-dependent photoluminescence spanning visible to near-infrared wavelengths. *Chem. Mater.*, **24**, 393–401.
14. Mariotti, D., Mitra. S., Švrček, V., (2013). Surface-engineered silicon nanocrystals, *Nanoscale*, **5**, 1385–1398.
15. Holmes, J.D., Ziegler, K.J., Doty, R.C., Pell, L.E., Johnston, K.P., Korgel, B. (2001). Highly luminescent silicon nanocrystals with discrete optical transitions, *J. Am. Chem. Soc.*, **123**, 3743–3745.
16. Švrček, V., Mariotti, D., Kondo, M. (2010). Microplasma-induced surface engineering of silicon nanocrystals in colloidal dispersion, *Appl. Phys. Lett.*, **97**, 161502–161505.

17. Švrček, V., Mariotti, D., Shibata, Y., Kondo, M. (2010). Hybrid heterojunction based on fullerenes and surfactant-free, self-assembled, closely-packed, silicon nanocrystals, *J. Phys. D: Appl. Phys.*, **43**, 415402–415407.
18. Švrček, V., Mariotti, D., Nagai, T., Shibata, Y., Turkevych, I., Kondo, M. (2011). Photovoltaic applications of silicon nanocrystal based nanostructures induced by nanosecond laser fragmentation in liquid media, *J. Phys. Chem. C*, **115**, 5084–5092.
19. Mariotti, D., Švrček, V., Hamilton, J.W.J., Schmidt, M., Kondo, M., (2012). Silicon nanocrystals in liquid media: optical properties and surface stabilization by microplasma induced non equilibrium liquid chemistry, *Adv. Funct. Mater.*, **22**, 954–964.
20. Iijima, S. (1991). Helical microtubules of graphitic carbon, *Nature*, **354**, 56–58.
21. Gabor, M., Zhong, Z., Bosnick, K., Park, J., McEuen. P.L. (2009). Extremely efficient multiple electron-hole pair generation in carbon nanotube photodiodesv, *Science*, **325**, 1367–1371.
22. Gonzalez Szwacki, N., Sadrzadeh, A., Yakobson, B. (2007). B_{80} fullerene: an ab initio prediction of geometry, stability, and electronic structure, *Phys. Rev. Lett.*, **98**, 166804–166808.
23. Brabec, C.J., Sariciftci, N.S., Hummelen, J.C., (2001). Plastic solar cells, *Adv. Funct. Mater.*, **11**, 15–26.
24. Li, F., Ruckenstein, E. (2004). Water-soluble poly(acrylic acid) grafted luminescent silicon nanoparticles and their use as fluorescent biological staining labels, *Nano Lett.*, **4**, 1463–1467.
25. Dohnalova, K., Poddubny, A., N Prokofiev, A. de Boer, A W., DAM., Umesh, Ch. P., Paulusse, J.M.J., Zuilhof H., Gregorkiewicz T. (2013). Surface brightens up Si quantum dots: direct bandgap-like size-tunable emission, *Light: Sci. Appl.*, **2**, e47.
26. Ueda, T., Matsuda, A., Tayagaki, T., Kanemitsu, Y. (2008). Carrier multiplication in carbon nanotubes studied by femtosecond pump-probe spectroscopy, *Appl. Phys. Lett.*, **92**, 233105(1)–233105(3).
27. Forrest, S.R. (2005). The limits to organic photovoltaic cell efficiency, *MRS Bull.*, **30**, 28–32.
28. Švrček, V., Cook, S., Kazaoui, S., Kondo M. (2011). Silicon nanocrystals and semiconducting single-walled nanotubes applied to photovoltaic cells, *J. Phys. Chem. Lett.*, **2**, 1646–1650.
29. Švrček, V., Normand, F.L., Pham-Huu, C., Ersen, O., Joulie, S., Ledoux, M.-J. (2006). Filling of single silicon nanocrystals within multi-walled carbon nanotubes, *Appl. Phys. Lett.*, **88**, 033112–033115.

30. Delerue, C., Lannoo, M., (2004). *Nanostructures: Theory and Modelling.* Berlin: Springer-Verlag.
31. Kanemitsu, Y., Uto, H., Masumoto, Y., Masumoto, T., Futagi, T., Mimura, H. (1993). Microstructure and optical properties of free-standing porous silicon films: size dependence of absorption spectra in Si nanometer-sized crystallites, *Phys. Rev. B*, **48**, 2827–2831.
32. Portoles, M.J.L., Diez, R.P., Dell'Arciprete, M.L., Caregnato, P., Romero, J.J., Martire, D.O., Azzaroni, O., Ceol'ın, M., Gonzalez, M., C. (2012). Understanding the parameters affecting the photoluminescence of silicon nanoparticles, *J. Phys. Chem. C*, **116**, 11315–11325.
33. Zou, J., Kauzlarich, S.M. (2008). Functionalization of silicon nanoparticles via silanization: alkyl, halide and ester, *J. Clust. Sci.*, **19**, 341–355.
34. Švrček, V., Kondo, M., Kalia, K., Mariotti, D. (2009). Photosensitive self-assembled nanoarchitectures containing surfactant-free Si nanoctystals produced by laser fragmentation in water, *Chem. Phys. Lett.*, **478**, 224–230.
35. Richmonds, C., Witzke, M., Bartling, B., Lee, S.W., Wainright, J., Liu, C.-C., Sankaran, R.M. (2011). Electron-transfer reactions at the plasma-liquid interface, *J. Am. Chem. Soc.*, **133**(44), 17582–17585.
36. Švrček, V., Slaoui, A., Muller, J.-C. (2004). Ex-situ prepared Si nanocrystals: their elaboration and characterization in embedded silica glass, *J. Appl. Phys.*, **95**, 3158–3163.
37. Švrček, V., Sasaki, T., Shimizu, Y., Koshizaki, N. (2006). Silicon nanocrystals formed by pulsed laser-induced fragmentation of electrochemically etched Si micrograi, *Chem. Phys. Lett.*, **429**, 483–488.
38. Guerra, R., Ossicini, S. (2010). High luminescence in small Si/SiO_2 nanocrystals: a theoretical study, *Phys. Rev. B*, **81**, 245307–245313.
39. Leitsmann, R., Panse, C., Küwen, F., Bechstedt, F. (2009). *Ab initio* characterization of transition-metal-doped Si nanocrystals, *Phys. Rev. B*, **80**, 104412–104421.
40. Gur, I., Fromer, N.A., Geier, M.L., Alivisatos, A.P. (2005). Air-stable all-inorganic nanocrystal solar cells processed from solution, *Science*, **310**, 462–465
41. Mangolini L, Thimsen E, Kortshagen U. (2005). High-yield plasma synthesis of luminescent silicon nanocrystals, *Nano Lett.*, **5**, 655–659.
42. Švrček, V., Fujiwara, H., Kondo, M. (2009). Luminsecent properties of doped freestanding silicon nanocrystals embedded in MEH-PPV, *Sol. Energy Mater. Sol. Cells*, **93**, 774–781.

43. Švrček, V., Fujiwara, H., Kondo, M., (2009). Top-down silicon nanocrystals and a conjugated polymer-based bulk heterojunction: optoelectronic and photovoltaic applications, *Acta Mater.*, **57**, 5986–5994.

44. Sauer R. (1973). Localized exciton bound to an isoelectronic trap in silicon, *Phys. Rev. Lett.*, **3**, 376–380.

45. Kelly, M.T., Bocarsly, A.B. (1997). Effects of SO_2 and I_2 on the photoluminescence of oxidized porous silicon, *Chem. Mater.*, **9**, 1659–1664.

46. Stutzmann, M., Biegelsen, D.K., Street, R. (1987). Detailed investigation of doping in hydrogenated amorphous silicon and germanium, *Phys. Rev. B*, **35**, 5666–5671.

47. Liu, X., Xu, G., Sui, Y., He, and Y., Bao, X. (2001). Fabrication and characterization of an organic-inorganic gradient surface made by polymethylsilsesquioxane, *Solid State Commun*, **19**, 397–402.

48. Fujii, M., Mimura, A., Hayashi, S., Yamamoto, Y., Murakami, K. (2002). Quantum confinement in phosphorus-doped silicon nanocrystals, *Phys. Rev. Lett*, **89**, 206805–206809.

49. Sanvito, S., Kwon, Y.K., Tomanek, D., Lambert, C.J. (2000). Fractional quantum conductance in carbon nanotubes, *Phys. Rev. Lett.*, **84**, 1974–1977.

50. Chen, Z.H., Appenzeller, J., Lin, Y.M., Sippel-Oakley, J., Rinzler, A.G., Tang, J.Y., Wind, S.J., Solomon, P.M., Avouris, P. (2006). An integrated logic circuit assembled on a single carbon nanotube, *Science*, **311**, 1735–1738.

51. Lin, Y.-M., Tsang, J.C., Freitag, M., Avouris, Ph. (2007). Impact of oxide substrate on electrical and optical properties of carbon nanotube devices, *Nanotechnology*, **18**, 295202–295208.

52. Tsang, S.C., Chen, Y.K., Harris, P.J.F., Green, M.L.H. (1994). A simple chemical method of opening and filling carbon nanotubes, *Nature*, **372**, 159–162.

53. Ajayan, P.M., Iijima, S. (1993). Capillarity-induced filling of carbon nanotubes, *Nature*, **361**, 333–819.

54. Monthioux, M. (2002). Filling single-wall carbon nanotubes, *Carbon*, **40**, 1809–1823.

55. Demoncy, N., Colliex, C., Brun, N., Stephan, O., Loiseau, A., Pascard, H. (1998). Filling carbon nanotubes with metals by the arc-discharge method: the key role of sulfur, *Eur. Phys. J. B*, **4**, 147–152.

56. Švrček, V., Sasaki, T., Shimizu, T., Koshizaki, N. (2006). Blue luminescent silicon nanocrystals prepared by ns pulsed laser ablation in water, *Appl. Phys. Lett.*, **89**, 213113 (1)–213113 (2).
57. Švrček, V., Mariotti, D., Kondo, M. (2009). Ambient-stable blue luminescent silicon nanocrystals prepared by nanosecond-pulsed laser ablation in water, *Opt. Express*, **17**, 520–527.
58. Lin, S., Burda, C., Mohamed, M.B., Nikoobakht, B., El-Sayed, M.A. (2000). Laser-induced shape changes of colloidal gold nanorods using femtosecond and nanosecond laser pulses, *J. Phys. Chem. B*, **104**, 6152–6163.
59. Mafune', F., Kohno, J., Takeda, Y., Kondow, T., Sawabe, H. (2000). Structure and stability of silver nanoparticles in aqueous solution produced by laser ablation, *J. Chem. Phys. B*, **104**, 8333–8337.
60. Švrček, V. (2008). Fabrication of filled carbon nanotubes with fresh silicon nanocrystals produced in-situ by nanosecond pulsed laser processing in environmentally friendly solutions, *J. Phys. Chem. C*, **112**, 13181–13188.
61. Mariotti, D., Švrček, V., Mathur, A., Dickinson, C., Matsubara, K., Kondo, M. (2013). Carbon nanotube growth activated by quantum-confined silicon nanocrystals, *J. Phys. D: Appl. Phys.*, **46**, 122001–122005.
62. Švrček, V., Ersen, O., Dintzer, T., Le Normand, F., Pham-Huu, C., Ledoux, M.J. (2006). Connection of silicon nanocrystals (Si-nc) with multi-walled carbon nanotubes, *Appl. Phys. A*, **83**, 153–157.
63. Kumar, S., Mehdipour, H., Ostrikov, K. (2013). Plasma-enabled graded nanotube biosensing arrays on a Si nanodevice platform: catalyst-free integration and in situ detection of nucleation events, *Adv. Mater.*, **25**, 69–74.
64. Page, A.J., Chandrakumar, K.R.S., Irle, S., Morokuma, K. (2011). SWNT nucleation from carbon-coated SiO_2 nanoparticles via a vapor-solid-solid mechanism, *J. Am. Chem. Soc.*, **133**, 621–628.
65. Hong, G., Chen, Y., Li, P., Zhang, J. (2012). Controlling the growth of single-walled carbon nanotubes on surfaces using metal and non-metal catalysts, *Carbon*, **50**, 2067–2082.

Chapter 11

Prototype PV Cells with Si Nanoclusters

Stefan Janz,[a] Philipp Löper,[b] and Manuel Schnabel[c]
[a] *Materials—Solar Cells and Technology, Fraunhofer Institute for Solar Energy Systems, Heidenhofstr. 2, 79110 Freiburg, Germany*
[b] *Institute of Microengineering, École Polytechnique Fédérale de Lausanne, Rue de la Maladière 71b, 2002 Neuchâtel 2, Switzerland*
[c] *Solar Cells—Development and Characterization, Fraunhofer Institute for Solar Energy Systems, Heidenhofstr. 2, 79110 Freiburg, Germany*
stefan.janz@ise.fraunhofer.de, philipp.loeper@epfl.ch, manuel.schnabel@ise.fraunhofer.de

11.1 Introduction

The impressive success of very high conversion efficiencies obtained with crystalline silicon (c-Si) has triggered the search for novel concepts which overcome the fundamental efficiency limits of c-Si solar cells. The market has seen a dramatic reduction of module costs due to up-scaled production. However, ongoing long-term cost reduction requires not only up-scaled production and increasingly sophisticated technologies but also the implementation of fundamentally new concepts which overcome the physical limitations of current technologies. One approach to overcoming the Shockley–Queisser limit [1] of c-Si is to introduce a second band gap

Nanotechnology and Photovoltaic Devices: Light Energy Harvesting with Group IV Nanostructures
Edited by Jan Valenta and Salvo Mirabella

ISBN 978-981-4463-63-8 (Hardcover), 978-981-4463-64-5 (eBook)
www.panstanford.com

which still essentially consists of c-Si [2, 3], which together with the first cell forms a tandem cell. This second band gap is provided by a dense array of silicon nanocrystals (Si NCs) embedded in a Si-based dielectric. The band gap of such a composite material can be tuned from that of bulk c-Si up to approximately 2 eV and is well suited for the top cell in the all c-Si tandem cell.

Evidence for quantum confined states in Si NCs embedded in silicon dioxide (SiO_2) has been given by photoluminescence [4], and luminescence quantum yields of up to 25% have been obtained (see also the chapter by Hiller et al.). As an alternative matrix to SiO_2, silicon carbide (SiC) is being investigated due to its superior transport properties. As a first step to the construction of a tandem solar cell, the benefit of the quantum confinement–based top cell must be demonstrated experimentally on an electrical level by investigating the top cell alone. Subsequently a suitable c-Si bottom solar cell and the interconnection of both cells have to be developed.

11.2 Motivation

The efficiency of a single-junction semiconductor solar cell is limited to 31% under 1 sun illumination [1, 3] (refer to the chapter by Valenta et al. for more details). The maximum efficiencies attained in practice for single-junction cells are 25.6% for silicon and 28.8% for gallium arsenide [5]. These results indicate that single-junction cells are already rather close to their theoretical efficiency limits and provide a strong incentive for research on solar cells which circumvent the Shockley–Queisser single-junction limit.

The most successful of these concepts is the tandem cell, which is the only concept to have surpassed the Shockley–Queisser limit in practice [5]. Tandem cells involve the use of more than one cell to gradually increase the efficiency limit. Adding a cell with a higher band gap allows high-energy photons to be converted with fewer thermalization losses, while adding a cell with a lower band gap lowers transmission losses. A tandem cell consisting of three cells is illustrated schematically in Fig. 11.1. Use of a tandem cell increases the maximum efficiencies to 42.9% for two cells and 49.3% for three cells [3]. Tandem cells have already

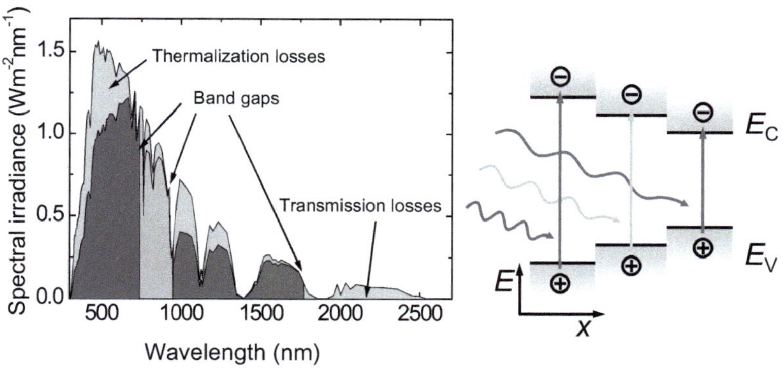

Figure 11.1 The AM1.5G solar spectrum and the maximum energy convertible to electricity by a triple-junction tandem cell (left). The increased conversion efficiency of a tandem cell arises from the matching of the different cells to different parts of the solar spectrum, as shown schematically (right). Reprinted from Ref. [6].

been produced in several materials systems. The most successful one consists of III–V semiconductors, in which triple-junction cells have achieved efficiencies of 37.9% under 1 sun illumination and 46% under concentration [5] (It is worth noting that the given efficiencies are laboratory efficiencies; significantly lower efficiencies are achieved in field tests due to solar spectrum fluctuations [7]). Another materials system within which tandem cells are commercially available is the amorphous/microcrystalline silicon (a-Si/μc-Si) system: a-Si has a band gap of 1.7 eV, which happens to be the optimum band gap for a top cell paired with c-Si [8], allowing an efficiency of 13.4% [5]. This system does not yet exceed the efficiency of conventional Si cells [5] but is significantly cheaper. Efficiency is limited mainly by the sub-band-gap absorption [9], low carrier mobility, and light-induced degradation of a-Si via the Staebler–Wronski effect [10, 11], which reduces the efficiency by up to 30%. The Staebler–Wronski effect has not been fully explained yet [12] but is frequently related to the structural disorder of a-Si. Finding an alternative to a-Si as the second cell on top of a c-Si cell is therefore of great interest.

11.3 Material Selection

Nanostructured semiconductors are promising candidates for the top cell of all-Si tandem cells. Their band gaps can be tuned by adjusting their sizes in one, two, or three dimensions [13–16], which means they can have the same band gap as a-Si while not exhibiting subband gap absorption and light-induced degradation, provided sufficiently defect-free layers are produced. Triple-junction cells with component cells of different nanostructure sizes are also conceivable.

Quantum wells and quantum dots (QDs) have been synthesized in most common semiconductors, such as silicon [4, 17–19], germanium [16], and III–V compound semiconductors [15]. The latter materials are without a doubt the most developed class, but they are also rather expensive due to the scarcity of some of the elements involved and the costs associated with processing the largely toxic materials. While they are well suited to the production of high-end semiconductor devices, these materials are less suitable for light-harvesting applications where a low production cost must be achievable. Silicon and germanium nanostructures both have their advantages and disadvantages; for example, in practice, quantum confinement is more easily obtained in Ge than in Si because the Bohr radius of Ge, an upper limit for strong quantum confinement, is almost three times larger than that of Si (14 and 5 nm, respectively [20]). However, Ge also has a lower band gap than Si (0.7 and 1.1 eV, respectively). As the goal is to produce a solar cell which will work in tandem with a c-Si solar cell, for which, as discussed above, a band gap of 1.7 eV would be ideal, the QD sizes required in both systems are comparable. Ultimately, there are other factors which suggest that Si nanostructures are more promising materials: their production will be much more cost efficient due to lower raw material costs and easier integration with the bottom component cell of a tandem cell, which would also be silicon based, and there is a wealth of information available on the properties of silicon interfaces with dielectrics such as SiO_2, Si_3N_4, and SiC, as well as with a-Si and a range of metals and transparent conducting oxides (TCOs), all of which can be important when designing a functional device.

Nanostructured silicon can be synthesized by a range of methods. Early research was conducted on electrochemically porosified Si [21], but fabrication methods have since matured; Si NCs are currently produced by laser pyrolysis [22], wet-chemical synthesis [23], plasma synthesis [24], direct deposition by plasma-enhanced chemical vapor deposition (PECVD) [25], or solid-phase crystallization from PECVD films [17, 26]. In this chapter, the discussion of synthetic routes is restricted to their viability for the production of Si NC solar cells. For more information on individual synthetic routes, the reader is referred to the chapter by Hiller et al. as well as to the review articles of Mangolini [19] and Janz et al. [18], which review direct Si NC synthesis and synthesis by solid phase crystallization of PECVD precursors, respectively.

For solar cell applications, the first design criterion is size control of the NCs to guarantee a spatially uniform band gap. Variation in Si NC size means the quantum confinement–induced energy levels of adjacent NCs may be at different energies. If the electronic wavefunctions are strongly confined, such as in an SiO_2 matrix or in vacuum, then the bandwidth of the energy levels becomes very narrow [27]; therefore, the energy bands of adjacent NCs would not line up, making direct tunneling transport impossible. The tunneling process would then need to be thermally activated, making transport less efficient. This issue can be mitigated by embedding NCs in a weakly confining matrix, such as SiC, in which the Si NC energy bands are broader. Further details on the carrier extraction can be found in the chapter by Garrido et al. Another effect of having a spread in Si NC size is that the absorption edge is different for every NC. The absorption of the sample is then the integral over the absorptions of the individual NCs, which leads to a smearing out of the absorption edge. However, for a tandem cell to work effectively, the solar spectrum must be divided as abruptly as possible between the top and bottom component cells, ideally involving a top cell made out of a direct-band-gap semiconductor; where this division is not possible, an attempt should nevertheless be made to keep the absorption edge of the top cell as abrupt as possible by ensuring that it is uniform across the Si NC material.

The second criterion is the conflicting requirement of having, on one hand, isolated NCs to ensure quantum confinement, and on the other hand, adjacent NCs to ensure electrical transport. This requirement can only be achieved by a uniform, high density of Si NCs embedded in a host matrix. The importance of this design criterion cannot be overstated; if the NCs are embedded in SiO_2, for example, then at least one atomic layer of SiO_2 is required between the Si NCs for confinement, but as few as four atomic layers do not permit efficient transport [28]. A high density of NCs also ensures improved absorption of incident light, which in a solar cell would lead to higher current as well as higher voltage output [27].

The final criterion is cost, including that of materials (process gases, chemicals), cooling water, electrical power, equipment, and maintenance. From this point of view, a low-temperature process utilizing cheap precursors would be desirable. However, for solar cell manufacturers, the real cost is often not so much the cost of one or two machines in the production line but the cost associated with developing a new process to the point at which it can be successfully integrated into a production line. Utmost reliability and reproducibility are required, as an increased number of bad cells which must be discarded quickly offsets the money saved from choosing to implement a cheaper process. Ultimately, even though the goal is a cheap process, the most important factor in making a process financially viable for a company is its reliability and compatibility with the existing process chain. The process chain for a "standard" Si solar cell includes inorganic wet chemistry, furnace diffusion, PECVD, screen printing, and fast firing. A synthetic route which relies on similar technology is more likely to be implemented industrially.

These criteria are difficult to meet with the direct synthesis methods introduced earlier. Laser pyrolysis is cheap but unlikely to produce monodisperse Si NCs. Organic wet chemistry will be difficult to integrate with standard solar cell processing. Laser pyrolysis, wet-chemical synthesis, and direct plasma synthesis all require additional processing to embed them inside a solid-state matrix in which they exhibit their quantum confinement–induced properties, and even then, it is doubtful whether the result will be a uniform, high density of Si NCs. The only synthetic route which has been shown to produce a uniform, high density of monodisperse Si NCs in a solid-state matrix and which is also compatible with

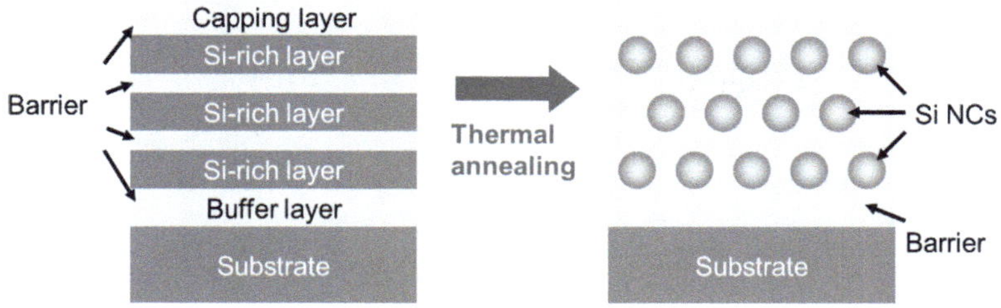

Figure 11.2 Fabrication of a Si NC superlattice by the multilayer approach [4]. Silicon-rich layers (SRLs) are sandwiched between stoichiometric layers (barrier). The barrier layers restrict the Si NC growth and thereby enable a narrow size distribution. Si NC formation is induced by thermal annealing at high temperatures, usually 1050°C–1150°C. Adapted from Ref. [6].

in-line solar cell manufacturing is the solid-phase crystallization of Si NCs from Si-rich multilayer precursors deposited by PECVD [4, 17]. The multilayer is prepared as a stack of alternating Si-rich and stoichiometric layers of a Si compound, as shown in Fig. 11.2. Upon annealing, the excess Si in the Si-rich layers precipitates and crystallizes at temperatures of 1050°C–1150°C. The stoichiometric layers act as barriers to crystallite growth and thus restrict the Si NC size distribution. Both PECVD and thermal annealing are in-line compatible processes already used in the manufacturing of Si solar cells (for details refer to the chapter by Hiller et al.).

The high thermal budget is of concern with regard to both cost and the cell structure which is produced. The latter point will be discussed later in this chapter. For now, we restrict ourselves to the point that Si NC layers, like any solar cell absorber, need to have a low defect density to avoid the recombination of photoexcited carriers before they can be collected as a photocurrent, and the observation that the passivation of such defects depends critically on the thermal budget employed in the growth process [29, 30]. The effect has been studied on Si NCs in SiO_2 grown by a superlattice approach using various annealing processes. All annealing processes led to the growth of Si NCs, as confirmed by high-resolution transmission electron microscopy (HRTEM); however, as shown in Fig. 11.3, the photoluminescence yield of the samples varied by a few orders of magnitude.

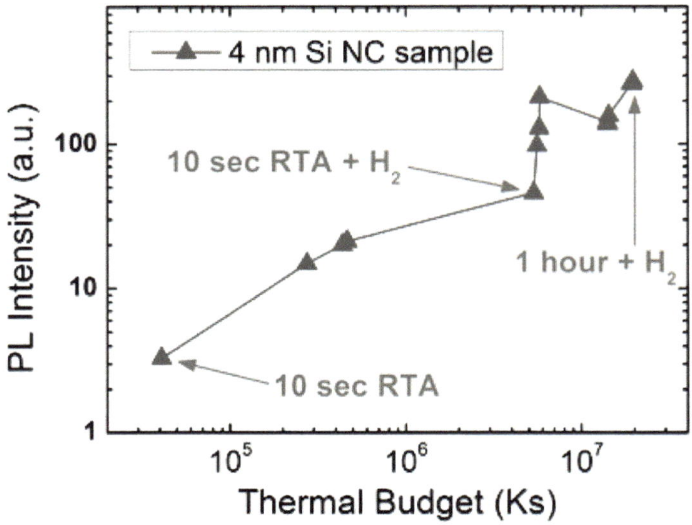

Figure 11.3 Photoluminescence intensity from identical samples of 4 nm Si NCs in SiO_2 for various annealing processes. Reprinted with permission from Ref. [30]. Copyright © 2013 WILEY-VCH Verlag GmbH & Co. KGaA, Weinheim. The photoluminescence intensity is a measure of the Si NC–SiO_2 interface quality. High thermal budgets are required to produce high-quality Si NCs. The thermal budget is plotted as the product of the annealing temperature of 1373 K [30] and annealing time.

Photoluminescence yield is a good measure of defect density because defects permit carriers to recombine nonradiatively and not contribute to the luminescence signal via radiative recombination. The main result, therefore, was that rapid thermal annealing (RTA) of the multilayer precursor leads to very defective NCs; nearly defect-free NCs could only be achieved by annealing for one hour (for details refer to the chapter by Hiller et al.).

High-temperature annealing over a long period of time seems to be indispensable for the production of defect-free Si NC layers. This approach has been studied intensely in three different materials systems: Si NCs in SiO_2 [17, 32–35], Si NCs in Si_3N_4 [36–38], and Si NCs in SiC [39–47]. The primary conclusion from all this work is that the matrix material has a strong effect on the overall properties of the Si NC film. The thermodynamics and kinetics of the solid-phase crystallization reaction, the defect concentration at the Si NC/matrix interface, the quantum confinement in the Si NCs and the ease of transport between Si NCs are all strongly dependent on the matrix. The overall properties of the Si NC films are determined much more by the nature of the matrix than by the exact sizes of the Si NCs. Therefore, we will hereafter refer to the materials as SiO_2/Si NCs, Si_3N_4/Si NCs, and SiC/Si NCs.

11.4 Current Collection

Having selected the most suitable type of Si NC film, we must now determine how to design a solar cell which will allow us to obtain the best photovoltaic performance from a given film.

In a semiconductor, photons are absorbed and generate electron–hole pairs. If nothing else were done, photoexcited electrons or holes would have no reason to flow in a particular direction and, after a certain time in the excited state, would recombine. To make a solar cell from a semiconductor which yields a current and a voltage, a structure is required which causes electrons to flow to one side of the device and holes to flow to the other side. If the device is disconnected, the electric field which results from having excess electrons at one terminal of the device and excess holes at the other gives rise to the open-circuit voltage of the solar cell. Connecting the two terminals with a wire causes electrons and holes to flow from one terminal to the other, giving rise to the short-circuit current and connecting them with an electrical load creates an intermediate situation in which both a voltage and a current are produced by the solar cell device.

In conventional solar cells, electrons and holes are forced to flow in opposite directions using *p*–*n* junctions. At the interface of *p*- and *n*-doped regions, a space charge region with a built-in voltage is formed. This voltage sweeps minority carriers which have reached the junction by diffusion across the junction, leading to an accumulation of electrons on one side and holes on the other. However, this sweep only works well if the minority carriers have a reasonable chance of diffusing to the junction in the first place; that is, the distance over which the carriers must diffuse, typically equal to the absorber thickness, must be lower than the minority carrier

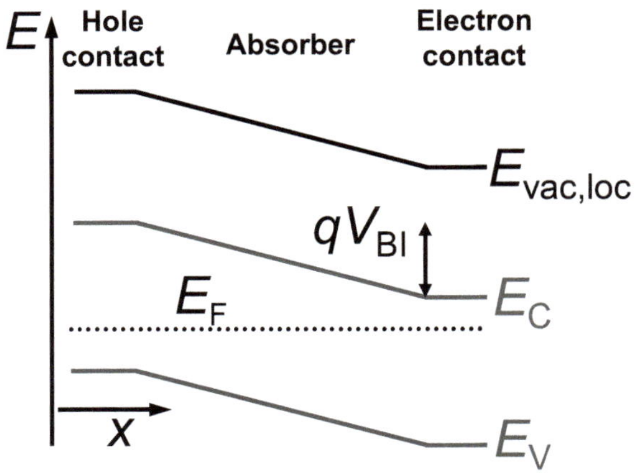

Figure 11.4 Band structure of a *p-i-n* solar cell [6]. E_c and E_v are the conduction and valence band edges, respectively, E_F is the Fermi level, and $E_{vac,loc}$ the local vacuum level. V_{bi} is the built-in voltage of the junction which is dropped across the entire absorber thickness, that is, the entire absorber is within the space–charge region of the junction. Reprinted from Ref. [6].

diffusion length L_{diff}. The latter value is given by $L_{diff} = \sqrt{(D\tau)}$, where D is the minority carrier diffusivity, and τ is the minority carrier lifetime.

In Si NC films, carrier diffusivity is expected to be rather low due to the insulating matrix and the many grain boundaries which act as scattering centers, while carrier lifetimes depend strongly on the defect density and cannot always be measured directly. For prototype devices, it is therefore reasonable to look to solar cell structures where the requirement of high diffusion length is relaxed, which can be accomplished with a *p-i-n* structure; the band structure of such a device is shown schematically in Fig. 11.4.

In this type of device, the absorber is intrinsic and sandwiched between *p*- and *n*-doped layers which act as the hole and electron contacts, respectively. This arrangement places the entire absorber within the space charge region of the *p–n* junction formed by the *p*- and *n*-doped layers. As all electron–hole pairs are now generated within the space charge region, there is no longer any need for them to diffuse to that region to be collected. Instead, we must now focus on whether the carriers can reach the electron and hole contacts by drift within the built-in electric field before recombination. The distance the carriers can drift before recombination is called the drift length L_{drift} and is the product of carrier lifetime τ and velocity v. The carrier velocity is defined as $v = \mu E$, where μ is the mobility and E is the built-in electric field of the p–i–n junction. The built-in potential V_{bi} shown in Fig 11.4 is the difference between the Fermi levels in the p- and n-doped layers, leading to the following overall expression for the drift length:

$$L_{\mathrm{drift}} = \mu\tau\frac{(E_{\mathrm{F,p}} - E_{\mathrm{F,n}})}{qd} \quad (11.1)$$

where $E_{\mathrm{F,p}}$ and $E_{\mathrm{F,n}}$ are the Fermi levels in the p- and n-doped layers, respectively, q is the elementary charge, and d is the absorber thickness. L_{drift} and L_{diff} both depend on the material parameter $\mu\tau$ (diffusivity and mobility are related via $D = \mu kT/q$, where k is the Boltzmann constant and T is the temperature). However, the drift length can be increased further for a given absorber material by having high doping levels in the p- and n-doped layers (hence increasing $E_{\mathrm{F,p}}$–$E_{\mathrm{F,n}}$) and by decreasing the absorber thickness d. From a theoretical standpoint, the p–i–n cell is preferable to the p–n cell but is also more difficult to implement as p- and n-doped layers must be placed on either side of the Si NC layer.

11.5 Doping

In practice, the viability of either structure also depends on how well the Si NC materials can be doped. Silicon is typically doped with boron (p-type doping) or phosphorus (n-type doping). Other dopants include aluminum and gallium (p-type) and arsenic and antimony (n-type). Dopants can be incorporated into the material in one of three ways: in the NCs, in the matrix, or at the NC/matrix interfaces.

If the dopant is incorporated into the Si NCs, then it can be expected to give rise to similar donor or acceptor levels as it would in bulk Si. However, because the Si NC is small, consisting of no more than a few thousand atoms, a single dopant atom is already equivalent to a doping level of $\sim 10^{19}$ cm^{-3}. Unless the overlap of electronic states between NCs is so high that the entire population of

NCs in a material can be described by a single density of states, it is not possible to achieve controlled doping levels below 10^{19} cm^{-3} or describe the behavior of the doped material with conventional Fermi statistics. A dopant atom can also cause significant strain of the NC, which further modifies the electronic states. This strain can be so high as to make incorporation of dopants into the Si NCs altogether energetically unfavorable, which has been shown theoretically for isolated, hydrogen-terminated Si NCs with diameters of 1–2.2 nm [48].

Dopant incorporation into the matrix can lead to conventional doping of the matrix. The matrix forms a continuous network throughout the film, so doping proceeds as it would in a bulk material. Free carriers thus supplied can be trapped by the Si NCs, leading to modulation doping [49, 50]. If the dopant selected only dopes Si but not the matrix material (be it SiC, Si_3N_4, or SiO_2), then conventional doping of the matrix will not occur. Instead, if the dopants are incorporated into the matrix but are close enough to a Si NC that they could tunnel to it and thereby lower their energy, then such a direct charge transfer process can occur [49]. More details on these processes are provided in the chapter by König et al. In both cases, doping of the Si NCs is achieved with the ionized dopant atoms remaining in the matrix.

The third possibility is that dopant atoms are located at the NC/matrix interface. As it is a region of high strain, incorporation of dopant atoms at this interface may be energetically favorable. In principle, such atoms could dope either the NCs or the matrix. However, dopant atoms are only electrically active if they form the same number of bonds as an NC or matrix atom (four for Si), leaving an excess electron or hole which becomes a free carrier. At the interface between a Si NC and the matrix, a dopant atom may form the number of bonds which it actually wants to form (for example, three for boron and five for phosphorus), and if it does so, then it no longer has an excess electron or hole and no longer behaves as a dopant. The same can occur for dopant atoms in the matrix when the matrix is amorphous.

Phosphorus doping of SiO_2/Si NCs has been attempted both by in situ doping of the multilayer precursor [51, 52] and in-diffusion of dopant after NC formation [53]. For both cases, the passivation of Si NC/matrix interface defects has been reported [51, 53], indicating that phosphorus is located at the interfaces. An increase in conductivity and Auger quenching of photoluminescence was also reported, suggesting that phosphorus also actively dopes the Si NCs [51, 52]; the same effects were observed upon boron doping [54, 55]. Boron doping of SiC/Si NCs showed that boron behaves as an active dopant [56]. However, it is unclear whether this is conventional doping of Si NCs or modulation doping, and there is evidence that initially boron compensates a background *n*-type doping in these films.

In summary, boron and phosphorus doping can clearly be used to enhance the conductivity of Si NC films. However, for SiO_2/Si NCs, the doping leads to increased nonradiative recombination which is counterproductive for solar cells. In SiC/Si NCs, the compensation of background doping by boron means that, depending on the exact processing parameters, the resulting film will be either *p*- or *n*-type. These effects complicate production of defined *p–n* junctions by direct incorporation of dopants into the Si NC material. A consideration of doping Si NC films therefore leads to the same conclusion as a theoretical consideration of minority carrier transport: a *p–i–n* cell is the most promising test device for a Si NC solar cell.

11.6 Device Concepts for Si NC Test Structures

The purpose of Si NC test structures is to show that Si NC materials are suitable as the top cell in an all-Si tandem cell which is more efficient than a single-junction silicon cell. Earlier in this chapter, we showed that, in a single-junction cell, transmission of low-energy photons leads to a loss in short-circuit current, while thermalization of electron–hole pairs excited by high-energy photons leads to a loss in open-circuit voltage. Si NCs have a higher band gap than a bulk Si cell, so neither absorption nor short-circuit current will be improved. However, due to their higher band gap, we expect an improvement in open-circuit voltage. Therefore, to show that a Si NC/bulk Si tandem cell can be more efficient than a conventional Si cell, one must demonstrate a higher open-circuit voltage than that possible with bulk Si.

This requirement means that a Si NC solar cell test device should be designed for maximum open-circuit voltage. Slow nonradiative carrier recombination, both within the Si NC film and at the interfaces within the device, is crucial. Maximizing the short-circuit current is not so important for the test devices so the use of very thick absorber layers or light trapping to maximise absorption is not necessary. Similarly, a minimization of the series resistance of the electron and hole contacts is not necessary. The materials for the electron and hole contacts should therefore be selected for their abilities to passivate the surface of the Si NC film and for their large inherent fields, rather than for their conductivities.

Unfortunately, the necessity of using temperatures up to 1150°C for the fabrication of the Si NC layer imposes severe requirements on the thermal stability and expansion coefficient of the substrate. Device fabrication is further complicated by the substrate hindering access to the layer backside. The simplest possibility is to use a conductive substrate, as shown in Fig. 11.5a, such as a Si wafer.

However, Si NCs in SiO_2 have penetration depths $1/\alpha$ of approximately 10 μm for 3 eV photons and 1 μm for 4 eV photons. Technologically viable Si NC layers meanwhile are typically less than 300 nm thick, so most of the solar spectrum is transmitted, generating charge carriers in the substrate wafer. This effect makes it rather difficult to differentiate between the photovoltaic properties of the Si NC film and those of the substrate wafer. The problem could be solved by using a substrate wafer with an optical band gap higher than that of the Si NC film, such as a silicon carbide wafer or epitaxial gallium nitride on a silicon carbide or sapphire wafer. However, the substrate must also have a suitable electron affinity and work function to establish a selective contact to electrons or holes.

This problem is somewhat mitigated by the concept depicted in Fig. 11.5b. A laterally conductive layer on an insulating substrate serves as the back contact to the Si NC layer. Unwanted light absorption can be suppressed more easily than in concept (a) because the conductive layer can be made as thin as series resistance considerations allow and can also be doped to give it sufficient

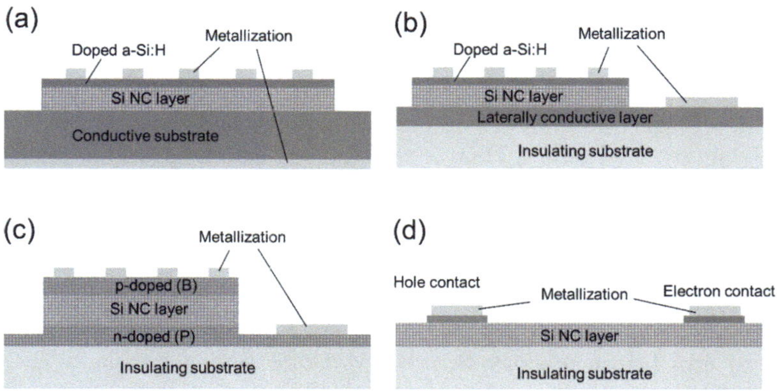

Figure 11.5 Device concepts for Si NC test structures. The contact can be established directly by a conductive substrate (a) or a doped layer on top of an insulating substrate (b). In a variant of (b), the Si NC layer itself is doped during its deposition, resulting in the structure shown in (c). For full flexibility in choosing the contact materials and processes, both contacts can be placed on top of the recrystallized layer (d). Reprinted with permission from Ref. [30]. Copyright © 2013 WILEY-VCH Verlag GmbH & Co. KGaA, Weinheim.

lateral conductivity. However, the high-temperature annealing will likely cause out-diffusion of dopants into the Si NC layer, blurring the *p*–*i*–*n* junction and possibly affecting Si NC formation. A variant of concept (b) is shown in Fig. 11.5c. In this variant, *n*-type, undoped, and *p*-type Si NC precursor layers are deposited and annealed together, resulting in a device that consists solely of Si NC material. The extent to which concept (c) is a classical *p*–*i*–*n* device is unclear due to the intricacies of doping NC films which were discussed previously. Nevertheless, this concept avoids having an additional material as a back contact in which unwanted absorption could occur.

The three device structures (a)–(c) discussed so far are all deficient, as the solid-phase crystallization process necessarily affects the physical properties of the electrical back contact. Any change of the thermal annealing step will change the electrical contact. Furthermore, the electrical properties of the Si NC layer are also affected by any changes in annealing, as the process controls the in-diffusion of dopants from the contact layer. Dopant diffusion could be avoided by incorporating a conductive diffusion barrier between the absorber and back contact, while a high-temperature stable back contact material would be required to provide uniform performance independent of annealing temperature.

Intermetallics and TCOs are certainly capable of fulfilling these requirements up to 900°C [57–67], so if the peak processing temperature for the solid-phase crystallization of Si NCs is at some stage reduced to this temperature range, then these structures will become viable. Limiting the processing temperature to 900°C would also permit the use of glass substrates for structures (b) and (c), in analogy to the crystalline silicon on glass (CSG) technology developed for single-junction Si cells [68]. However, these processes are not yet sufficiently developed to permit the reliable characterization of Si NC films annealed at temperatures approaching 1150°C.

Implementing both contacts on the front side of the recrystallized layer circumvents this problem (Fig. 11.5d). Si NCs are first formed by thermal annealing; selective electron and hole contacts are both established afterwards. However, much higher carrier diffusion or drift lengths are required for this approach than for any vertical structure because the technologically feasible separation of the electron and hole contacts is in the μm range. In addition, the large width of the *p–i–n* junction reduces the drift length as *d* in Eq. 11.1 is now the lateral separation of electron and hole contacts. Device performance is likely to be limited by the small fraction of the cell area from which carriers will be successfully collected. The device structure shown in Fig. 11.5d was realized by Rölver et al. and Berghoff et al. with a 50 nm thick Si film instead of Si NCs. The authors achieved a lateral *p–i–n* structure with Schottky contacts [69] or doped regions formed by ion implantations [70]. These devices did not contain any Si NCs but are included here as technological information.

To meet all the device design requirements outlined above, a novel membrane-based *p–i–n* solar cell was developed at Fraunhofer ISE. The membrane cell shown in Fig. 11.6 is the most promising device concept for the reliable photovoltaic characterization of Si NC films. Its key feature is the local removal and encapsulation of the substrate after solid-phase crystallization of

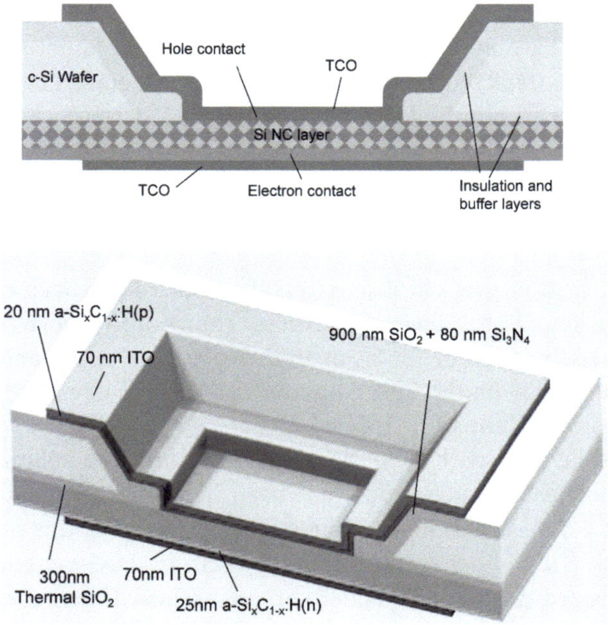

Figure 11.6 Schematic cross section (top). Reprinted from Ref. [6]. 3D view (bottom) [43] of a membrane-based *p–i–n* Si NC solar cell. Reprinted with permission from [P. Löper, M. Canino, D. Qazzazie, M. Schnabel, M. Allegrezza, C. Summonte, S.W. Glunz, S. Janz, M. Zacharias, Silicon nanocrystals embedded in silicon carbide: investigation of charge carrier transport and recombination, *Applied Physics Letters*, 102 (2013) 033507]. Copyright [2013], AIP Publishing LLC.

the Si NCs. The electrically insulating encapsulation is opened above the Si NC layer such that it can be accessed freely from both sides. This device structure can be used for any thin films which involve high-temperature processing, such as silicon or germanium NCs in a dielectric matrix or thin c-Si or c-SiC bulk films. A Si wafer can be used as the substrate and is insulated and structured using Si-based dielectric layers deposited by PECVD. By this method, the selective contact materials and interface pretreatment can be chosen independently of the solid-phase crystallization parameters. Materials with a band gap comparable to or higher than that of the Si NC film can be chosen for the electrically active components, permitting an unambiguous characterization of the Si NC material.

Furthermore, compatibility with standard Si process technology is assured. The only drawbacks of this cell are the large number of processing steps required and their fragility. A detailed overview of the processing required is given in reference [71].

11.7 Device Results

In this section, we will discuss results which have been obtained with various Si NC device structures. The results reported will be reviewed in relation to the material synthetic route and device structure. The limitations and potentials of given device structures will again be addressed.

A solar cell with a Si NC absorber fabricated following the wet-chemical synthetic route has been reported by Liu et al. [72]; Fig. 11.7 depicts this device. The major advantage of this process is the lack of high-temperature steps. The authors used wet-chemically synthesized Si NCs embedded in poly-3(hexylthiophene) as an absorber, in a superstrate configuration with indium tin oxide (ITO) on glass as a substrate and an Al back contact. The most critical point in the process chain seems to be the transfer of the synthesized particles into the solid absorber matrix, as the NC can even oxidize in solution. The oxide surrounding the Si NCs plays a dominant role in terms of passivation and transport, which means that control of this process, from solution to the encapsulated state in the final absorber

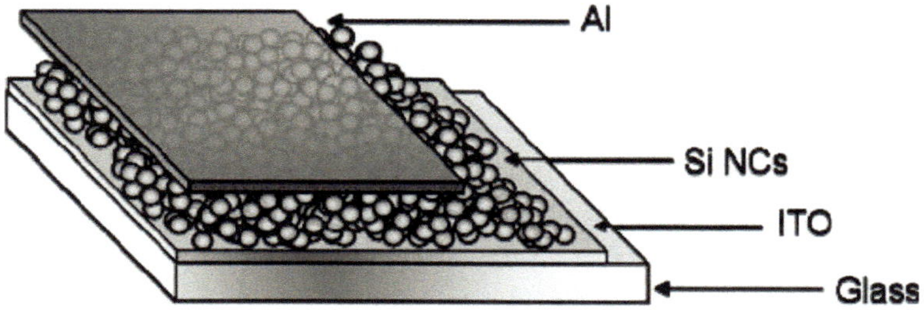

Figure 11.7 Solar cell device with Si NCs from wet-chemical synthesis inside a poly-3(hexylthiophene) matrix. The charge-separating junction is formed by the work function difference between Al and ITO. Reprinted with permission from Ref. [72].

matrix, is of the utmost importance. However, 510 mV open-circuit voltage and 0.148 mA/cm^2 short-circuit current have been reported. These results are included here for completeness, even though the rest of this chapter is restricted to devices from Si NCs prepared by high-temperature annealing.

The most obvious conductive substrate for the solar cell structure shown in Fig. 11.5a is a Si wafer. However, as the incident light is not absorbed completely in the NC film, the wafer also absorbs part of the light and contributes to the photovoltaic effect. The Si wafer is of excellent electronic quality and can mask any effects of the Si NC layer. However, Si wafer-based solar cells can still be used to characterize the Si NC layer in an indirect way as the heteroemitter of the Si wafer solar cell. This approach was followed by Cho et al. [73], who implemented a phosphorous-doped SiO_2/Si NC multilayer as the heteroemitter. The device and the illuminated I–V curves are shown in Fig. 11.8. The authors used a multilayer consisting of 15 or 25 bilayers with very thin SiO_2 barriers (1 and 2 nm thick) and silicon-rich oxide (SRO) layers with the composition $SiO_{0.89}$. The SRO layers were doped with 0.23 at% phosphorous, and Si NC formation was provoked by tube furnace annealing at 1100°C for 90 minutes. The short-circuit current decreases with increased

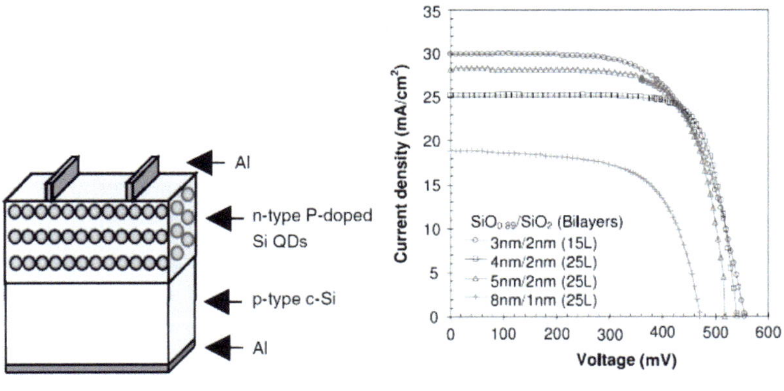

Figure 11.8 Schematic diagram of an implementation of the device structure in Fig. 11.5a (left), and corresponding I–V curves (right). As most of the incident light is absorbed in the wafer, the device must be considered to be a Si wafer solar cell. The Si NC layer is heavily doped with phosphorous. Reprinted with permission from Ref. [73].

Si NC layer thickness, but quite high fill factors are obtained. As the multilayer is heavily doped, this result might be attributed to free carrier or defect absorption. However, the device structure is not suited to investigate quantum confinement effects or to assess the potential of the Si NC layer as a solar cell absorber.

When an insulating substrate is used, highly doped layers are implemented as the back contact to the NC layer (the solar cell base), as shown in Fig. 11.5b. This approach was pursued by Kurokawa et al. [74] and Yamada et al. [75], who realized SiC/Si NC *p*–*i*–*n* structures on quartz glass with heavily doped bulk Si as a back contact. The drawback of this approach is diffusion of dopants into the Si NC film during high-temperature treatments, which might adversely affect crystallization and the electronic quality of the film. Furthermore, clear separation of the photovoltaic activity of the NC layer from that of the rather thick poly-Si film is quite challenging. The devices had an active area of 0.00785 cm^2, consisting of 40 bilayers, and they were crystallized in forming gas at 900°C for 30 minutes, followed by a hydrogen plasma treatment at 340°C. The target layer thicknesses were 5 nm for the Si-rich layer and 1 nm [74] or 2 nm [75] for the barrier. Both works sought to avoid the crystallization of the SiC matrix and pursued this goal by introducing nitrogen or oxygen into the SiC/Si NC layer. The effect of adding nitrogen was presented by Kurokawa et al. [74], using the structure in Fig. 11.5b with a 100 nm p^{++} poly-Si back contact. The open-circuit voltage of a nitrogen-containing SiC/Si NC superlattice was 289 mV, compared to 165 mV of the reference SiC/Si NC material without nitrogen (see Fig. 11.9). Yamada et al. [75] incorporated oxygen into the SiC/Si NC layer and reported an open-circuit voltage of 518 mV using the same device structure but with a 530 nm n^{++} poly-Si layer as the back contact (see Fig. 11.10). The authors ascribe the improved device performance to reduced leakage currents through the amorphous SiC matrix compared to the original case of the partially crystallized and therefore more conductive SiC. However, diffusion of dopants into the Si NC film during the thermal treatments might adversely affect crystallization and the electronic quality of the film. Furthermore, clear separation of the photovoltaic activity of the NC layer from that of the poly-Si bulk film is difficult. The poly-Si thicknesses used by Kurokawa and

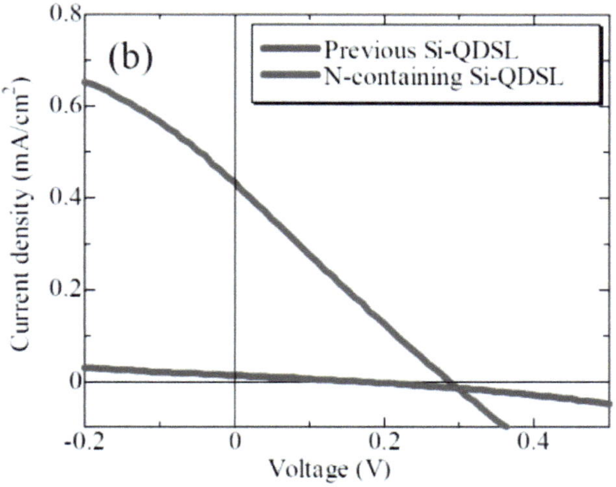

Figure 11.9 I–V curves under illumination of the SiC/Si NC solar cell with and without nitrogen incorporation. Reprinted with permission from Ref. [74].

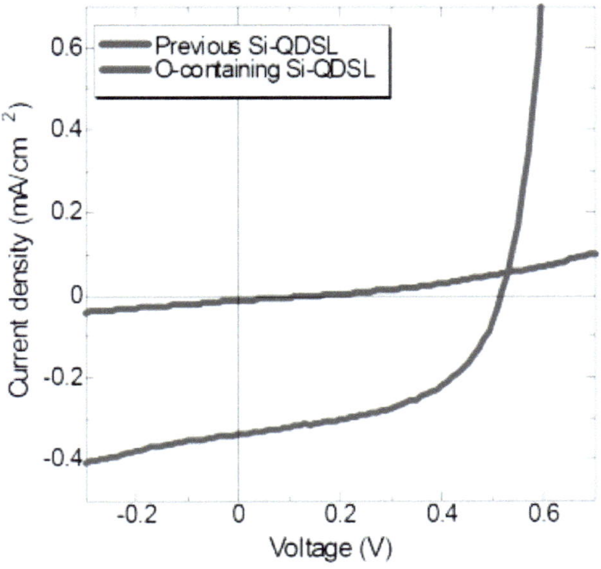

Figure 11.10 I–V curves under illumination of the SiC/Si NC solar cell with and without oxygen incorporation. Reprinted with permission from Ref. [75].

Yamada were 100 and 500 nm, respectively, which are in the range of the penetration depth of green light (250 nm for a wavelength of 514 nm). The devices were illuminated from the poly-Si side, so considerable charge carrier generation occurred in the poly-Si layer. However, the influence of poly-Si on the photovoltaic performance was not analyzed further.

In a variant of (b), the Si NC layer itself can be doped during its deposition, resulting in the structure shown in Fig. 11.5c. This approach in principle allows measurement of the nanocrystal layer independently from any bulk Si contributions. A realization of this structure was presented by Perez-Wurfl et al. [76], who fabricated SiO_2/Si-rich oxide multilayers on quartz glass and doped in situ the lowermost bilayers with phosphorus and the top bilayers with boron. After annealing to form Si NCs, a mesa etch was performed to contact the doped bottom layers. The device yielded an open-circuit voltage of 492 mV. The smeared-out doping profile reported by the authors clearly limits device performance, and the high impurity concentration very likely affects NC formation. A more detailed device analysis in that work came to the conclusion that the device was limited by the high series resistance of 28 kΩcm (see Fig. 11.11). Both approaches, the laterally conductive layer and the conductive substrate, suffer from the fact that the physical properties of the electrical contact cannot be independently influenced but are determined by the material and the recrystallization process. To suppress out-diffusion of dopants from the back contact layer, implementation of a diffusion barrier would be required.

One method of separating contact formation from the recrystallization process and substrate properties is placement of both contacts on the front side of the recrystallized layer, as in Fig. 11.5d. However, this approach requires much higher carrier diffusion lengths than any vertical structure. The measurement length in this device is higher than in the vertical *p–i–n* device of Perez-Wurfl et al. [76], increasing the likelihood of performance limitations from series resistances which are even more severe than those observed by those authors (28 kΩcm, see Fig. 11.11).

For full flexibility in tuning the physical properties of both contacts, our group at Fraunhofer ISE has developed the solar cell test structure shown in Fig. 11.6; the substrate is locally removed by

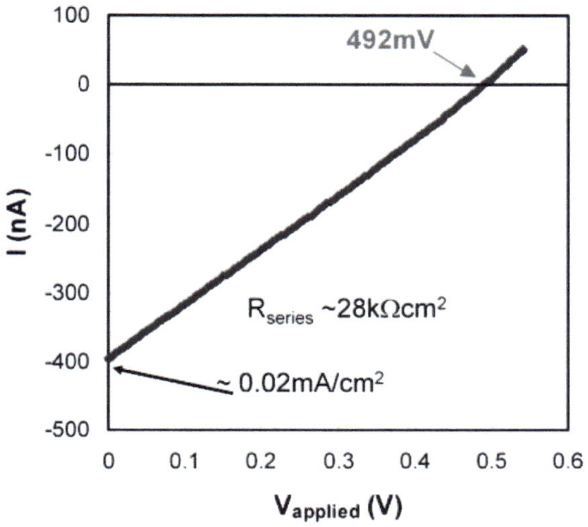

Figure 11.11 Measured *I–V* curves of a 2.2 mm^2 SiO_2/Si NC *p–i–n* diode under 1 sun illumination. Reprinted from *Solar Energy Materials and Solar Cells*, **100**, I. Perez-Wurfl, L. Ma, D. Lin, X. Hao, M.A. Green, G. Conibeer, Silicon nanocrystals in an oxide matrix for thin film solar cells with 492 mV open circuit voltage, 65–68, Copyright (2012), with permission from Elsevier.

chemical etching which facilitates large-area rear-side access to the NC layer [71].

High-temperature annealing is performed before structuring, and deposition of the contact layer and the Si NC film is separated from the substrate by an insulation layer (see Fig. 11.6); thus, there is no out-diffusion of impurities to influence Si NC formation. The electrically active parts are vertically stacked and as such can be described by one-dimensional device physics. No lateral conductivity of any semiconductor device part is required. Figure 11.12 depicts the current–voltage curve of a membrane-based *p–i–n* solar cell with a SiC/Si NC multilayer as the absorber which was reported in Ref. [44]. This cell has an open-circuit voltage of 320 mV and a short-circuit current of 0.35 mA/cm^2. The drawback of this device is the high technological complexity. The membranes must withstand numerous structuring steps, and the device functionality severely depends on the insulating layers shown in Fig. 11.6 (bottom). To optimize the device functionality,

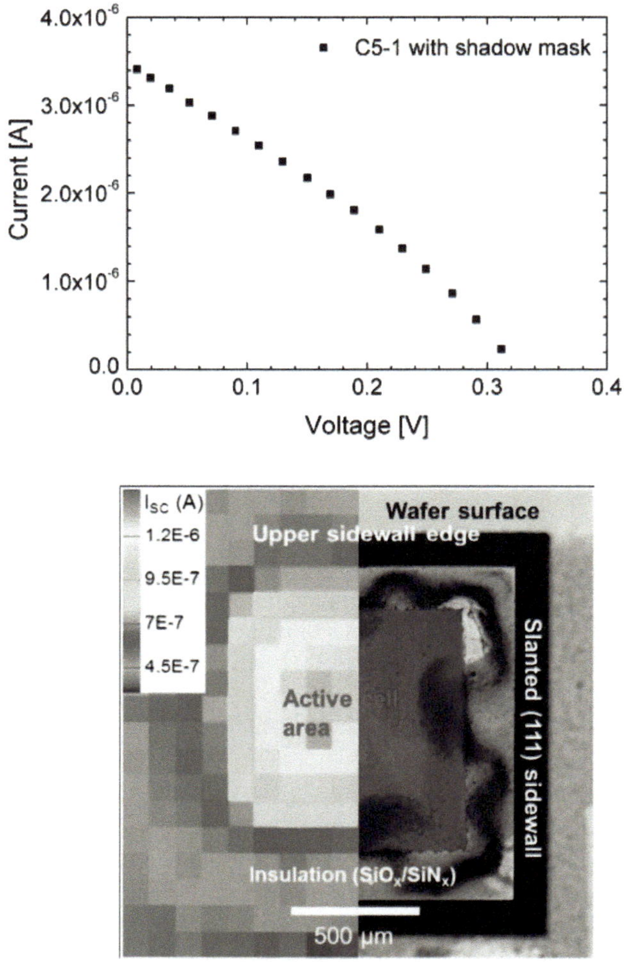

Figure 11.12 (Top) Current–voltage curve of a membrane-based *p–i–n* solar cell with a SiC/Si NC multilayer absorber and doped a-Si_xC_{1-x}:H selective contacts. An open-circuit voltage of 320 mV was achieved. The best cell showed 370 mV open-circuit voltage. Reprinted with permission from Ref. [44]. (Bottom) light beam–induced current map from a similar device, showing that any eventual wafer contribution is effectively suppressed by the insulating layers (SiO_x/SiN_x stack). The right-hand side is a light microscopy image of the same cell included to highlight the correspondence between the location of the light-beam induced current signal and the membrane. Reprinted with permission from Ref. [30]. Copyright © 2013 WILEY-VCH Verlag GmbH & Co. KGaA, Weinheim.

extensive precharacterization is necessary. A powerful tool to test the suppression of the wafer signal is light beam–induced current (LBIC) mapping, as shown in Fig. 11.12 (bottom).

After proving that photocurrent and photovoltage indeed stem from the active cell area and Si NC layer, solar cell device characterization can be conducted to assess the photovoltaic properties of the Si NC layer. Illumination-dependent device measurements were employed to gain insight into the transport and recombination properties of the SiC/Si NC layer [43].

In that work, I–V curves in dark and under illumination between 1 and 20 suns were recorded and then modeled to extract the effective mobility lifetime product, $(\mu\tau)_{\text{eff}}$. The illumination-dependent I–V curves are reprinted in Fig. 11.13. The open-circuit voltage and short-circuit current at 1 sun are 282 mV

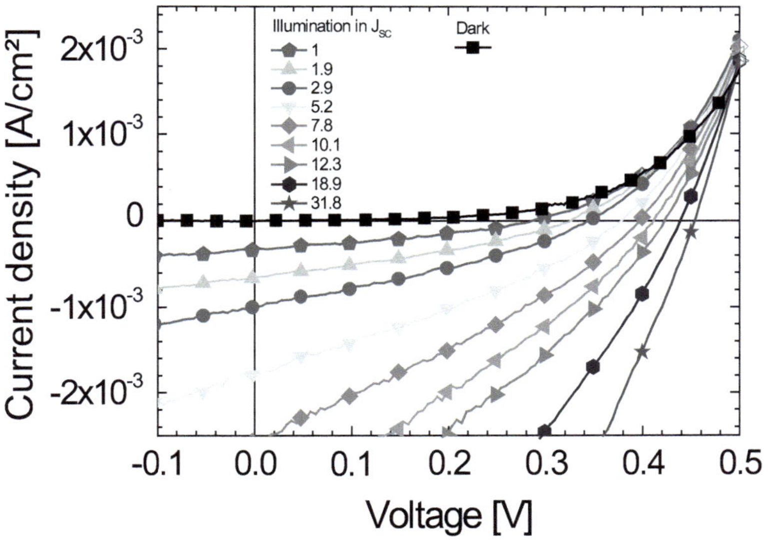

Figure 11.13 Current–voltage curves of a SiC/Si NC membrane cell at varying illumination levels between 1 and 32 suns. Reprinted with permission from [P. Löper, M. Canino, D. Qazzazie, M. Schnabel, M. Allegrezza, C. Summonte, S.W. Glunz, S. Janz, M. Zacharias, Silicon nanocrystals embedded in silicon carbide: investigation of charge carrier transport and recombination, *Applied Physics Letters*, 102 (2013) 033507]. Copyright [2013], AIP Publishing LLC.

and 0.339 mA/cm^2, respectively. Unlike in dark conditions, the illuminated I–V curve shows a pronounced slope at 0 V, indicating a shunt current path which is only present under illumination. A detailed device analysis was carried out [43], assuming a voltage-dependent photocurrent collection function in the approximation of a uniform field, as originally proposed by Crandall [43]. This analysis was governed by the idea of a thin, undoped, and highly recombinative absorber within the electrical field given by highly doped electron and hole contacts. For this case, the approximations are reasonable and allow for an analytical approach. For devices with voltage-dependent photocurrent collection, the photocurrent can be expressed as the product of a voltage-dependent current collection function and the optically generated current:

$$\begin{aligned} J_{\text{light}} &= J_0 \left(\exp\left(\frac{qU}{nk_{\text{B}}T}\right) - 1 \right) + \frac{U}{R_{\text{p}}} - J_{\text{gen}}\chi(U) \\ &= J_{\text{dark}} - J_{\text{gen}}\chi(U) \end{aligned} \tag{11.2}$$

where U denotes the junction voltage, which is the externally applied voltage U_{ext} corrected for series resistance, $U = U_{\text{ext}} - J_{\text{light}}$; J_0 is the dark saturation current density; J_{gen} is the photogenerated current, equal to the number of electron–hole pairs excited per unit time; $\chi(U)$ is the voltage-dependent current collection function; n is the ideality factor; and R_{p} is the parallel or shunt resistance. Recombination in thin film p–i–n solar cells can be described by the ratio of drift length to absorber thickness. Under the assumption of a uniform electrical field over the absorber, Crandall [43] derived the following collection function:

$$\chi(U) = L_{\text{drift}}/d(l - \exp(-d/L_{\text{drift}})), \tag{11.3}$$

where d is the thickness of the intrinsic absorber and L_{drift} is the drift length,

$$L_{\text{drift}} = (\mu\tau)_{\text{eff}}(U_{\text{FB}} - U)/d. \tag{11.4}$$

Equation 11.4 is a reformulation of Eq. 11.1 in terms of parameters that are more accessible experimentally. The electrical field results from the potential difference of the flat-band voltage U_{FB} and the applied voltage U, which drops over the absorber thickness d. $(\mu\tau)_{\text{eff}}$ is the product of the effective charge carrier lifetime and mobility. When the externally applied voltage equals the flat-band voltage, the drift length is zero, and the photogenerated current is no longer collected. Hence, the flat-band voltage U_{FB} can be obtained from the experimentally accessible intersection of dark and light curves at $\chi(U_{FB}) = 0$.

The collection function derived by Crandall refers to the entire voltage range between 0 V and U_{oc} and provides a powerful tool for analyzing the entire I–V curve [77]. However, the function requires knowledge of the diode parameters to reduce the number of unknown parameters. Therefore, the series resistance, ideality factor, and shunt resistance were extracted from the dark I–V data and then used to fit the illumination-dependent I–V curves. The illumination was set to $C \cdot J_{gen}$ where C is the light concentration calculated from the ratio of the respective short-circuit current to that under 1 sun illumination. For J_{gen}, an optical limit at 1 sun illumination of 5.97 mA was used. The flat-band voltage was set to $1.2 \cdot U_{FB}$, where U_{FB} was obtained from the intersection point with the dark curve (Fig. 11.13). Thus, the only free fit parameters used are $(\mu\tau)_{eff}$ and J_0. Figure 11.14 shows measurements under illumination by 1, 1.9, 2.9, and 5.2 suns along with fits with the parameters $(\mu\tau)_{eff} = 2.6 \times 10^{-11}$ cm^2/V and $J_0 = 5.2 \times 10^{-6}$ A/cm^2. Excellent fits are clearly obtained for a wide range of illumination intensities with the same parameter set. The value of $(\mu\tau)_{eff} = 2.6 \times 10^{-11}$ cm^2/V must be regarded as a lower limit, as it combines all recombination in the entire device.

Even though the membrane-based device shown in Fig. 11.6 has so far only been realized with Si NCs embedded in SiC as the absorber, it can also be applied to oxide- or nitride-based NC materials. The implementation of Si NCs in SiO_2 as the absorber layer can be realized by preparing an additional intermediate layer between the Si NC layer and the Si wafer to etch the insulation layer selectively to the SiO_2 with Si NCs. However, the practical realization is more complex than with a SiC matrix due to prolonged KOH etching. KOH attacks the SiO_2 matrix and mechanically destabilizes the entire system. Furthermore, device characterization is facilitated by the superior conductivity of SiC with NCs with respect to SiO_2.

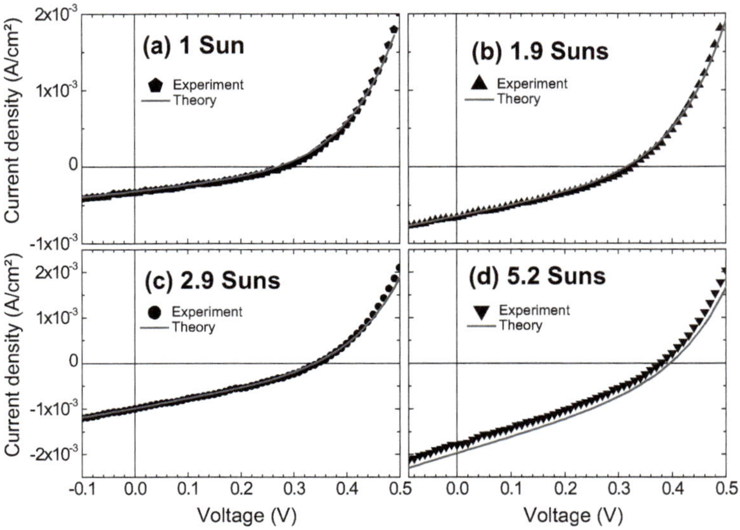

Figure 11.14 Current–voltage curves of membrane-based *p–i–n* devices and the respective fits to the data of the one-diode model with a voltage-dependent current collection function in the uniform field approximation. Series resistance, parallel resistance, and ideality factor were fixed at the dark values. Reprinted with permission from [P. Löper, M. Canino, D. Qazzazie, M. Schnabel, M. Allegrezza, C. Summonte, S.W. Glunz, S. Janz, M. Zacharias, Silicon nanocrystals embedded in silicon carbide: investigation of charge carrier transport and recombination, *Applied Physics Letters*, 102 (2013) 033507]. Copyright [2013], AIP Publishing LLC.

About a decade after the advent of Si NC PV devices which was induced by the work of Green [2], several device varieties have been pursued. Table 11.1 gives an overview of the main achievements on the device level. The development of Si NC photovoltaic devices clearly remains an emerging field. Open-circuit voltages are mostly around 400 mV and approach 500 mV but do not yet exceed typical values obtained with Si wafers (700 mV). However, appropriate device concepts and characterization methods are available now due to recent progress [43]. After the Si NC material has been developed with a well-defined and tunable band gap, the next step is optimization of the transport and recombination parameters of the Si NC material to attain improved photovoltaic performance.

Table 11.1 Overview of Si NC photovoltaic devices without wafer contribution. Devices were realized on quartz glass (QZ) substrates, doped poly-Si on QZ substrates, oxidized wafers which were later etched back, and ITO-covered glass

Material	Fabrication method	Device	V_{oc} (mV)	J_{sc} (mA/cm^2)	Reference
Si NC in SiO_2	Cosputtering, SPC	p–i–n, in situ doped SiO_2/Si NC on QZ, Al contacts	492, (349)	0.02	[76, 78]
Si NC in SiC Si NC in N-doped SiC	PECVD, SPC	n–i–p, a-Si(n), SiC/Si NC (i), p-type poly-Si on QZ	165, 289	0.013, 0.433	[74]
Si NC in SiC Si NC in O-doped SiC	PECVD, SPC	n–i–p a-Si(p), SiC/Si NC (i), n–type poly-Si on QZ	165, 518	0.013, 0.35	[75]
Si NC in SiC	PECVD, SPC	p–i–n, a-Si_xC_{1-x}(p), SiC/Si NC (i), a- Si_xC_{1-x}(n), substrate-free	370	0.35	[71]
Colloidal Si NC	Plasma	ITO/Si NC in/Al	510	0.148	[72]

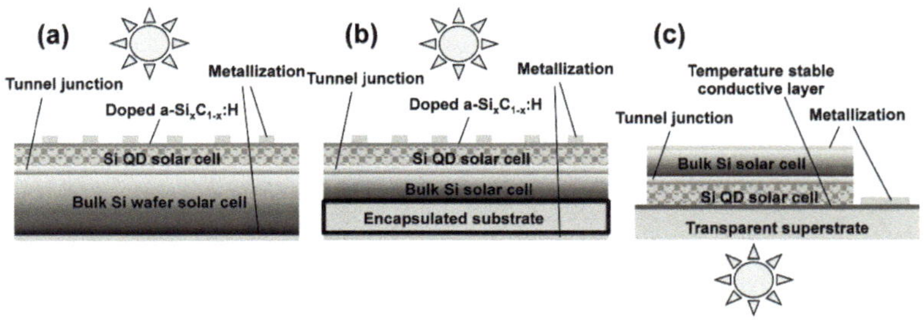

Figure 11.15 Tandem solar cell structures for the c-Si tandem concept: (a) high-efficiency concept with a wafer-based bottom solar cell, (b) thin-film approach using an encapsulated low-cost substrate, and (c) thin-film superstrate approach for a reduced thermal budget on the bottom solar cell [6].

11.8 Tandem Solar Cell Development

The newly developed Si-based absorber with a band gap larger than that of c-Si is just one of many pieces in the overall tandem solar cell device. Depending on the final device structure and process chains, even well-known parts such as the Si bulk bottom solar cell may require significant adaptations. To date, most publications have focused on the monolithic approach, in which both solar cell absorbers are interconnected with a tunnel contact material. Provided the shortcomings of the different Si NC materials are remedied, various monolithic tandem solar cell device concepts are possible. Three possible tandem device structures are presented in Fig. 11.15: (a) a high-efficiency approach with a monocrystalline Si wafer bottom solar cell, (b) a low-cost approach based on an encapsulated foreign substrate and (c) a low-cost approach in superstrate configuration. The three device structures are conceptually guided by (a) high-efficiency solar cells such as passivated emitter and rear contact (PERC) solar cells [79], (b) recrystallized wafer equivalent (RexWE) solar cells [80], and (c) the c-Si on glass (CSG) approach [81]. The implications of the three structures for the Si NC material and the feasibility of a tandem device shall be briefly described here.

Structure (a) must compete with commercialized wafer solar cells. Industrial high-efficiency wafer solar cells already exceed 20%

efficiency, and it might be inferred that the tandem concept can only be competitive if the tandem efficiency clearly exceeds the wafer "technological limit" of 26% [82]. However, if one additional PECVD plus an annealing step (for QD cell fabrication) is sufficient to increase the efficiency of a simple standard wafer solar cell (Al back surface field) by 2% absolute, this concept would be still very attractive. In other words, because the addition of a Si NC top solar cell can be achieved with relatively few processing steps, the cost–benefit relationship is similar to that involved in opting for slightly more expensive bottom cell designs such as the HIT cell [83]: if the increase in cost is low, only a moderate efficiency increase of a few % absolute is required. The more important problem is whether the bottom solar cell and tunnel junction are able to withstand the thermal budget needed to fabricate the Si QD top solar cell. Structure (b) is subject to the same requirements, although it is less severely applied because the bottom cell in structure (b) is not as good as the wafer cell in structure (a) to begin with, which makes it less sensitive to damage from the processing of the top cell. Furthermore, if the thin-film Si bottom cell in structure (b) is produced by gas-phase deposition of Si followed by annealing, the addition of the tunnel junction and Si QD top cell would not even require any additional process steps, merely a slightly longer gas-phase deposition step. This would require harmonization of the gas-phase deposition parameters for the two cells, and of the anneal used, but would greatly decrease the cost of adding a QD solar cell to the thin-film Si cell, which in turn decreases the efficiency increase which must be attained for equal or better cost efficiency. However, texturing of structure (b) will be challenging if a superlattice QD cell is to be grown on top. The usual textures for such thin-film devices show very small structures on which pinhole-free PECVD layer growth is challenging, and even untextured Si films deposited on foreign substrate can exhibit appreciable roughness. Consequently, the development of a Si QD material without front-side texture and scattering structures on the rear side of the cell may be desirable for approach (b). On the other hand, structure (a) is usually textured with random pyramids. As their period and height are on the order of several micrometers and their surface consists of flat (111) planes, pinhole-free PECVD multilayer growth may well be possible. This could even be advantageous to increase the effective light path through the QD layer. Structure (c) is distinguished by a flat surface for Si NC growth and lack of thermal impact to the bulk Si solar cell, but there is a need for a transparent conductive layer. The challenges are thus shifted away from the bulk Si cell which must withstand the thermal budget in structures (a) and (b) to the requirements of the transparent conductive layer. Realistic options include tungsten silicide, which provides a stable contact up to 900°C. Unless novel materials permit higher temperatures, the feasibility of structure (c) critically depends on reduction of the Si QD thermal budget. In summary, each tandem structure implies different requirements for the Si QD properties and those of other device components. Structure (a) is compatible with the superlattice approach but imposes the highest requirements for the absorption as well as recombination and transport properties. In structure (b), the multilayer approach interferes with the surface roughness of the bulk Si solar cell. However, for Si QD materials beyond the multilayer approach, this structure appears to be the most feasible. Structure (c) is viable only for peak temperatures up to 900°C involved in Si QD fabrication.

11.8.1 *Current Matching*

In each of the three structures, the Si QD cell must deliver and conduct a current of 15 to 20 mA/cm^2 to be considered current-matched to the bottom solar cell. Such high currents strongly emphasize the need for efficient electric transport through the QD material. The thickness of the Si QD solar cell required to achieve current matching with the c-Si bottom cell was calculated for different material systems by Summonte et al. [84, 85]. Thicknesses of approximately 5 μm for Si NCs in SiO_2 [85] and 500 nm for Si NCs in SiC [84] were found to be sufficient for this purpose. While the exact numbers are subject to the Si NC density, the general trend that Si NCs in SiC absorb more strongly can be explained with the additional absorption by the matrix.

11.9 Future Trends

11.9.1 *Thermal Budget–Compatible Processing*

Some methods exist for integrating the NC absorbers with a bulk Si bottom cell without exposing the latter to the thermal budgets required for production of the former, such as lift-off and bonding or spectral splitting of the incoming sunlight to direct short- and long-wavelength light to separate Si NC and bulk Si cells, respectively. However, the most economical and standard procedure so far is crystallization of the NC absorber on top of the bottom c-Si cell which already includes the tunnel contact. Solid-phase crystallization for 30 minutes at 1100°C will clearly have an impact on dopant diffusion and smearing of *p–n* junctions. In superstrate configurations where transparent substrates are needed, high thermal loads could also lead to severe problems. In recent publications, Canino et al. [86] and Hiller et al. [87] reported the first successful experiments in which the thermal budget could be reduced significantly for Si NCs in SiC and SiN_xO_y matrices, respectively. This success was mainly achieved by initiating the crystallization with RTA and terminating it at much lower temperatures. Alternative strategies could include application of high-temperature stable collectors based on semi-insulating polycrystalline silicon (SIPOS) technology [88] and diffusion barrier layers acting as a tunnel contact, as in metal-insulator-semiconductor (MIS) solar cells [89].

11.9.2 *Increased Conductivity of the Si NC Material*

The approach of amorphous multilayer deposition and subsequent high-temperature thermal annealing has resulted in excellent optical properties for the SiO_2-based material and a comparatively narrow Si NC size distribution, and it is compatible with large-area production. At the present stage of material development, wavefunction coupling of Si NCs in SiO_2 is not sufficiently strong to provide electrical transport at low electric fields. Such a material is not suitable as a solar cell absorber due to the immense series resistance losses. Luo et al. [90] noted that the NC size variation is extremely critical for NC–NC interaction and thus miniband formation. This effect might be a fundamental problem arising from the disorder introduced by the NC size variation. By partially abandoning the multilayer structure, however, electrical transport was enhanced by 10 orders of magnitude. In this specific case, quantum confinement was present only in one dimension (1D, Si quantum wells) [91]. As the NCs merged across the barriers, the quantum confinement was also lost in this direction. Evidently, the technological limits of arranging Si NCs in SiO_2 and achieving a higher volume and areal NC density are far from exhausted. A careful exploitation of the apparent tradeoff between quantum confinement and coalescence remains to be conducted to achieve optimal material properties. The first encouraging results have been reported by Gutsch et al. [28] with enhanced areal density of the NCs, and additional doping [51] led to significantly enhanced conductivity.

11.9.3 *Reduction of Electronic Defects*

With regard to the SiC matrix, a reduction of the defect density is the prerequisite for an investigation of quantum confinement effects. The major focus is thus not maximization of the Si NC density but high-quality growth of Si NCs, ensuring a low defect density within the SiC matrix and at the SiC/Si NC interface. A ternary SiC/SiO_x superlattice has already been employed to tackle this challenge, but the SiC matrix has been found to be extremely defect rich [92, 93]. The addition of O-[75] or N-[74] impurities to the SiC matrix proved to suppress SiC crystallization and resulted in enhanced device results. However, the supposed defect reduction was not investigated in detail. Finally, a combination of direct Si NC synthesis with matrix-embedding techniques is a candidate for size control in conjunction with electrical transport. For example, plasma-synthesized or wet-chemically synthesized Si NCs could be deposited on a substrate and covered by a monolayer of silicon oxide using atomic layer deposition (ALD).

11.10 Conclusion

All-Si-based tandem solar cells are a quite promising solution to the efficiency limitations of conventional c-Si solar cells. Si QDs which are NCs with a diameter below 10 nm are viable materials for the top cell, as they have a tunable band gap, are made of abundant and safe materials, and are less likely to suffer from light-induced degradation than the currently used amorphous Si. Furthermore, multiple-exciton generation per photon is possible in Si QDs, which improves the conversion of highly energetic photons (see also the chapter by Marri et al.). In this chapter we focus on the most common method at present for fabrication of monodisperse Si NCs which is precipitation from multilayers of a Si-based dielectric material. Reasons for this trend seem to be good size control of the NCs, a high degree of freedom in matrix materials, dopant incorporation, potentially low costs, and perfect compatibility with solar cell processing. We present different solar cell structures which have been used to prove the quantum confinement effect in Si QDs and discuss the assets and drawbacks. Several of them are using existing solar cell technologies but also a membrane cell device which has been especially developed to work best under open-circuit conditions is presented. The comparison of solar cell performance (mainly V_{oc} values) and a close examination of the results are another topic in this chapter. So far, solar cells have been produced with Si QD absorbers in SiO_2 with V_{oc} as high as 490 mV. In addition to disturbing influences from the substrate or dopants, the major material drawback seems to be the high barrier for minority carriers which hinders extraction from the QD. Devices with Si NCs in a SiC:O matrix implementing a highly doped poly-Si layer led to V_{oc} as high as 518 mV. Furthermore we discuss major technology issues which have to be tackled until a working tandem solar cell device can be realized such as, for example, tunnel contacts for monolithic interconnection. Finally, we discuss future trends and necessary improvements like reduction of thermal budget during processing, enhanced electrical conductivity and reduction of electronic defects in the new absorber materials which should enable penetration of this exciting technology into the photovoltaic market in a midterm perspective.

References

1. W. Shockley, H.J. Queisser, Detailed balance limit of efficiency of p-n junction solar cells, *Journal of Applied Physics*, 32 (1961) 510–519.
2. M.A. Green, Potential for low dimensional structures in photovoltaics, *Materials Science and Engineering B*, 74 (2000) 118–124.
3. M.A. Green, *Third Generation Photovoltaics: Advanced Solar Energy Conversion*, Springer, Berlin, Heidelberg, New York, 2003.
4. M. Zacharias, J. Heitmann, R. Scholz, U. Kahler, M. Schmidt, J. Bläsing, Size-controlled highly luminescent silicon nanocrystals: A SiO/SiO_2 superlattice approach, *Applied Physics Letters*, 80 (2002) 661–663.
5. M.A. Green, K. Emery, Y. Hishikawa, W. Warta, E.D. Dunlop, Solar cell efficiency tables (Version 45). Progress in Photovoltaics: Research and Applications, 23(1) (2015) 1–9.
6. P. Löper, *Silicon Nanostructures for Photovoltaics*, 2013, Dissertation, Technical Faculty, University of Freiburg, Reiner Lemoine-Stiftung, Neuss, ISBN, 978-3-8440-2866-9
7. A.W. Bett, F. Dimroth, G. Stollwerck, O.V. Sulima, III-V compounds for solar cell applications, *Applied Physics A: Materials Science and Processing*, A69 (1999) 119–129.
8. F. Meillaud, A. Shah, C. Droz, E. Vail at-Sau vain, C. Miazza, Efficiency limits for single-junction and tandem solar cells, *Solar Energy Materials and Solar Cells*, 90 (2006) 2952–2959.
9. W. Ma, T. Horiuchi, C.C. Lim, H. Okamoto, Y. Hamakawa, Optimum design and its experimental approach of a-Si/ /poly-Si tandem solar cell, *Solar Energy Materials and Solar Cells*, 32 (1994) 351–368.
10. D.L. Staebler, C.R. Wronski, Reversible conductivity changes in discharge-produced amorphous Si, *Applied Physics Letters*, 31 (1977) 292–294.
11. D.L. Staebler, C.R. Wronski, Optically induced conductivity changes in discharge-produced hydrogenated amorphous silicon, *Journal of Applied Physics*, 51 (1980) 3262–3268.
12. R.A. Street, *Hydrogenated Amorphous Silicon*, Cambridge University Press, Cambridge, 1991.
13. R. Guerra, E. Degoli, S. Ossicini, Size, oxidation, and strain in small Si/SiO_2 nanocrystals, *Physical Review B*, 80 (2009) 1–5.
14. C.-W. Jiang, M.A. Green, Silicon quantum dot superlattices: modeling of energy bands, densities of states, and mobilities for silicon tandem solar cell applications, *Journal of Applied Physics*, 99 (2006) 114902.

15. O.I. Micic, A.J. Nozik, Synthesis and characterization of binary and ternary III-V quantum dots, *Journal of Luminescence*, 70 (1996) 95–107.
16. S. Cosentino, M. Miritello, I. Crupi, G. Nicotra, F. Simone, C. Spinella, A. Terrasi, S. Mirabella, Room-temperature efficient light detection by amorphous Ge quantum wells, *Nanoscale Research Letters*, 8 (2013) 128.
17. A.M. Hartel, D. Hiller, S. Gutsch, P. Löper, S. Estradé, F. Peiró, B. Garrido, M. Zacharias, Formation of size-controlled silicon nanocrystals in plasma enhanced chemical vapor deposition grown SiO_xN_y/SiO_2 superlattices, *Thin Solid Films*, 520 (2011) 121–125.
18. S. Janz, P. Löper, M. Schnabel, Silicon nanocrystals produced by solid phase crystallisation of superlattices for photovoltaic applications, *Materials Science and Engineering B*, 178 (2013) 542–550.
19. L. Mangolini, Synthesis, properties, and applications of silicon nanocrystals, *Journal of Vacuum Science and Technology B: Microelectronics and Nanometer Structures*, 31 (2013) 020801-020801–020829.
20. J. Singh, Radiative lifetime and spontaneous emission in amorphous semiconductors, *Journal of Optoelectronics and Advanced Materials*, 7 (2005) 129–136.
21. L.T. Canham, Silicon quantum wire array fabrication by electrochemical and chemical dissolution of wafers, *Applied Physics Letters*, 57 (1990) 1046–1048.
22. F. Huisken, G. Ledoux, O. Guillois, C. Reynaud, Light-emitting silicon nanocrystals from laser pyrolysis, *Advanced Materials*, 14 (2002) 1861–1865.
23. J.D. Holmes, K.J. Ziegler, R.C. Doty, L.E. Pell, K.P. Johnston, B.A. Korgel, Highly luminescent silicon nanocrystals with discrete optical transitions, *Journal of the American Chemical Society*, 123 (2001) 3743–3748.
24. D. Jurbergs, E. Rogojina, L. Mangolini, U. Kortshagen, Silicon nanocrystals with ensemble quantum yields exceeding 60%, *Applied Physics Letters*, 88 (2006) 233116/233111–233113.
25. T.-Y. Kim, N.-M. Park, K.-H. Kim, G.Y. Sung, Y.-W. Ok, T.-Y. Seong, C.-J. Choi, Quantum confinement effect of silicon nanocrystals in situ grown in silicon nitride films, *Applied Physics Letters*, 85 (2004) 5355–5357.
26. J. López-Vidrier, S. Hernández, A.M. Hartel, D. Hiller, S. Gutsch, P. Löper, L. López-Conesa, S. Estradé, F. Peiró, M. Zacharias, B. Garrido, Structural

and optical characterization of size controlled silicon nanocrystals in SiO_2/SiO_xN_y multilayers, *Energy Procedia*, 10 (2011) 43–48.

27. P. Löper, R. Müller, D. Hiller, T. Barthel, E. Malguth, S. Janz, J.C. Goldschmidt, M. Hermle, M. Zacharias, Quasi-Fermi-level splitting in ideal silicon nanocrystal superlattices, *Physical Review B*, 84 (2011) 195317.
28. S. Gutsch, J. Laube, A.M. Hartel, D. Hiller, N. Zakharov, P. Werner, M. Zacharias, Charge transport in Si nanocrystal/SiO_2 superlattices, *Journal of Applied Physics*, 113 (2013) 133703.
29. D. Hiller, S. Goetze, M. Zacharias, Rapid thermal annealing of size-controlled Si nanocrystals: dependence of interface defect density on thermal budget, *Journal of Applied Physics*, 109 (2011) 1–5.
30. P. Löper, M. Canino, J. López-Vidrier, M. Schnabel, F. Schindler, F. Heinz, A. Witzky, M. Bellettato, M. Allegrezza, D. Hiller, A. Hartel, S. Gutsch, S. Hernández, R. Guerra, S. Ossicini, B. Garrido, S. Janz, M. Zacharias, Silicon nanocrystals from high-temperature annealing: characterization on device level, *Physica Status Solidi A*, 210 (2013) 669–675.
31. D. Hiller, S. Goetze, M. Zacharias, Rapid thermal annealing of size-controlled Si nanocrystals: dependence of interface defect density on thermal budget, *Journal of Applied Physics*, 109 (2011) 054308.
32. D. Hiller, M. Jivanescu, A. Stesmans, M. Zacharias, $P_{b(0)}$ centers at the Si-nanocrystal/SiO_2 interface as the dominant photo luminescence quenching defect, *Journal of Applied Physics*, 107 (2010) 084309.
33. G. Seguini, S. Schamm-Chardon, P. Pellegrino, M. Perego, The energy band alignment of Si nanocrystals in SiO_2, *Applied Physics Letters*, 99 (2011) 082107.
34. D. Timmerman, J. Valenta, K. Dohnalova, W.D.A.M. de Boer, T. Gregorkiewicz, Step-like enhancement of luminescence quantum yield of silicon nanocrystals, *Nature Nanotechnology*, 6 (2011) 710–713.
35. V. Vinciguerra, G. Franzo, F. Priolo, F. lacona, C. Spinella, Quantum confinement and recombination dynamics in silicon nanocrystals embedded in Si/SiO_2 superlattices, *Journal of Applied Physics*, 87 (2000) 8165–8173.
36. N.-M. Park, C.-J. Choi, T.-Y. Seong, S.-J. Park, Quantum confinement in amorphous silicon quantum dots embedded in silicon nitride, *Physical Review Letters*, 86 (2001) 1355–1357.
37. B. Rezgui, A. Sibai, T. Nychyporuk, M. Lemiti, G. Bremond, Luminescence mechanisms in Si quantum dots-SiN_x nanocomposite structures, *Journal of Vacuum Science and Technology, B*, 27 (2009) 2238–2241.

38. Y.-H. So, S. Huang, G. Conibeer, M.A. Green, Formation and photoluminescence of Si nanocrystals in controlled multilayer structure comprising of Si-rich nitride and ultrathin silicon nitride barrier layers, *Thin Solid Films*, 519 (2011) 5408–5412.
39. M. Künle, S. Janz, K.G. Nickel, O. Eibl, Changes in chemical composition and nanostructure of SiC thin films prepared by PECVD during thermal annealing, *Physica Status Solidi A*, 208 (2011) 1885–1895.
40. M. Künle, S. Janz, K.G. Nickel, A. Heidt, M. Luysberg, O. Eibl, Annealing of nm-thin $Si_{1-x}C_x$/SiC multilayers, *Solar Energy Materials and Solar Cells*, 115 (2013) 11–20.
41. M. Künle, T. Kaltenbach, P. Löper, A. Hartel, S. Janz, O. Eibl, K.-G. Nickel, Si-rich a-SiC:H thin films: structural and optical transformations during thermal annealing, *Thin Solid Films*, 519 (2010) 151–157.
42. Y. Kurokawa, S. Miyajima, A. Yamada, M. Konagai, Preparation of nanocrystalline silicon in amorphous silicon carbide matrix, Japanese *Journal of Applied Physics, Part 1*, 45 (2006) 1064–1066.
43. P. Löper, M. Canino, D. Qazzazie, M. Schnabel, M. Allegrezza, C. Summonte, S.W. Glunz, S. Janz, M. Zacharias, Silicon nanocrystals embedded in silicon carbide: investigation of charge carrier transport and recombination, *Applied Physics Letters*, 102 (2013) 033507.
44. P. Löper, A. Witzky, A. Hartel, S. Gutsch, D. Hiller, J. C. Goldschmidt, S. Janz, S. W. Glunz, M. Zacharias, *Photovoltaic Properties of Silicon Nanocrystals in Silicon Carbide*, et al. in Physics, Simulation, and Photonic Engineering of Photovoltaic Devices, Proceedings of SPIE vol 8256. (eds A. Freundlich and J.-F. F. Guillemoles) 82560G-82561-82566, SPIE, San Francisco, 2012, doi: 10.1117/12.906669.
45. D. Song, E.-C. Cho, G. Conibeer, C. Flynn, Y. Huang, M.A. Green, Structural, electrical and photovoltaic characterization of Si nanocrystals embedded SiC matrix and Si nanocrystals/c-Si heterojunction devices, *Solar Energy Materials and Solar Cells*, 92 (2008) 474–481.
46. D. Song, E.-C. Cho, G. Conibeer, Y. Huang, C. Flynn, M.A. Green, Structural characterization of annealed $Si_{1-x}C_x$/SiC multilayers targeting formation of Si nanocrystals in a SiC matrix, *Journal of Applied Physics*, 103 (2008) 083544.
47. C. Summonte, S. Mirabella, R. Balboni, A. Desalvo, I. Crupi, F. Simone, A. Terrasi, in *Proceedings of the 23rd European Photovoltaic Solar Energy Conference*, Valencia, Spain, 2008, pp. 730–733.
48. F. lori, E. Degoli, E. Luppi, R. Magri, I. Marri, G. Cantele, D. Ninno, F. Trani, S. Ossicini, Doping in silicon nanocrystals: an ab initio study of the

structural, electronic and optical properties, *Journal of Luminescence*, 121 (2006) 335–339.

49. D. König, J. Rudd, Formation of Si or Ge nanodots in Si_3N_4 with in-situ donor modulation doping of adjacent barrier material, *AIP Advances*, 3 (2013) 012109.
50. D. König, J. Rudd, S. Shrestha, G. Conibeer, M.A. Green, in *Proceedings of the 24th European Photovoltaic Solar Energy Conference*, Hamburg, Germany, 2009, pp. 575–578.
51. S. Gutsch, A.M. Hartel, D. Hiller, N. Zakharov, P. Werner, M. Zacharias, Doping efficiency of phosphorus doped silicon nanocrystals embedded in a SiO_2 matrix, *Applied Physics Letters*, 100 (2012) 233115–233115–233114.
52. X.J. Hao, E.-C. Cho, G. Scardera, Y.S. Shen, E. Bellet-Amalric, D. Bellet, G. Conibeer, M.A. Green, Phosphorus-doped silicon quantum dots for all-silicon quantum dot tandem solar cells, *Solar Energy Materials and Solar Cells*, 93 (2009) 1524–1530.
53. B.J. Simonds, I. Perez-Wurfl, Y.-H. So, A.S. Wan, S. McMurray, P.C. Taylor, Phosphorous doping of superlattice silicon quantum dots in silicon dioxide, *Physica Status Solidi C*, 9 (2012) 1908–1911.
54. X.J. Hao, E.-C. Cho, C. Flynn, Y.S. Shen, S.C. Park, G. Conibeer, M.A. Green, Synthesis and characterization of boron-doped Si quantum dots for all-Si quantum dot tandem solar cells, *Solar Energy Materials and Solar Cells*, 93 (2009) 273–279.
55. A. Mimura, M. Fujii, S. Hayashi, K. Yamamoto, Quenching of photoluminescence from Si nanocrystals caused by boron doping, *Solid-State Communications*, 109 (1999) 561–565.
56. C. Summonte, M. Canino, M. Allegrezza, M. Bellettato, A. Desalvo, R. Shukla, I.P. Jain, I. Crupi, S. Milita, L. Ortolani, L. López-Coneza, S. Estradé, F. Peiró, B. Garrido, Boron doping of silicon rich carbides: electrical properties, *Materials Science and Engineering B*, 178 (2013) 551–558.
57. B.S. Richards, J.E. Cotter, C.B. Honsberg, S.R. Wenham, in *Proceedings of the 28th IEEE Photovoltaic Specialists Conference*, Anchorage, Alaska, 2000, pp. 1–4.
58. B.S. Richards, Comparison of TiO_2 and other dielectric coatings for buried-contact solar cells: a review, *Progress in Photovoltaics: Research and Applications*, 12 (2004) 253–281.
59. P. Alen, M. Vehkamaki, M. Ritala, M. Leskela, Diffusion barrier properties of atomic layer deposited ultrathin Ta_2O_5 and TiO_2 films, *Journal of the Electrochemical Society*, 153 (2006) G304-G308.

60. A.M. Dhote, O. Auciello, D.M. Gruen, R. Ramesh, Studies of thin film growth and oxidation processes for conductive Ti-Al diffusion barrier layers via in situ surface sensitive analytical techniques, *Applied Physics Letters*, 79 (2001) 800–802.
61. O. Auciello, Science and technology of thin films and interfacial layers in ferroelectric and high-dielectric constant heterestructures and application to devices, *Journal of Applied Physics*, 100 (2006) 051614–051614–051615.
62. T.P. Chow, A.J. Steckl, Refractory metal silicides: thin-film properties and processing technology, *IEEE Transactions on Electron Devices*, ED-30 (1983) 1480–1497.
63. S.P. Murarka, Silicide thin-films and their applications in microelectronics, *Intermetallics*, 3 (1995) 173–186.
64. S.H. Luo, W.L. Liu, M. Zhang, Z.F. Di, S.Y. Wang, Z.T. Song, C.L. Lin, S.C. Zou, Buried tungsten silicide layer in silicon on insulator substrate by Smart-Cut®, *Semiconductor Science and Technology*, 19(2004) 1329–1332.
65. C.S. Park, B.J. Cho, D.-L. Kwong, Thermally stable fully silicided Hf-silicide metal-gate electrode, *IEEE Electron Device Letters*, 25 (2004) 372–374.
66. T. Minami, T. Miyata, T. Yamamoto, Stability of transparent conducting oxide films for use at high temperatures, *Journal of Vacuum Science & Technology A*, 17(4) (1999) 1822–1826.
67. O.J. Gregory, X. Chen, E.E. Crisman, Strain and temperature effects in indium-tin-oxide sensors, *Thin Solid Films*, 518 (2010) 5622–5625.
68. M.A. Green, P.A. Basore, N. Chang, D. Clugston, R. Egan, R. Evans, D. Hogg, S. Jarnason, M. Keevers, P. Lasswell, J. O'Sullivan, U. Schubert, A. Turner, S.R. Wenham, T. Young, Crystalline silicon on glass (CSG) thin-film solar cell modules, *Solar Energy*, 77 (2004) 857–863.
69. R. Rolver, B. Berghoff, D.L. Batzner, B. Spangenberg, H. Kurz, Lateral Si/SiO_2 quantum well solar cells, *Applied Physics Letters*, 92 (2008) 212108.
70. B. Berghoff, S. Suckow, R. Rolver, N. Wilck, B. Spangenberg, H. Kurz, in *Proceedings of the 24th European Photovoltaic Solar Energy Conference and Exhibition*, Hamburg, Germany, 2009, pp. 388–390.
71. P. Löper, D. Stüwe, M. Künle, M. Bivour, C. Reichel, R. Neubauer, M. Schnabel, M. Hermle, O. Eibl, S. Janz, M. Zacharias, S.W. Glunz, A membrane device for substrate-free photovoltaic characterization of quantum dot based p-i-n solar cells, *Advanced Materials*, 24 (2012) 3124–3129.

72. C.-Y. Liu, U. Kortshagen, A silicon nanocrystal schottky junction solar cell produced from colloidal silicon nanocrystals, *Nanoscale Research Letters*, 5 (2010) 1253–1256.
73. E.-C. Cho, S. Park, X. Hao, D. Song, G. Conibeer, S.-C. Park, M.A. Green, Silicon quantum dot/crystalline silicon solar cells, *Nanotechnology*, 19 (2008) 245201.
74. Y. Kurokawa, S. Tomita, S. Miyajima, A. Yamada, M. Konagai, in *Proceedings of the 24th European Photovoltaic Solar Energy Conference*, Hamburg, Germany, 2009, pp. 398–403.
75. S. Yamada, Y. Kurokawa, S. Miyajima, A. Yamada, M. Konagai, in *Proceedings of the 35th IEEE Photovoltaic Specialists Conference*, Honolulu, Hawaii, USA, 2010, pp. 766–769.
76. I. Perez-Wurfl, L. Ma, D. Lin, X. Hao, M.A. Green, G. Conibeer, Silicon nanocrystals in an oxide matrix for thin film solar cells with 492 mV open circuit voltage, *Solar Energy Materials and Solar Cells*, 100 (2012) 65–68.
77. S.S. Hegedus, Current-voltage analysis of a-Si and a-SiGe solar cells including voltage-dependent photocurrent collection, *Progress in Photovoltaics*, 5 (1997) 151–168.
78. I. Perez-Wurfl, X. Hao, A. Gentle, D.-H. Kim, G. Conibeer, M.A. Green, Si nanocrystal p-i-n diodes fabricated on quartz substrates for third generation solar cell applications, *Applied Physics Letters*, 95(2009) 153506.
79. A. Mohr, P. Engelhart, C. Klenke, S. Wanka, A.A. Stekolnikov, M. Scherff, R. Seguin, S. Tardon, T. Rudolph, M. Hofmann, F. Stenzel, J.Y. Lee, S. Diez, J. Wendt, W. Brendle, S. Schmidt, J.W. Miiller, P. Wawer, M. Hofmann, P. Saint-Cast, J. Nekarda, D. Erath, J. Rentsch, R. Preu, in *Proceedings of the 26th European Photovoltaic Solar Energy Conference and Exhibition*, Hamburg, Germany, 2011, pp. 2150–2153.
80. I. Kuzma-Filipek, K.V. Nieuwenhuysen, J.V. Hoeymissen, M.R. Payo, E.V. Kerschaver, J. Poortmans, R. Mertens, G. Beaucarne, E. Schmich, S. Lindekugel, S. Reber, Efficiency (>15%) for thin-film epitaxial silicon solar cells on 70 cm^2 area offspec silicon substrate using porous silicon segmented mirrors, *Progress in Photovoltaics: Research and Applications*, 18 (2010) 137–143.
81. Z. Ouyang, O. Kunz, A.B. Sproul, S. Varlamov, Influence of the absorber doping for p-type polycrystalline silicon thin-film solar cells on glass prepared by electron beam evaporation and solid-phase crystallization, *Journal of Applied Physics*, 109 (2011) 054510.

82. R.M. Swanson, in *Proceedings of the 31st IEEE Photovoltaic Specialists Conference*, Orlando, USA, 2005, pp. 889–894.

83. S. De Wolf, A. Descoeudres, Z. C. Holman, C. Ballif, High-efficiency Silicon Heterojunction Solar Cells: A Review, in *Green.* (2012) 7.

84. C. Summonte, A. Canino, M. Allegrezza, M. Bellettato, A. Desalvo, S. Mirabella, A. Terrasi, in *Proceedings of the 26th European Photovoltaic Solar Energy Conference and Exhibition*, Hamburg, Germany, 2011, pp. 361–366.

85. C. Summonte, E. Centurioni, A. Desalvo, A. Canino, S. Mirabella, R. Agosta, F. Simone, A. Terrasi, M.A. Di Stefano, S. Di Marco, S. Ravesi, S. Lombardo, in *Proceedings of the 24th European Photovoltaic Solar Energy Conference*, Hamburg, Germany, 2009, pp. 317–321.

86. M. Canino, M. Allegrezza, M. Bellettato, C. Summonte, A. Desalvo, S. Mirabella, P. Löper, M. Schnabel, J.L. Vidrier, S. Hernandez, L. Lopez, S. Estrade, M. Perani, D. Cavalcoli, F. Peiro, B. Garrido, S. Janz, in *Proceedings of the E-MRS Spring Meeting*, Strasbourg, France, 2013.

87. S. Janz, P. Löper, M. Schnabel, M. Zacharias, D. Hiller, S. Gutsch, A. Hartel, C. Summonte, A. Canino, M. Allegrezza, S. Ossicini, R. Guerra, I. Marri, B. Garrido, S. Hernandez, J. Lopez-Vidrier, J. Valenta, T. Kubera, M. Foti, C. Gerardi, in *Proceedings of the 26th European Photovoltaic* Solar Energy Conference and Exhibition, Hamburg, Germany, 2011, pp. 22–27.

88. Y.H. Kwark, R.M. Swanson, N-type SIPOS and poly-silicon emitters, *Solid-State Electronics*, 30 (1987) 1121–1125.

89. M.A. Green, R.B. Godfrey, MIS solar cell - general theory and new experimental results for silicon, *Applied Physics Letters*, 29 (1976) 610–612.

90. J.-W. Luo, P. Stradins, A. Zunger, Matrix-embedded silicon quantum dots for photovoltaic applications: a theoretical study of critical factors, *Energy and Environmental Science*, 4 (2011) 2546–2557.

91. B. Berghoff, S. Suckow, R. Rolver, B. Spangenberg, H. Kurz, A. Sologubenko, J. Mayer, Quantum wells based on Si/SiO_x stacks for nanostructured absorbers, *Solar Energy Materials and Solar Cells*, 94 (2010) 1893–1896.

92. K. Ding, U. Aeberhard, W. Beyer, O. Astakhov, F. Kohler, U. Breuer, F. Finger, R. Carius, U. Rau, Annealing induced defects in SiC, SiO_x single layers, and SiC/SiO_x hetero-superlattices, *Physica Status Solidi A*, (2012) published online.

93. K. Ding, U. Aeberhard, O. Astakhov, U. Breuer, M. Beigmohamadi, S. Suckow, B. Berghoff, W. Beyer, F. Finger, R. Carius, U. Rau, Defect passivation by hydrogen reincorporation for silicon quantum dots in SiC/SiO_x hetero-superlattice, *Journal of Non-Crystalline Solids*, (2012) published online.

Index